Introduction

to

Scientific Computing

THE MATLAB CURRICULUM SERIES

To facilitate the use of MATLAB throughout the curriculum, Prentice Hall has developed the following series of MATLAB-related titles:

Marcus, *Matrices and* MATLAB: *A Tutorial*
0-13-562901-2   ©1993

Etter, *Engineering Problem Solving with* MATLAB
0-13-293069-2   ©1993

Ogata, *Solving Control Engineering Problems with* MATLAB
0-13-182213-6   ©1994

Hanselman/Kuo, MATLAB *Tools for Control Systems Analysis and Design*
Second Edition
0-13-202293-1   ©1995

Garcia, *Numerical Methods for Physics*
0-13-151986-7   ©1994

Burrus et al, *Computer-Based Exercises for Signal Processing*
0-13-219825-8   ©1994

Polking, MATLAB *Manual for Ordinary Differential Equations*
0-13-133944-3   ©1995

Hanselman/Littlefield, *Mastering* MATLAB: *A Comprehensive Tutorial and Reference*
0-13-191594-0   ©1996

Van Loan, *Introduction to Scientific Computing: A Matrix-Vector Approach Using* MATLAB
0-13-125444-8   ©1997

Introduction
to
Scientific Computing

A Matrix-Vector Approach Using MATLAB®

Charles F. Van Loan
Department of Computer Science
Cornell University

The MATLAB Curriculum Series

Prentice Hall   *Upper Saddle River, New Jersey 07458*

Library of Congress Cataloging-in-Publication Data

Van Loan, Charles F.
    Introduction to scientific computing : a matrix-vector approach
  using MATLAB / Charles F. Van Loan.
      p.   cm. -- (The MATLAB curriculum series)
    Includes bibliographical references (p.   -   ) and index.
    ISBN 0-13-125444-8
    1. Computer Science--Mathematics. 2. Electronic digital
  computers--Programming. 3. MATLAB. I. Title. II. Series.
  QA76.9.M35V375   1997
    510'.285'53--dc20   96-31544
                                                         CIP

Acquisitions Editor: Tom Robbins
Production Editor: Joseph Scordato
Director Prod. & Mfg.: David Riccardi
Production Manager: Bayani Mendoza DeLeon
Cover Designer: Bruce Kenselaar
Copy Editor: Patricia Daly
Buyer: Donna Sullivan
Editorial Assistant : Nancy Garcia

 ©1997 by Prentice-Hall, Inc.
Simon & Schuster/A Viacom Company
Upper Saddle River, NJ 07458

MATLAB is a registered trademark of
The MathWorks, Inc.

The author and publisher of this book have used their best efforts in preparing this book. These efforts
include the development, research, and testing of the theories and programs to determine their effectiveness.
The author and publisher make no warranty of any kind, expressed or implied, with regard to these programs
or the documentation contained in this book. The author and publisher shall not be liable in any event for
incidental or consequential damages in connection with, or arising out of, the furnishing, performance, or use
of these programs.

Printed in the United States of America

10 9 8 7 6 5 4 3 2 1

ISBN   0-13-125444-8

Prentice-Hall International (UK) Limited, London
Prentice-Hall of Australia Pty. Limited, Sydney
Prentice-Hall Canada Inc., Toronto
Prentice-Hall Hispanoamericana, S.A., Mexico
Prentice-Hall of India Private Limited, New Delhi
Prentice-Hall of Japan, Inc., Tokyo
Simon & Schuster Asia Pte. Ltd., Singapore
Editora Prentice-Hall do Brasil, Ltda., Rio de Janeiro

Dedicated to

Cleve B. Moler

Contents

Preface   ix
Software   xi

1   Power Tools of the Trade   1

1.1   Vectors and Plotting   2
1.2   More Vectors, More Plotting, and Now Matrices   15
1.3   Building Exploratory Environments   29
1.4   Error   44
1.5   Designing Functions   52
      M-Files and References   66

2   Polynomial Interpolation   68

2.1   The Vandermonde Approach   70
2.2   The Newton Representation   76
2.3   Properties   82
2.4   Special Topics   86
      M-Files and References   93

3   Piecewise Polynomial Interpolation   94

3.1   Piecewise Linear Interpolation   94
3.2   Piecewise Cubic Hermite Interpolation   105
3.3   Cubic Splines   112
      M-Files and References   123

4   Numerical Integration   124

4.1   The Newton-Cotes Rules   125
4.2   Composite Rules   133
4.3   Adaptive Quadrature   138
4.4   Special Topics   143
4.5   Shared Memory Adaptive Quadrature   148
      M-Files and References   154

5 Matrix Computations   155

5.1   Setting Up Matrix Problems   155
5.2   Matrix Operations   164
5.3   Once Again, Setting Up Matrix Problems   173
5.4   Recursive Matrix Operations   179
5.5   Distributed Memory Matrix Multiplication   186
      M-Files and References   192

6 Linear Systems   194

6.1   Triangular Problems   194
6.2   Banded Problems   200
6.3   Full Problems   207
6.4   Analysis   218
      M-Files and References   223

7 The QR and Cholesky Factorizations   224

7.1   Least Squares Fitting   224
7.2   The QR factorization   229
7.3   The Cholesky Factorization   236
7.4   High-Performance Cholesky   250
      M-Files and References   257

8 Nonlinear Equations and Optimization   258

8.1   Finding Roots   259
8.2   Minimization a Function of One Variable   277
8.3   Minimizing Multivariate Functions   287
8.4   Solving Systems of Nonlinear Equations   295
      M-Files and References   306

9 The Initial Value Problem   308

9.1   Basic Concepts   309
9.2   The Runge-Kutta Methods   319
9.3   The Adams Methods   328
      M-Files and References   339

Bibliography   341
Index   343

Preface

MATLAB affects the way we do research in scientific computing because it encourages experimentation with interesting mathematical ideas. Visualization and vector-level thinking are supported in a way that permits focus on high-level issues. It is by clearing such a wide path from research to applications that MATLAB has been such an uplifting force in computational science.

For exactly the same reasons, MATLAB can uplift the teaching of introductory scientific computing. Students need to *play* with the mathematics that stands behind each and every new method that they learn. They need graphics to appreciate convergence and error. They need a matrix-vector programming language to solidify their understanding of linear algebra and to prepare for a world of advanced array-level computing. They need a total problem solving environment tapping into the very latest algorithmic research that has a bearing on science and engineering In short, they need MATLAB.

In this textbook I present all the topics that are usually covered in a one-semester introduction to scientific computing. But graphics and matrix-vector manipulation have been folded into the presentation in a way that gets students to appreciate the connection between continuous mathematics and computing. Each of the nine chapters comes equipped with a theorem. Analysis is complemented with computational experiments that are designed to bolster intuition. Indeed, the text revolves around examples that are packaged in 200+ **m**-files. These codes are critical to the overall presentation. Collectively they communicate all the key mathematical ideas and an appreciation for the subtleties of numerical computing. They also illustrate many features of MATLAB that are likely to be useful later on in the student's computational career.

Snapshots of advanced computing are given in sections that deal with parallel adaptive quadrature and parallel matrix computations. Our treatment of recursion includes divided differences, adaptive approximation, quadrature, the fast Fourier transform, Strassen matrix multiplication, and the Cholesky factorization.

Numerical linear algebra is not confined to the matrix computation units. Because of the graphics thread throughout the text, it permeates the entire presentation beginning in Chapter 1. That first chapter is yet another get-started-with-MATLAB tutorial, but it is driven by examples that set the stage for the numerical algorithms that follow.

I want to thank the students of CS 222 at Cornell University who inspired me to write this book. My colleagues Yuying Li and Steve Vavasis were of immense help during the revision process.

Cindy Robinson has been my administrative assistant since 1987 and has seen me through the production of five textbooks. Cindy's thoughtful support was essential during this period.

Finally, Cleve Moler has played a very critical role in my academic career ever since I first walked into his office while an undergraduate at the University of Michigan. As teacher, Ph.D. advisor, and force behind MATLAB, Cleve has defined the way I look at mathematics and computing. I am extremely happy to dedicate this textbook to him.

Software

A wide range of software for all the problems considered in this book is available via Netlib:

World Wide Web: `http://www.netlib.org/index.html`
Anonymous ftp: `ftp://ftp.netlib.org`

Via email, send a one-line message:

```
mail netlib@ornl.gov
send index
```

to get started.

The m-files mentioned at the end of each chapter are available through

`ftp://ftp.cs.cornell.edu/pub/cv`

Chapter 1

Power Tools of the Trade

§**1.1** Vectors and Plotting

§**1.2** More Vectors, More Plotting, and Now Matrices

§**1.3** Building Exploratory Environments

§**1.4** Error

§**1.5** Designing Functions

MATLAB is a matrix-vector-oriented system that supports a wide range of activity that is crucial to the computational scientist. In this chapter we get acquainted with this system through a collection of examples that sets the stage for the proper study of numerical computation. The MATLAB environment is very easy to use and you might start right now by running the overview files `intro` and `demo`. There is an excellent tutorial in the *Student Edition to Matlab*. Our introduction is similar in spirit but also previews the central themes that occur with regularity in the following chapters.

We start with the exercise of plotting. MATLAB has an extensive array of visualization tools. But even the simplest plot requires setting up a vector of function values, and so very quickly we are led to the many vector-level operations that MATLAB supports. Our mission is to build up a linear algebra sense to the extent that vector-level thinking becomes as natural as scalar-level thinking. MATLAB encourages this in many ways, and plotting is the perfect start-up topic. The treatment is spread over two sections.

Building environments that can be used to explore mathematical and algorithmic ideas is the theme of §1.3. A pair of random simulations is used to illustrate how MATLAB can be used in this capacity.

In §1.4 we learn how to think and reason about error. Error is a fact of life in computational science, and our examples are designed to build an appreciation for two very important types of error. Mathematical errors result when we take what is infinite or continuous and make it finite or discrete. Rounding errors arise because floating-point representation and arithmetic is inexact.

The last section is devoted to the art of designing effective functions. The user-defined function is the fundamental building block in scientific computation. The design and analysis of MATLAB functions are detailed through examples in the final section.

1.1 Vectors and Plotting

Suppose we want to plot the function $f(x) = \sin(2\pi x)$ across the interval $[0, 1]$. In MATLAB there are three components to this task:

- A vector of x-values that range across the interval must be set up:

$$0 = x_1 < x_2 < \cdots < x_n = 1.$$

- The function must be evaluated at each x-value:

$$y_k = f(x_k) \qquad k = 1, \ldots, n$$

- A polygonal line that connects the points $(x_1, y_1), \ldots, (x_n, y_n)$ must be displayed.

If we take 21 equally spaced x-values, then the result might look like the plot shown in Fig 1.1. The

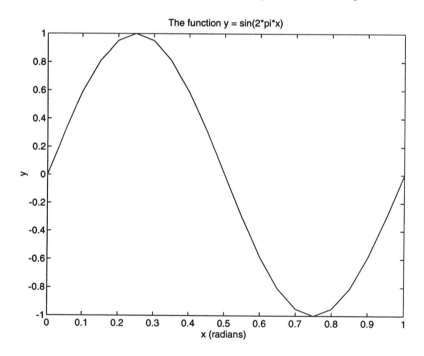

FIGURE 1.1 A crude plot of $\sin(2\pi x)$

plot is "crude" because the polygonal effect is noticeable in regions where the function is changing rapidly. But otherwise the graph looks quite good. Our introduction to MATLAB begins with the details of the plotting process and the vector computations that go along with it. The $\sin(2\pi x)$ example is used throughout because it is simple and structured. Exploiting that structure leads naturally to some vector operations that are well supported in the MATLAB environment.

1.1.1 Setting Up Vectors

When you invoke the MATLAB system, you enter the *command window* and are prompted to enter commands with the symbol ">>". For example,

```
>> x = [10.1 20.2 30.3]
```

MATLAB is an interactive environment and it responds with

```
x =

    10.1000    20.2000    30.3000

>>
```

This establishes x as a length-3 row vector. Square brackets delineate the vector and spaces separate the components. On the other hand, the exchange

```
>> x = [ 10.1; 20.2; 30.3]

x =

    10.1000
    20.2000
    30.3000
```

establishes x as a length-3 column vector. Again, square brackets define the vector being set up. But this time semicolons separate the component entries and a column vector is produced.

In general, MATLAB displays the consequence of a command unless it is terminated with a semicolon. Thus

```
>> x = [ 10.1; 20.2; 30.3];
```

sets up the same column 3-vector as in the previous example, but there is no echo that displays the result. However, the dialog

```
x = [10.1; 20.2; 30.3];
x

x =

    10.1000
    20.2000
    30.3000
```

shows that the contents of a vector can be displayed merely by entering the name of the vector. Even if one component in a vector is changed with no terminating semicolon, MATLAB displays the whole vector:

```
x = [10.1; 20.2; 30.3];
x(2) = 21


x =

    10.1000
    21.0000
    30.3000
```

It is clear that when dealing with large vectors, a single forgotten semicolon can result in a deluge of displayed output.

To change the orientation of a vector from row to column or column to row, use an apostrophe. Thus

```
x = [10.1 20.2 30.3]'
```

establishes x as a length-3 column vector. Placing an apostrophe after a vector effectively takes its transpose.

The plot shown in Fig 1.1 involves the equal spacing of $n = 21$ x-values across $[0, 1]$; that is

```
x = [0 .05 .10 .15 .20 .25 .30 .35 .40 .45 .50 ...
        .55 .60 .65 .70 .75 .80 .85 .90 .95 1.0 ]
```

The ellipsis symbol "..." permits the entry of commands that occupy more than one line.

It is clear that for even modest values of n, we need other mechanisms for setting up vectors. Naturally enough, a for-loop can be used:

```
n = 21;
h = 2*pi/(n-1);
for k=1:n
    x(k) = (k-1)*h;
end
```

This is a MATLAB *script*. It assigns the same length-21 vector to x as before and it brings up an important point.

> *In MATLAB, variables are not declared by the user but are created on a need-to-use basis by a memory manager. Moreover, from MATLAB's point of view, every variable is a complex matrix indexed from unity.*

Scalars are 1-by-1 matrices. Vectors are "skinny" matrices with either one row or one column. We have much more to say about "genuine" matrices later. Complex computations will also be covered. Our initial focus is on real vectors and scalars.

In the preceding script, n, h, k, and x are active variables. It is instructive to trace how x "turns into" a vector during the execution of the for-loop. After one pass through the loop, x is a length-1 vector (i.e., a scalar). During the second pass, the reference x(2) prompts the memory manager to make x a 2-vector. During the third pass, the reference x(3) prompts the memory manager to make x a 3-vector. And so it goes until by the end of the loop, x has length 21. It is a convention in MATLAB that this kind of vector construction yields row vectors.

The MATLAB `zeros` function is handy for setting up the shape and size of a vector prior to a loop that assigns it values. Thus

```
n = 21;
h = 2*pi/(n-1);
x = zeros(1,n);
for k=1:n;
    x(k) = (k-1)*h;
end
```

computes x as row vector of length 21 and initializes the values to zero. It then proceeds to assign the appropriate value to each of the 21 components. Replacing x = `zeros(1,n)` with the command x = `zeros(n,1)` sets up a length 21 column vector. This style of vector set-up is recommended for two reasons. First, it forces you to think explicitly about the orientation and length of the vectors that you are working with. This reduces the chance for "dimension mismatch" errors when vectors are combined. Second, it is more efficient because the memory manager does not have to "work" so hard with each pass through the loop.

MATLAB supplies a `length` function that can be used to probe the length of any active vector. To illustrate its use, the script

```
u = [10 20 30];
n = length(u);
v = [10;20;30;40];
m = length(v);
u = [50 60];
p = length(u);
```

assigns the values of 3, 4, and 2 to n, m, and p, respectively.

This brings up another important feature of MATLAB. It supports a very extensive **help** facility. For example, if we enter

```
help length
```

then MATLAB responds with

```
LENGTH Number of components of a vector.
 LENGTH(X) returns the length of vector X.  It is equivalent
 to MAX(SIZE(X)).
```

So extensive and well structured is the help facility that it obviates the need for us to go into excessive detail when discussing many of MATLAB's capabilities. Get in the habit of playing around with each new MATLAB feature that you learn, exploring the details via the **help** facility. Start right now by trying

```
help help
```

Here in Chapter 1 there are many occasions to use the help facility as we proceed to acquire enough familiarity with the system to get started. Before continuing, you are well advised to try

```
help who
help whos
help clear
```

to learn more about the management of memory. We have already met a number of MATLAB language features and functions you can organize your own mini-review by entering

```
help for
help zeros
help ;
help []
```

1.1.2   Regular Vectors

Regular vectors arise so frequently that MATLAB has a number of features that support their construction. With the *colon notation* it is possible to establish row vectors whose components are equally spaced. The command

```
x = 20:24
```

is equivalent to

```
x = [ 20 21 22 23 24]
```

The spacing between the component values is called the *stride* and the vector x has unit stride. Nonunit strides can also be specified. For example,

```
x = 20:2:29;
```

This stride-2 vector is the same as

```
x = [20 22 24 26 28]
```

Negative strides are also permissible. The assignment

```
x = 10:-1:1
```

is equivalent to

```
x = [ 10 9 8 7 6 5 4 3 2 1]
```

As seen from the examples, the general use of the colon notation has the form

$$\langle Starting\ Index\rangle : \langle Stride\rangle : \langle Bounding\ Index\rangle$$

If the starting index is beyond the bounding index, then the *empty vector* is produced:

```
x = 3:2

x =

    []
```

The empty vector has length zero and is denoted with a square bracket pair with nothing in between.

The colon notation also works with nonintegral values. The command

```
x = 0:.05:1
```

sets up a length-21 row vector with the property that $x_i = (i-1)/20$, $i = 1, \ldots, 9$. Alternatively, we could multiply the vector 0:20 by the scalar .05:

```
x = .05*(0:20)
```

However, if nonintegral strides are involved, then it is preferable to use the `linspace` function. If a and b are real scalars, then

```
x = linspace(a,b,n)
```

returns a row vector of length n whose kth entry is given by

$$x_k = a + (k-1)*(b-a)/(n-1)$$

For example,

```
x = linspace(0,1,21)
```

is equivalent to

```
x = [0 .05 .10 .15 .20 .25 .30 .35 .40 .45 .50 ...
        .55 .60 .65 .70 .75 .80 .85 .90 .95 1.0 ]
```

In general, a reference to `linspace` has the form

```
linspace(⟨Left Endpoint⟩,⟨Right Endpoint⟩,⟨Number of Points⟩)
```

Logarithmic spacing is also possible. The assignment

```
x = logspace(-2,2,6);
```

is the same as x = [.01 .1 1 10 100 1000]. More generally, x = logspace(a,b,n) sets

$$x_k = 10^{a+(b-a)(k-1)/(n-1)} \qquad k = 1, \ldots, n$$

and is equivalent to

```
m = linspace(a,b,n);
for k=1:n
    x(k) = 10^m(k);
end
```

The `linspace` and `logspace` functions bring up an important detail. Many of MATLAB's functions can be called with a reduced parameter list that is often useful in simple, canonical situations. For example, `linspace(a,b)` is equivalent to `linspace(a,b,100)` and `logspace(a,b)` is equivalent to `logspace(a,b,50)`. Make a note of these shortcuts as you become acquainted with MATLAB's many features.

So far we have not talked about how MATLAB displays results except to say that if a semicolon is left off the end of a statement, then the consequences of that statement are displayed. Thus if we enter

```
x = .123456789012345*logspace(1,5,5)'
```

then the vector `x` is displayed according to the active *format*. For example,

```
x =

   1.0e+04 *

    0.0001
    0.0012
    0.0123
    0.1235
    1.2346
```

The preceding is the `short` format. The `long`, `short e`, and `long e` formats are also handy as depicted in Fig 1.2. The `short` format is active when you first enter MATLAB . The `format`

short	long	short e	long e
1.0e+14 *	1.0e+14 *		
0.0000	0.00000000000001	1.2346e+00	1.234567890123450e+00
0.0000	0.00000000000012	1.2346e+01	1.234567890123450e+01
0.0000	0.00000000000123	1.2346e+02	1.234567890123450e+02
0.0000	0.00000000001235	1.2346e+03	1.234567890123450e+03
0.0000	0.00000000012346	1.2346e+04	1.234567890123450e+04
0.0000	0.00000000123457	1.2346e+05	1.234567890123450e+05
0.0000	0.00000001234568	1.2346e+06	1.234567890123450e+06
0.0000	0.00000012345679	1.2346e+07	1.234567890123450e+07
0.0000	0.00000123456789	1.2346e+08	1.234567890123450e+08
0.0000	0.00001234567890	1.2346e+09	1.234567890123450e+09
0.0001	0.00012345678901	1.2346e+10	1.234567890123450e+10
0.0012	0.00123456789012	1.2346e+11	1.234567890123450e+11
0.0123	0.01234567890123	1.2346e+12	1.234567890123450e+12
0.1235	0.12345678901234	1.2346e+13	1.234567890123450e+13
1.2346	1.23456789012345	1.2346e+14	1.234567890123450e+14

FIGURE 1.2 The display of .123456789012345*logspace(1,15,15)

command is used to switch formats. For example,

```
format long
```

It is important to remember that the display of a vector is independent of its internal floating point representation, something that we discuss in §1.4.4.

1.1.3 Evaluating Functions

We return to the task of plotting $\sin(2\pi x)$. MATLAB comes equipped with a host of built-in functions including sin. (Enter `help elfun` to see the available elementary functions.) The script

```
n = 21;
x = linspace(0,1,n);
y = zeros(1,n);
for k=1:n
    y(k) = sin(x(k));
end
```

sets up a vector of sine values that correspond to the values in x. But many of the built-in functions like sin accept vector arguments, and the preceding loop can be replaced with a single reference as follows:

```
n = 21;
x = linspace(0,1,n);
y = sin(2*pi*x);
```

The act of replacing a loop in MATLAB with a single vector-level operation will be referred to as *vectorization* and has three fringe benefits:

- *Speed.* Many of the built-in MATLAB functions execute faster when they operate on a vector of arguments.

- *Clarity.* It is often easier to read a vectorized MATLAB script than its scalar-level counterpart.

- *Education.* Scientific computing on advanced machines requires that one be able to think at the vector level. MATLAB encourages this and as the title of this book indicates, we have every intention of fostering this style of algorithmic thinking.

As a demonstration of the vector-level manipulation that MATLAB supports, we dissect the following script:

```
m = 5;
n = 4*m+1;
x = linspace(0,1,n);
y = zeros(1,n);
a = x(1:m+1);
y(1:m+1)     =  sin(2*pi*a);
y(2*m+1:-1:m+2) =  y(1:m);
y(2*m+2:n) = -y(2:2*m);
```

which sets up the same vector y as before but with one-fourth the number of scalar sine evaluations. The idea is to exploit symmetries in the table shown in Fig 1.3.

The script starts by assigning to a a *subvector* of x. In particular, the assignment to a is equivalent to

k	x_k	$\sin(x_k)$
1	0	0.000
2	18	0.309
3	36	0.588
4	54	0.809
5	72	0.951
6	90	1.000
7	108	0.951
8	126	0.809
9	144	0.588
10	162	0.309
11	180	0.000
12	198	-0.309
13	216	-0.588
14	234	-0.809
15	252	-0.951
16	270	-1.000
17	288	-0.951
18	306	-0.809
19	324	-0.588
20	342	-0.309
21	360	-0.000

FIGURE 1.3 Selected Values of the sine function (x_k in degrees)

```
a = [0.00   0.05   0.10   0.15   0.20   0.25]
```

In general, if v is a vector of integers that are valid subscripts for a row vector z, then

```
w = z(v);
```

is equivalent to

```
for k=1:length(v)
    w(k) = z(v(k));
end
```

The same idea applies to column vectors. Extracted subvectors have the same orientation as the parent vector.

Assignment to a subvector is also legal provided the named subscript range is valid. Thus

```
y(1:m+1) = sin(2*pi*a);
```

is equivalent to

```
for k=1:m+1
    y(k) = sin(2*pi*a(k));
end
```

Now comes the first of two mathematical exploitations. The sine function has the property that

$$\sin\left(\frac{\pi}{2} + x\right) = \sin\left(\frac{\pi}{2} - x\right)$$

Thus

$$\begin{bmatrix} \sin(10h) \\ \sin(9h) \\ \sin(8h) \\ \sin(7h) \\ \sin(6h) \end{bmatrix} = \begin{bmatrix} \sin(0h) \\ \sin(h) \\ \sin(2h) \\ \sin(3h) \\ \sin(4h) \end{bmatrix} \qquad h = 2\pi/20$$

Note that the components on the left should be stored in reverse order in y(7:11) while the components on the right have already been computed and are housed in y(1:5). (See Fig 1.3.) The assignment

```
y(m+1:2*m+1)  =  y(m:-1:1);
```

establishes the necessary values in y(7:11).

At this stage, y(1:2*m+1) contains the sine values from $[0, \pi]$ that are required. To obtain the remaining values, we exploit a second trigonometric identity:

$$\sin(\pi + x) = -\sin(x)$$

We see that this implies

$$\begin{bmatrix} \sin(11h) \\ \sin(12h) \\ \sin(13h) \\ \sin(14h) \\ \sin(15h) \\ \sin(16h) \\ \sin(17h) \\ \sin(18h) \\ \sin(19h) \\ \sin(20h) \end{bmatrix} = - \begin{bmatrix} \sin(h) \\ \sin(2h) \\ \sin(3h) \\ \sin(4h) \\ \sin(5h) \\ \sin(6h) \\ \sin(7h) \\ \sin(8h) \\ \sin(9h) \\ \sin(10h) \end{bmatrix} \qquad h = 2\pi/20$$

The sine values on the left belong (in reverse order) in y(12:21) while those on the right have already been computed and occupy y(2:11). Hence the construction of y(1:21) is completed with the assignment

```
y(2*m+2:n)  =  -y(2:2*m+1);
```

(See Fig 1.3.)

Why go though such contortions when y = sin(2*pi*linspace(0,1,21)) is so much simpler? The reason is that more often than not, function evaluations are expensive and one should always be searching for relationships that reduce their number. Of course, sin is not expensive. But the vector computations detailed in this subsection, above are instructive because we must learn to be sparing when it comes to the evaluation of functions.

1.1.4   Displaying Tables

Any vector can be displayed by merely typing its name and leaving off the semicolon. However, sometimes a more customized output is preferred, and for that a facility with the `disp` and `sprintf` functions is required.

But before we can go any further we must introduce the concept of a *script file*. Already, our scripts are getting too long and too complicated to assemble line-by-line in the command window. The time has come to enlist the services of a text editor and to store the command sequence in a file that can then be executed.

To illustrate the idea, we set up a script file that can be used to display the table in Fig 1.3. We start by entering the following into a file named `SineTable.m`:

```
% Script File: SineTable
%
% Prints a short table of sine evaluations.
%
    n = 21;
    x = linspace(0,1,n);
    y = sin(2*pi*x);
    disp(' ')
    disp('  k      x(k)    sin(x(k))')
    disp('------------------------')
    for k=1:21
       degrees = (k-1)*360/(n-1);
       disp(sprintf(' %2.0f      %3.0f      %6.3f    ',k,degrees,y(k)));
    end
    disp( ' ');
    disp('x(k) is given in degrees.')
    disp(sprintf('One Degree = %5.3e Radians',pi/180))
```

The .m suffix is crucial, for then the preceding command sequence is executed merely by entering `SineTable` at the prompt:

```
>> SineTable
```

This displays the table shown in Fig 1.3 assuming that MATLAB can find `SineTable.m`. This is assured if the file is in the current working directory or if `path` is properly set. Review what you must know about key file organization by entering `help dir cd ls lookfor`.

Focusing on `SineTable` itself, there are a number of new features that we must explain. The script begins with a sequence of *comments* indicating what happens when it is run. Comments in MATLAB begin with the percent symbol "%". Aside from enhancing readability, the lead comments are important because they are displayed in response to a `help` enquiry. That is,

```
help SineTable
```

Use `type` to list the entire contents of a file (e.g., `type SineTable`).

The `disp` command has the form

```
disp(⟨string⟩)
```

Strings in MATLAB are enclosed by single quotes. (We have much more to say about strings in §1.5.3.) The commands

```
disp(' ')
disp('  k      x(k)    sin(x(k))')
disp('----------------------')
```

are used to print a blank line, a heading, and a dashed line.

The `sprintf` command is used to produce a string that includes the values of named variables. It has the form

sprintf(⟨*String with Format Specifications*⟩,⟨*List-of-Variables*⟩)

A variable must be listed for each format. Sample format insertions include `%5.0f`, `%8.3f`, and `%10.6e`. The first integer in a format specification is the total width of the field. The second number specifies how many places are allocated to the fractional part. In the script, the command

```
disp(sprintf(' %2.0f     %3.0f     %6.3f     ',k,degrees,y(k)));
```

prints a line with three numbers. The three numbers are stored in `k`, `degrees`, and `y(k)`. The values in `k` and `degrees` are printed as integers because 0 places are allocated for the decimal portion. On the other hand, `y(k)` is printed with three decimal places. The `e` format is used to specify mantissa/exponent style. For example,

```
disp(sprintf('One Degree = %5.3e Radians',pi/180))
```

This produces the output

```
One Degree = 1.745e-02 Radians
```

1.1.5   A Simple Plot

We are now in a position to solve the plotting problem posed at the beginning of this section. The script

```
n = 21;
x = linspace(0,1,n);
y = sin(2*pi*x);
plot(x,y)
title('The Function  y = sin(2*pi*x)')
xlabel('x (in radians)')
ylabel('y')
```

reproduces Fig 1.1. It draws a polygonal line in a *figure window* that connects the vertices (x_k, y_k), $k = 1:21$ in order. In its most simple form, `plot` takes two vectors of equal size and plots the second versus the first. The scaling of the axes is done automatically. The `title`, `xlabel`, and `ylabel` functions enable us to "comment" the plot. Each requires a string argument.

To produce a better plot with no "corners," we increase n so that the line segments that make up the graph are sufficiently short, thereby rendering the impression of a genuine curve. For example,

```
n = 200;
x = linspace(0,1,n);
y = sin(2*pi*x);
plot(x,y)
title('The function y = sin(2*pi*x)')
xlabel('x (in radians)')
ylabel('y')
```

(See Fig 1.4.) The smoothness of the displayed curve depends on the spacing of the underlying

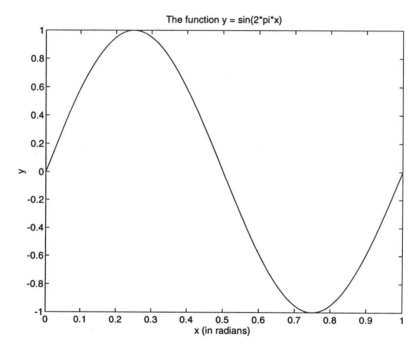

FIGURE 1.4 A smooth plot of $\sin(2\pi x)$

sample points. Here is a script file that produces a sequence of increasingly refined plots:

```
% Script File: SinePlot
%
% Displays increasingly smooth plots of sin(2*pi*x).

    for n = [4 8 12 16 20 50 100 200]
        x = linspace(0,1,n);
        y = sin(2*pi*x);
        plot(x,y)
        title(sprintf('Plot based upon n = %3.0f points.',n))
        pause(1);
    end
```

There are three new features to discuss. First, notice the use of a general vector in the specification of the `for`-loop. The count variable `n` takes on the values in the specified vector one at a time. Second, observe the use of `sprintf` in the reference to `title`. This enables us to report the number of points associated with each plot. Finally, the fragment makes use of the `pause` function. In general, a reference of the form `pause(s)` holds up execution for approximately `s` seconds. Because a sequence of plots is produced in the preceding example, the `pause(1)` command permits a 1-second viewing of each plot.

Problems

P1.1.1 The built-in functions like `sin` accept vector arguments and return vectors of values. If `x` is an n vector, then

$$y = \begin{cases} \texttt{abs(x)} \\ \texttt{sqrt(x)} \\ \texttt{exp(x)} \\ \texttt{log(x)} \\ \texttt{sin(x)} \\ \texttt{cos(x)} \\ \texttt{asin(x)} \\ \texttt{acos(x)} \\ \texttt{atan(x)} \end{cases} \Rightarrow y_i = \begin{cases} |x_i| \\ \sqrt{x_i}, \ x_i \ge 0 \\ e^{x_i} \\ \log(x_i), \ x_i > 0 \\ \sin(x_i) \\ \cos(x_i) \\ \arcsin(x_i), \ -1 \le x_i \le 1 \\ \arccos(x_i), \ -1 \le x_i \le 1 \\ \arctan(x_i) \end{cases} \ i = 1{:}n$$

The vector `x` can be either a row vector or a column vector and `y` has the same shape. Write a script file that plots these functions in succession with appropriate pauses in between the plots.

P1.1.2 Define the function

$$f(x) = \begin{cases} \sqrt{1 - (x-1)^2} & 0 \le x \le 2 \\ \sqrt{1 - (x-3)^2} & 2 < x \le 4 \\ \sqrt{1 - (x-5)^2} & 4 < x \le 6 \\ \sqrt{1 - (x-7)^2} & 6 < x \le 8 \end{cases}$$

Set up a length-201 vector y with the property that $y_i = f(8 * (i-1)/201)$.

1.2 More Vectors, More Plotting, and Now Matrices

We continue to refine our vector intuition by considering several additional plotting situations. New control structures are introduced and some of Matlab's matrix algebra capabilities are presented.

1.2.1 Vectorizing Function Evaluations

Consider the problem of plotting the rational function

$$f(x) = \left(\frac{1 - \dfrac{x}{24}}{1 + \dfrac{x}{24} + \dfrac{x^2}{384}} \right)^8$$

across the interval $[0, 1]$. (This happens to be an approximation to the function e^{-x}.) Here is a scalar approach:

```
n = 200;
x = linspace(0,1,n);
y = zeros(1,n);
for k=1:n
    y(k) = ((1 - x(k)/24)/(1 + x(k)/24 + (x/384)*x))^8
end
plot(x,y)
```

However, by using vector operations that are available in MATLAB, it is possible to replace the loop with a single, vector-level command:

```
% Script File: ExpPlot
%
% Plots the function
%
%      f(x) = ((1 - x/24)/(1 + x/24 + x^2/384))^8
%
% across [0,1].
%
    x  = linspace(0,1,200);
    y = ((1 - (x/24))./(1 + (x/24) + (x/384).*x)).^8
    plot(x,y);
    title('(1 - x/24)/(1 + x/24 + x^2/384))^8')
```

The assignment to y involves the familiar operations of *vector scale*, *vector add*, and *vector subtract*, and the not-so-familiar operations of *pointwise vector multiply*, *pointwise vector divide*, and *pointwise vector exponentiation*. To clarify each of these operations, we break the one-liner down into a sequence of operations:

```
z = (1/24)*x;
num = 1 - z;
w = x/384;
q = w.*x;
denom = 1 + z + q;
quotient = num./denom;
y = quotient.^8
```

MATLAB supports scalar-vector multiplication. The command

```
z = (1/24)*x;
```

multiplies every component in x by $(1/24)$ and stores the result in z. The vector z has exactly the same length and orientation as x. The command

```
num = 1 + z;
```

adds 1 to every component of z and stores the result in w. Thus w = 1 + [20 30 40] and w = [21 31 41] are equivalent. Strictly speaking, scalar-plus-vector is not a legal vector space operation, but it is a very handy MATLAB feature.

Now let us produce the vector of denominator values. The command w = x/384 is equivalent to w = (1/384)*x. It is also the same as w = x*(1/384). The command

```
q = w.*x
```

makes use of pointwise vector multiplication and produces a vector q with the property that each component is equal to the product of the corresponding components in w and x. Thus

```
q = [2 3 4].*[20 30 50]
```

is equivalent to

```
q = [40 90 200]
```

The same rules apply when the two operands are column vectors. The key is that both vectors that make up the multiply have to be identical in length and orientation.

The command denom = 1 + z + q sets denom(i) to 1 + z(i) + q(i) for all i. Vector addition, like vector subtraction, requires both operands to have the same length and orientation.

The pointwise division quotient = num./denom performs as expected. The ith component of quotient is set to num(i)/denom(i). Lastly, the command

```
y = quotient.^8
```

raises each component in quotient to the 8th power and assembles the results in the vector y.

1.2.2 Scaling and Superpositioning

Consider the plotting of the function $\tan(x) = \sin(x)/\cos(x)$ across the interval $[-\pi/2, 11\pi/2]$. This is interesting because the function has poles at points where the cosine is zero. The script

```
x = linspace(-pi/2,11*pi/2,200);
y = tan(x);
plot(x,y);
```

produces a plot with minimum information because the autoscaling feature of the plot function must deal with an essentially infinite range of y-values. This can be corrected by using the axis function:

```
x = linspace(-pi/2,11*pi/2,200);
y = tan(x);
plot(x,y);
axis([-pi/2 9*pi/2 -10 10]);
```

This axis function is used to scale manually the axes in the current plot, and it requires a 4-vector whose values define the x and y ranges. In particular,

```
axis([xmin xmax ymin ymax])
```

imposes the x-axis range xmin $\leq x \leq$ xmax and a y-axis range ymin $\leq y \leq$ ymax. In our example, the $[-10, 10]$ range in the y-direction is somewhat arbitrary. Other values would work. The idea is to choose the range so that the function's poles are dramatized without sacrificing the quality of the plot in domains where it is nicely behaved. (See Fig 1.5.) We mention that the command axis by itself returns the system to the original autoscaling mode.

Another way to produce the same graph is to plot the first branch and then to reuse the function evaluations for the remaining branches:

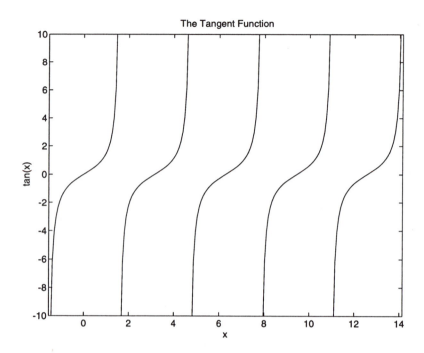

FIGURE 1.5 A plot of $\tan(x)$

```
% Script File: TangentPlot
%
% Plots the function tan(x), -pi/2 <= x <= 11pi/2

    ymax = 10;
    x = linspace(-pi/2,pi/2,40);
    y = tan(x);
    plot(x,y)
    axis([-pi/2 9*pi/2 -ymax ymax])
    title('The Tangent Function')
    xlabel('x')
    ylabel('tan(x)')
    hold on
    for k=1:4
        xnew = x+ k*pi;
        plot(xnew,y);
    end
    hold off
```

This script has a number of new features that require explanation. The `hold on` command effectively tells MATLAB to superimpose all subsequent plots on the current figure window. Each time through the `for`-loop, a different branch is plotted. The axis scaling is frozen during these computations. The `xnew` calculation produces the required x-domain for each branch plot. During the kth pass through the loop, the expression `xnew + k*pi` establishes a vector of equally spaced values across the interval

$$[-\pi/2 + k\pi, -\pi/2 + (k+1)\pi]$$

The same vector of tan-evaluations is used in each branch plot. Observe that with superpositioning we produce a plot with only one-fifth the number of `tan` evaluations that our initial solution required.

The `hold off` command shuts down the superpositioning feature and sets the stage for "normal" plotting thereafter.

Another way that different graphs can be superimposed in the same plot is by calling `plot` with an extended parameter list. Suppose we want to plot the functions $\sin(2\pi x)$ and $\cos(2\pi x)$ across the interval $[0, 1]$ and to mark the point where they intersect. The script

```
x = linspace(0,1,200);
y1 = sin(2*pi*x);
y2 = cos(2*pi*x);
plot(x,y1);
hold on;
plot(x,y2,'-');
plot([1/8 5/8],[1/sqrt(2) -1/sqrt(2)],'*')
hold off
```

accomplishes this task. (See Fig 1.6.) The first three-argument call to `plot` uses a dashed line to produce the graph of $\cos(2\pi x)$. Other line designations are possible (e.g., '–','-.'). The second three-argument call to plot places an asterisk at the intersection points $(1/8, 1/\sqrt{2})$ and $(5/8, -1/\sqrt{2})$. Other point designations are possible (e.g., '+', '.', 'o'.) The key idea is that when plot is used to draw a graph, an optional third parameter can be included. This parameter is a string that specifies the "nature of the pen" that is doing the drawing. Colors may also be specified. The superpositioning can also be achieved as follows:

```
% Script File: SineAndCosPlot
%
% Plots the functions sin(2*pi*x) and cos(2*pi*x) across [0,1]
% and marks their intersection.

    x = linspace(0,1,200);
    y1 = sin(2*pi*x);
    y2 = cos(2*pi*x);
    plot(x,y1,x,y2,[1/8 5/8],[1/sqrt(2) -1/sqrt(2)],'*');
```

This illustrates plot's "multigraph" capability. The syntax is as follows:

```
plot(⟨First Graph Specification⟩,...,⟨Last Graph Specification⟩)
```

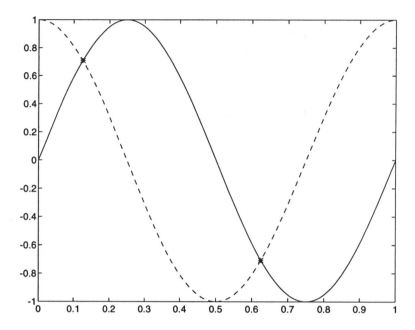

FIGURE 1.6 Superpositioning

where each graph specification has the form

$\langle Vector \rangle, \langle Vector \rangle, \langle String\ (optional) \rangle$

If some of the string arguments are missing, then MATLAB chooses them in a way that fosters clarity in the overall plot.

1.2.3   Polygons

Suppose that we have a polygon with n vertices. If x and y are column vectors that contain the coordinate values, then

```
plot(x,y)
```

does *not* display the polygon because (x_n, y_n) is not connected to (x_1, y_1). To rectify this we merely "tack on" an extra copy of the first point:

```
x = [x;x(1)];
y = [y;y(1)];
plot(x,y)
```

Thus, the three points $(1, 2)$, $(4, -2)$, and $(3, 7)$ could be represented with the three-vectors x = [1 4 3] and y=[2 -2 7]. The x and y updates produce x = [1 4 3 1] and y = [2 -2 7 2]. Plotting the revised y against the revised x displays the triangle with the designated vertices.

The preceding "concatenation" of a component to a vector is a special case of a general operation whereby vectors can be glued together. If r1, r2,...,rm are row vectors, then

```
v = [ r1 r2 ...  rm]
```

is also a row vector obtained by placing the component vectors r1,...,rm side by side. For example,

```
v = [linspace(1,10,10) linspace(20,100,9)];
```

is equivalent to

```
v = [ 1  2  3  4  5  6  7  8  9  10  20  30  40  50  60  70  80  90  100];
```

Similarly, if c1, c2,..., cm are column vectors, then

```
v = [ c1 ; c2 ; ...  ; cm]
```

is also a column vector, obtained by stacking c1,...,cm.

Continuing with our polygon discussion, assume that we have executed the commands

```
t = linspace(0,2*pi,361);
c = cos(t);
s = sin(t);
plot(c,s);
axis('square')
```

The object displayed is a regular 360-gon with "radius" 1. The command axis('square') ensures that the same scales are used in both the x and y directions. This is important in this application because a regular polygon would not look regular if the two scales were different.

With the preceding sine/cosine vectors computed, it is possible to display various other regular n-gons simply by connecting appropriate subsets of points. For example,

```
x = [c(1) c(121) c(241) c(361)];
y = [s(1) s(121) s(241) s(361)];
plot(x,y)
```

plots the equilateral triangle whose vertices are at the $0°$, $120°$, and $240°$ points along the unit circle. This kind of non-unit stride subvector extraction can be elegantly handled in MATLAB using the colon notation. The preceding triplet of commands is equivalent to

```
x = c(1:120:361);
y = s(1:120:361);
plot(x,y)
```

More generally, if sides is a positive integer that is a divisor of 360, then

```
x = c(1:(360/sides):361);
y = s(1:(360/sides):361);
plot(x,y)
```

plots a regular polygon with that number of sides. Here is a script that displays nine regular polygons in nine separate subwindows:

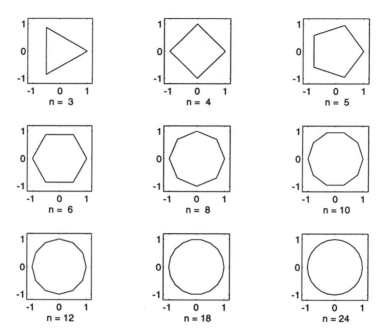

FIGURE 1.7 Regular Polygons

```
% Script File Polygons
%
% Plots selected regular polygons.

    close all
    clc
    theta = linspace(0,2*pi,361);
    c = cos(theta);
    s = sin(theta);
    k=0;
    for sides = [3 4 5 6 8 10 12 18 24]
        stride = 360/sides;
        k=k+1;
        subplot(3,3,k)
        plot(c(1:stride:361),s(1:stride:361))
        xlabel(sprintf(' n = %2.0f',sides));
        axis([-1.2 1.2 -1.2 1.2])
        axis('square')
    end
```

FIGURE 1.7 shows what is produced when this script is executed. The `close all` command closes
all windows and `clc` clears the command window and places the cursor in the home position. It
is a good idea to begin script files with these commands to work with a "clean slate."

The key new feature in `Polygons` is `subplot`. The command `subplot(3,3,k)` says "break up the current figure window into a 3-by-3 array of subwindows, and place the next plot in the kth one of these." The subwindows are indexed as follows:

$$\boxed{1}\quad\boxed{2}\quad\boxed{3}$$
$$\boxed{4}\quad\boxed{5}\quad\boxed{6}$$
$$\boxed{7}\quad\boxed{8}\quad\boxed{9}$$

In general, `subplot(m,n,k)` splits the current figure into an m-row by n-column array of subwindows that are indexed left to right, top to bottom.

1.2.4   Some Matrix Computations

Let's consider the problem of plotting the function

$$f(x) = 2\sin(x) + 3\sin(2x) + 7\sin(3x) + 5\sin(4x)$$

across the interval $[-10, 10]$. The scalar-level script

```
n = 200;
x = linspace(-10,10,n)';
y = zeros(n,1);
for k=1:n
    y(k) = 2*sin(x(k)) + 3*sin(2*x(k)) + 7*sin(3*x(k)) + 5*sin(4*x(k));
end
plot(x,y)
title('f(x) = 2sin(x) + 3sin(2x) + 7sin(3x)  +5sin(4x)')
```

does the trick. (See Fig 1.8.) Notice that `x` and `y` are column vectors. The `sin` evaluations can be vectorized giving this superior alternative:

```
n = 200;
x = linspace(-10,10,n)';
y = 2*sin(x) + 3*sin(2*x) + 7*sin(3*x) + 5*sin(4*x);
plot(x,y)
title('f(x) = 2sin(x) + 3sin(2x) + 7sin(3x)  +5sin(4x)')
```

But any linear combination of vectors is "secretly" a matrix-vector product. That is,

$$
2\begin{bmatrix} 3 \\ 1 \\ 4 \\ 7 \\ 2 \\ 8 \end{bmatrix}
+ 3\begin{bmatrix} 5 \\ 0 \\ 3 \\ 8 \\ 4 \\ 2 \end{bmatrix}
+ 7\begin{bmatrix} 8 \\ 3 \\ 3 \\ 1 \\ 1 \\ 1 \end{bmatrix}
+ 5\begin{bmatrix} 1 \\ 6 \\ 8 \\ 7 \\ 0 \\ 9 \end{bmatrix}
=
\begin{bmatrix} 3 & 5 & 8 & 1 \\ 1 & 0 & 3 & 6 \\ 4 & 3 & 3 & 8 \\ 7 & 8 & 1 & 7 \\ 2 & 4 & 1 & 0 \\ 8 & 2 & 1 & 9 \end{bmatrix}
\begin{bmatrix} 2 \\ 3 \\ 7 \\ 5 \end{bmatrix}
$$

Matlab supports matrix-vector multiplication, and the script

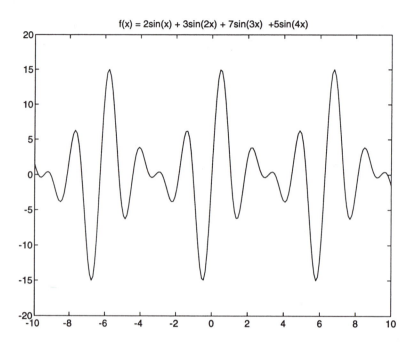

FIGURE 1.8 A sum of sines

```
A = [3 5 8 1; 1 0 3 6; 4 3 3 8; 7 8 1 7; 2 4 1 0; 8 2 1 9];
y = A*[2;3;7;5];
```

shows how to initialize a small matrix and engage it in a matrix-vector product. Note that the matrix is assembled row by row with semicolons separating the rows. Spaces separate the entries within a row. An ellipses (...) can be used to spread a long commands over more than one line, which is sometimes useful for clarity:

```
A = [3 5 8 1;...
     1 0 3 6;...
     4 3 3 8;...
     7 8 1 7;...
     2 4 1 0;...
     8 2 1 9];
y = A*[2;3;7;5];
```

In the sum-of-sines plotting problem, the vector y can also be constructed as follows:

```
n = 200;
x = linspace(-10,10,n)';
y = zeros(n,1);
m = 4
```

```
A = zeros(n,m)
for j=1:m
   for k=1:n
      A(k,j) = sin(j*x(k));
   end
end
y = A*[2;3;7;5];
plot(x,y)
title('f(x) = 2sin(x) + 3sin(2x) + 7sin(3x)  +5sin(4x)')
```

This illustrates how a matrix can be initialized at the scalar level. But a matrix is just an aggregation of its columns, and MATLAB permits a column-by-column synthesis, bringing us to the final version of our script:

```
% Script File: SumOfSines
%
% Plots f(x) = 2sin(x) + 3sin(2x) + 7sin(3x) + 5sin(4x)
% across the interval [-10,10].
   close all; clc
   n = 200;
   x = linspace(-10,10,n)';
   A = [sin(x) sin(2*x) sin(3*x) sin(4*x)];
   y = A*[2;3;7;5];
   plot(x,y)
   title('f(x) = 2sin(x) + 3sin(2x) + 7sin(3x)  +5sin(4x)')
```

An expression of the form

$$[\ \langle Column\ 1 \rangle \ \ \langle Column\ 2 \rangle \ \ldots \ \ \langle Column\ m \rangle \ ]$$

is a matrix with m columns. Of course, the participating column vectors must have the same length.

Another way to initialize A is to use a single loop whereby each pass sets up a single column:

```
n = 200;
m = 4;
A = zeros(n,m);
for j=1:m
   A(:,j) = sin(j*x);
end
```

The notation $A(:,j)$ names the jth column of A. Notice that the size of A is established with a call to `zeros`. The `size` function can be used to determine the dimensions of any active variable. (Recall that all variables are treated as matrices.) Thus the script

```
A = [1 2 3;4 5 6];
[r,c] = size(A);
```

assigns 2, the row dimension, to r and 3, and the column dimension to c. Many MATLAB functions return more than one value and `size` is our first exposure to this. Note that the output values are enclosed with square brackets.

Matrices can also be built up by row. In SumOfSines, the kth row of A is given by `sin(k*(1:4))` so we also initialize A as follows:

```
n = 200;
m = 4;
A = zeros(n,m);
for k=1:n
    A(k,:) = sin(k*(1:m));
end
```

The notation `A(k,:)` identifies the kth row of A.

As a final example, suppose that we want to plot *both* of the functions

$$f(x) = 2\sin(x) + 3\sin(2x) + 7\sin(3x) + 5\sin(4x)$$
$$g(x) = 8\sin(x) + 2\sin(2x) + 6\sin(3x) + 9\sin(4x)$$

in the same window. Obviously, a double application of the preceding ideas solves the problem:

```
n = 200;
x = linspace(-10,10,n)';
A = [sin(x) sin(2*x) sin(3*x) sin(4*x)];
y1 = A*[2;3;7;5];
y2 = A*[8;2;6;9];
plot(x,y1,x,y2)
```

But a set of matrix-vector products that involve the same matrix is "secretly" a single matrix-matrix product:

$$\left. \begin{array}{c} \begin{bmatrix} 1 & 2 \\ 3 & 4 \end{bmatrix} \begin{bmatrix} 5 \\ 7 \end{bmatrix} = \begin{bmatrix} 19 \\ 43 \end{bmatrix} \\ \\ \begin{bmatrix} 1 & 2 \\ 3 & 4 \end{bmatrix} \begin{bmatrix} 6 \\ 8 \end{bmatrix} = \begin{bmatrix} 22 \\ 50 \end{bmatrix} \end{array} \right\} \equiv \begin{bmatrix} 1 & 2 \\ 3 & 4 \end{bmatrix} \begin{bmatrix} 5 & 6 \\ 7 & 8 \end{bmatrix} = \begin{bmatrix} 19 & 22 \\ 43 & 50 \end{bmatrix}$$

Since MATLAB supports matrix-matrix multiplication, our script transforms to

```
n = 200;
x = linspace(-10,10,n)';
A = [sin(x) sin(2*x) sin(3*x) sin(4*x)];
y = A*[2 8;3 2;7 6;5 9];
plot(x,y(:,1),x,y(:,2))
```

But the `plot` function can accept matrix arguments. The command

```
plot(x,y(:,1),x,y(:,2))
```

is equivalent to

```
plot(x,y)
```

and so we obtain

```
% Script File: SumOfSines2
%
% Plots the functions
%           f(x) = 2sin(x) + 3sin(2x) + 7sin(3x) + 5sin(4x)
%           g(x) = 8sin(x) + 2sin(2x) + 6sin(3x) + 9sin(4x)

% across the interval [-10,10].
   close all; clc
   n = 200;
   x = linspace(-10,10,n)';
   A = [sin(x) sin(2*x) sin(3*x) sin(4*x)];
   y = A*[2 8;3 2;7 6;5 9];
   plot(x,y)
```

In general, plotting a matrix against a vector is the same thing as plotting each of the matrix columns against the vector. Of course, the row dimension of the matrix must equal the length of the vector.

It is also possible to plot one matrix against another. If X and Y have the same size, then the corresponding columns will be plotted against each other with the command plot(X,Y).

Finally, we mention the "backslash" operator that can be invoked whenever the solution to a linear system of algebraic equations is required. For example, suppose we want to find scalars $\alpha_1, \ldots, \alpha_4$ so that if

$$f(x) = \alpha_1 \sin(x) + \alpha_2 \sin(2x) + \alpha_3 \sin(3x) + \alpha_4 \sin(4x)$$

then $f(1) = -2$, $f(2) = 0$, $f(3) = 1$, and $f(4) = 5$. These four stipulations imply

$$\begin{array}{ccccccccc}
\alpha_1 \sin(1) & + & \alpha_2 \sin(2) & + & \alpha_3 \sin(3) & + & \alpha_4 \sin(4) & = & -2 \\
\alpha_1 \sin(2) & + & \alpha_2 \sin(4) & + & \alpha_3 \sin(6) & + & \alpha_4 \sin(8) & = & 0 \\
\alpha_1 \sin(3) & + & \alpha_2 \sin(6) & + & \alpha_3 \sin(9) & + & \alpha_4 \sin(12) & = & 1 \\
\alpha_1 \sin(4) & + & \alpha_2 \sin(8) & + & \alpha_3 \sin(12) & + & \alpha_4 \sin(16) & = & 5
\end{array}$$

That is,

$$\begin{bmatrix} \sin(1) & \sin(2) & \sin(3) & \sin(4) \\ \sin(2) & \sin(4) & \sin(6) & \sin(8) \\ \sin(3) & \sin(6) & \sin(9) & \sin(12) \\ \sin(4) & \sin(8) & \sin(12) & \sin(16) \end{bmatrix} \begin{bmatrix} \alpha_1 \\ \alpha_2 \\ \alpha_3 \\ \alpha_4 \end{bmatrix} = \begin{bmatrix} -2 \\ 0 \\ 1 \\ 5 \end{bmatrix}$$

Here is how to set up and solve this 4-by-4 linear system:

```
X = [1 2 3 4 ; 2 4 6 8 ; 3 6 9 12 ; 4 8 12 16];
Z = sin(X);
f = [-2; 0; 1; 5]
alpha = Z\f
```

Observe that `sin` applied to a matrix returns the matrix of corresponding sine evaluations. This is typical of many of MATLAB's built-in functions. For linear system solving, the backslash operator requires the matrix of coefficients on the left and the right hand side vector (as a column) on the right. The solution to the preceding example is

$$\alpha = \begin{bmatrix} -0.2914 \\ -8.8454 \\ -18.8706 \\ -11.8279 \end{bmatrix}$$

Problems

P1.2.1 Suppose z = [10 40 20 80 30 70 60 90]. Indicate the vectors that are specified by z(1:2:7), z(7:-2:1), and z([3 1 4 8 1]).

P1.2.2 Suppose z = [10 40 20 80 30 70 60 90]. What does this vector look like after each of these commands? then

```
z(1:2:7) = zeros(1,4)
z(7:-2:1) = zeros(1,4)
z([3 4 8 1]) = zeros(1,4)
```

P1.2.3 Given that the commands

```
x = linspace(0,1,200);
y = sqrt(1-x.^2);
```

have been carried out, show how to produce a plot of the circle $x^2 + y^2 = 1$ without any additional square roots or trigonometric evaluations.

P1.2.4 Produce a single plot that displays the graphs of the functions $\sin(kx)$ across $[0, 2\pi]$, $k = 1{:}5$.

P1.2.5 Write a MATLAB script that plots the functions x, x^2, x^3, ..., x^m across the interval $[0, 1]$. All the plots should appear in the same window. Assume that m is a positive integer.

P1.2.6 Assume that x is an initialized MATLAB array and that m is a positive integer. Using the `ones` function, the pointwise array multiply operator `.*`, and MATLAB's ability to scale and add arrays, write a fragment that computes an array y with the property that the ith component of y has the following value:

$$y_i = \sum_{k=0}^{n} \frac{x_i^k}{k!}$$

P1.2.7 Write a MATLAB fragment to plot the following ellipses in the same plot window:

$$\text{Ellipse 1:} \quad x_1(t) = 3 + 6\cos(t) \quad y_1(t) = -2 + 9\sin(t)$$
$$\text{Ellipse 2:} \quad x_2(t) = 7 + 2\cos(t) \quad y_2(t) = 8 + 6\sin(t)$$

P1.2.8 The sine function has period 2π. Consider the following MATLAB script:

```
x = linspace(0,2*pi);
y = sin(x);
plot(x/2,y)
hold on
for k=1:3
   plot((k*pi)+x/2,y)
end
hold off
```

What function does this script plot? Over what range of x-values does the plot range?

P1.2.9 Assume that x, y, and z are MATLAB arrays initialized as follows:

```
x = linspace(0,2*pi,100);
y = sin(x);
z = exp(-x);
```

Write a MATLAB fragment that plots the function $e^{-x}\sin(x)$ across the interval $[0, 4\pi]$. To receive full credit, the fragment should not involve any additional calls to sin or exp. Exploit the fact that sin has period 2π and that the exponential function satisfies $e^{a+b} = e^a e^b$.

P1.2.10 Consider the following grid:

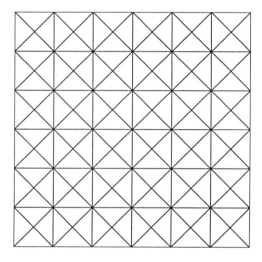

Assume that the MATLAB variables x0, y0, s, and n are initialized as follows:

x0	x coordinate of the lower left corner
y0	y coordinate of the lower left corner
s	the spacing between the horizontal (or vertical) lines
n	the size of the grid (This is 6 in the preceding example.)

Assume that the hold toggle is on and ignore axis settings. **(a)** Write a MATLAB script that draws the outer perimeter of the grid (i.e., the square boundary). **(b)** Write a MATLAB script that draws all the vertical line segments in the grid. **(c)** Write a MATLAB script that draws all the line segments in the grid that have negative slope.

1.3 Building Exploratory Environments

A consequence of MATLAB's ease of use and versatility is that it encourages the exploration of mathematical and algorithmic ideas. Many computational scientists like to precede the rigorous analysis of a problem with MATLAB-based experimentation. We use three examples to show this, learning many new features of the system as we go along.

1.3.1   The Up/Down Sequence

Suppose x_1 is a given positive integer and that for $k \geq 1$ we define the sequence $x_1, x_2, \ldots$ as follows:

$$x_{k+1} = \begin{cases} x_k/2 & \text{if } x_k \text{ is even} \\ 3x_k + 1 & \text{if } x_k \text{ is odd} \end{cases}$$

Thus, if $x_1 = 7$, then the following sequence unfolds:

$$7, 22, 11, 34, 17, 52, 26, 13, 40, 20, 10, 5, 16, 8, 4, 2, 1, 4, 2, 1, 4, 2, 1, \ldots$$

We will call this the *up/down* sequence for obvious reasons. Note that it cycles once the value of 1 is reached. A number of interesting questions are suggested:

- Does the sequence always reach the cycling stage?

- Let n be the smallest index for which $x_n = 1$. How does n behave as a function of the initial value x_1?

- Are there any systematic patterns in the sequence worth noting?

Our goal is to develop a script file that can be used to shed light on these and related issues.

We start with a script that solicits a starting value and then generates the sequence, assembling the values in a vector x:

```
x(1) = input('Enter initial positive integer:');
k = 1;
while (x(k) ~= 1)
   if rem(x(k),2) == 0
      x(k+1) = x(k)/2;
   else
      x(k+1) = 3*x(k)+1;
   end
   k = k+1;
end
```

The input command is used to set up x(1). It has the form

$$\text{input}(\langle string\ message \rangle)$$

and prompts for keyboard input. For example,

```
Enter initial positive integer:
```

Whatever number you type, it is assigned to x(1).

After x(1) is initialized, the generation of the sequence takes place under the auspices of a while-loop. Each pass through the loop requires a test of the current x(k) in accordance with the rule for x(k+1) given earlier. This is handled by an if-then-else.

Let's look at the details. In MATLAB, a test of the form x(k) == 1 renders a zero if it is true and a 1 if it is false.[1] All the usual comparisons are supported:

---

[1] Remember, MATLAB v.4 has only one type, the complex matrix. There is no Boolean type.

Notation	Meaning
<	less than
<=	less than or equal
==	equal
>=	greater than or equal
>	greater than
~=	not equal

A while-loop has the form

```
while (⟨Condition⟩)
    ⟨Statements⟩
end
```

An if-then-else is structured as follows:

```
if ⟨Condition⟩
    ⟨Statements⟩
else
    ⟨Statements⟩
end
```

Both of these control structures operate in the usual way. The condition is numerically valued, and is interpreted as *true* if it is nonzero.

The remainder function rem is used to check whether or not x(k) is even. Assuming that a and b are positive integers, a call of the form rem(a,b) returns the remainder when b is divided into a.

Now one of the things we do not know is whether or not the up/down sequence reaches 1. To guard against the production of an unacceptably large x-vector, we can put a limit on how many terms to generate. Setting that limit to 500 and presizing x to that length, we obtain

```
x = zeros(500,1);
x(1) = input('Enter initial positive integer:');
k = 1;
while ((x(k) ~= 1) & (k < 500))
    if rem(x(k),2) == 0
        x(k+1) = x(k)/2;
    else
        x(k+1) = 3*x(k)+1;
    end
    k = k+1;
end
n = k;
x = x(1:n);
```

The index of the first sequence member that equals 1 is assigned to n and x is "trimmed" to that length with the assignment x = x(1:n). Notice the use of the *and* operator & in the while-loop condition. The and, or, and not operations are all possible in MATLAB :

Notation	Meaning
&	and
|	or
~	not
xor	exclusive or

The usual definitions apply with the understanding that 1 and 0 are used for true and false respectively. Thus (x(k) == 1) & (k < 500)) has the value of 1 if x(k) equals 1 and k is strictly less than 500. If neither of these conditions are satisfied, then the logical expression equals 0.

Computing x(1:n) brings us to the stage where we must decide how to display it and its properties. Of course, we could display the vector simply by leaving off the semicolon in x = x(1:n);. Alternatively, we can make use of sprintf's vectorizing capability:

```
disp(sprintf('%-5.0f',x))
```

When a vector like x is passed to sprintf, all its values are printed using the designated formats. (It just keeps cycling through the format string until every vector component is processed. The minus sign in front of the 5.0f left justifies the display of the value in the 5-character wide field.

Among the numerical properties of x that are interesting are the maximum value and the number of integers $\leq x_1$ that are "hit" by the up/down process:

```
[xmax,imax] = max(x);
disp(sprintf('\n x(%1.0f) = %1.0f is the max.',imax,xmax))
density = sum(x<=x(1))/x(1);
disp(sprintf(' The density is %5.3f.',density))
```

When the max function is applied to a vector, it returns the maximum value and the index where it occurs. It is also possible to use max in an expression. For example,

```
GrowthFactor = max(x)/x(1)
```

assigns to GrowthFactor the ratio of the largest value in x to x(1). Notice the use of the 1.0f format. For integers greater than one digit in length, extra space is accorded as necessary. This ensures that there is no gap between the displayed subscript and the right parenthesis, a small aesthetic point.

The assignment to density requires two explanations. First, it is legal to compare vectors in MATLAB . The comparison x<=x(1) returns a vector of 0's and 1's that is the same size as x. If x(k) <= x(1) is true, then the kth component of this vector is one. The sum function applied to a vector sums its entries. Thus sum(x<=x(1)) is precisely the number of components in x that are less than or equal to x(1).

Graphical display is also in order and can help us appreciate the "flow of events" as the sequence winds its way to unity:

```
close all
figure
plot(x)
title(sprintf('x(1) = %1.0f, n = %1.0f',x(1),n));
figure
```

```
plot(-sort(-x))
title('Sequence values sorted.')
I = find(rem(x(1:n-1),2));
if length(I)>1
   figure
   plot((1:n),zeros(1,n),I+1,x(I+1),I+1,x(I+1),'*')
   title('Local Maxima')
end
```

This script involves a number of new features. First, the reference `plot(x)` plots the components of x against their indices. It is equivalent to `plot((1:n)',x)`.

Second, the `sort` function is used to produce a plot of the sequence with its values ordered from large to small. If v is a vector with length m, then u = `sort(v)` permutes the values in v and assigns them to u so that

$$u_1 \leq u_2 \leq u_3 \leq \cdots \leq u_m$$

With the minus signs in `-sort(-x)` we produce a "big-to-little" sort.

Third, the expression `rem(x(1:n-1),2) == 1` returns a 0-1 vector that designates which components of `x(1:n-1)` are odd. The function `rem`, like many of MATLAB's built-in functions, accepts vector arguments and merely returns a vector of the function applied to each of the components. The `find` function returns a vector of subscripts that designate which entries in a vector are nonzero. Here are some tables that should clarify these issues:

k	x(1:n-1)	rem(x(1:n-1))
1	17	1
2	52	0
3	26	0
4	13	1
5	40	0
6	20	0
7	10	0
8	5	1
9	16	0
10	8	0
11	4	0
12	2	0

k	I = find(rem(x(1:n-1)))
1	1
2	4
3	8

If the vector I is nonempty, then a plot of I+1 is produced showing the pattern of the sequence's "local maxima." (The vector I+1 contains the indices of values in `x(1:n-1)` that are produced by the "up operation" $3x_k + 1$.)

The last thing to discuss is `figure`. In all prior examples, our plots have appeared in a single window. New plots erase old ones. But with each reference to `figure`, a new window is opened and these are indexed from 1. Thus `figure(1)` refers to a plot of x, `figure(2)` designates the plot of x sorted, and if I is nonempty, then `figure(3)` contains a plot of its local maxima. The `close all` statement clears all windows and ensures that the figure indexing starts at 1.

Putting it all together, here is a script that can be used to explore a single up/down sequence with a user-specified starting value:

```
% Script File: UpDown
%
% Generates a column vector x(1:n) of positive integers
% where x(1) is solicited and
%
%            x(k+1) = x(k)/2    if x(k) is even.
%            x(k+1) = 3x(k)+1   if x(k) is odd.
%
% The value of n is either 500 or the first index with the
% property that x(n) = 1, whichever comes first.

    x = zeros(500,1);
    x(1) = input('Enter initial positive integer:');
    k = 1;
    while ((x(k) ~= 1) & (k < 500))
        if rem(x(k),2) == 0
            x(k+1) = x(k)/2;
        else
            x(k+1) = 3*x(k)+1;
        end
        k = k+1;
    end
    n = k;
    x = x(1:n);

    clc
    disp(sprintf('x(1:%1.0f) = \n',n))
    disp(sprintf('%-8.0f',x))
    [xmax,imax] = max(x);
    disp(sprintf('\n x(%1.0f) = %1.0f is the max.',imax,xmax))
    density = sum(x<=x(1))/x(1);
    disp(sprintf(' The density is %5.3f.',density))

    close all
    figure
    plot(x)
    title(sprintf('x(1) = %1.0f, n = %1.0f',x(1),n));
    figure
    plot(-sort(-x))
    title('Sequence values sorted.')
    I = find(rem(x(1:n-1),2));
    if length(I)>1
        figure
        plot((1:n),zeros(1,n),I+1,x(I+1),I+1,x(I+1),'*')
        title('Local Maxima')
    end
```

Intuition about the up/down sequence could be acquired by repeatedly running this script file. However, to make this enterprise more convenient, we write another script file that invokes UpDown:

```
% Script File: RunUpDown
%
% Environment for studying the up/down sequence.
% Stores selected results in file UpDownOutput.

    while(input('Another Example? (1=yes, 0=no)'))
       diary UpDownOutput
       UpDown
       diary off
       if (input('Keep Output? (1=yes, 0=no)')~=1)
          delete UpDownOutput
       end
    end
```

By using this script we can keep trying new starting values until one of special interest is found. The while-loop keeps running as long as you want to test another starting value. Before UpDown is run, the diary UpDownOutout command creates a file called UpDownOutput. Everything that is now written to the command window during the execution of UpDown is now also written to UpDownOutPut. After UpDown is run, we turn off this feature with diary off. The script then asks if the output should be kept. If not, then the file UpDownOutput is deleted. Note that it is possible to record several possible runs of UpDown, but as soon as the if condition is true, everything is erased. The advantage of writing output to a file is that it can then be edited to make it look nice. For example,

```
For the starting value x(1) = 293, the UpDown sequence is

  293   880   440   220   110    55   166    83   250
  125   376   188    94    47   142    71   214   107
  322   161   484   242   121   364   182    91   274
  137   412   206   103   310   155   466   233   700
  350   175   526   263   790   395  1186   593  1780
  890   445  1336   668   334   167   502   251   754
  377  1132   566   283   850   425  1276   638   319
  958   479  1438   719  2158  1079  3238  1619  4858
 2429  7288  3644  1822   911  2734  1367  4102  2051
 6154  3077  9232  4616  2308  1154   577  1732   866
  433  1300   650   325   976   488   244   122    61
  184    92    46    23    70    35   106    53   160
   80    40    20    10     5    16     8     4     2
    1

x(84) = 9232 is the max and the density is 0.181.
```

The figures from the final UpDown run are available for printing as well.

1.3.2   Random Processes

Many simulations performed by computational scientists involve random processes. In order to implement these on a computer, it is necessary to be able to generate sequences of random numbers. In MATLAB this is done with the built-in functions **rand** and **randn**. The command x = **rand(1000,1)** creates a length-1000 column vector of real numbers chosen randomly from the interval $(0,1)$. The uniform$(0,1)$ distribution is used, meaning that if $0 < a < b < 1$, then the fraction of values that fall in the range $[a,b]$ will be about $b - a$. The **randn** function should be used if a sequence of normally distributed random numbers is desired. The underlying probability distribution is the normal$(0,1)$ distribution. A brief, graphically oriented description of these functions should clarify their statistical properties.

Histograms are a common way of presenting statistical data. Here is a script that illustrates **rand** and **randn** using this display technique:

```
% Script File: Histograms
%
% Histograms of rand(1000,1) and randn(1000,1).
%
    close all
    clc

    subplot(2,1,1)
    x = rand(1000,1);
    hist(x,30)
    axis([-1 2 0 60])
    title('Distribution of Values in rand(1000,1)')
    xlabel(sprintf('Mean = %5.3f. Median = %5.3f.',mean(x),median(x)))

    subplot(2,1,2)
    x = randn(1000,1);
    hist(x,linspace(-2.9,2.9,100))
    title('Distribution of Values in randn(1000,1)')
    xlabel(sprintf('Mean = %5.3f. Standard Deviation = %5.3f',mean(x),std(x)))
```

(See Fig 1.9.) Notice that **rand** picks values uniformly from $[0,1]$ while the distribution of values in **randn(1000,1)** follows the familiar "bell shaped curve." The mean, median, and standard deviation functions **mean**, **median**, and **std** are referenced. The histogram function **hist** can be used in several ways and the script shows two of the possibilities. A reference like **hist(x,30)** reports the distribution of the x-values according to where they "belong" with respect to 30 equally spaced bins spread across the interval $[\min(x), \max(x)]$. The bin locations can also be specified by passing **hist** a vector in the second parameter position (e.g., **hist(x,linspace(-2.9,2.9,100))**). This is done for the histogram of the normally distributed data.

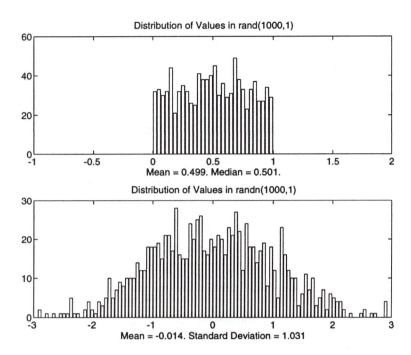

FIGURE 1.9 The uniform and normal distributions

To appreciate further the distinction between **rand** and **randn**, we run the following script which displays 10,000 points that are randomly generated by the two functions:

```
% Script File: Clouds
%
% 2-dimensional pictures of the uniform and normal distributions.

   close all; clc
   Points = rand(1000,2);
   subplot(1,2,1)
   plot(Points(:,1),Points(:,2),'.')
   title('Uniform distribution.')
   axis([0 1 0 1])
   axis('square')

   Points = randn(1000,2);
   subplot(1,2,2)
   plot(Points(:,1),Points(:,2),'.')
   title('Normal distribution.')
   axis([-3 3 -3 3])
   axis('square')
```

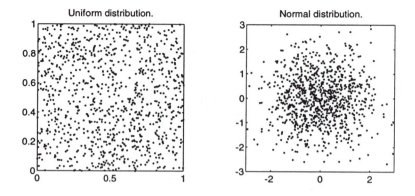

FIGURE 1.10 Two-dimensional (2D) uniform and normal distributions

(See Fig 1.10.) In this example, the calls to **rand** and **randn** are used to produce two-column matrices of random numbers.

Building on **rand** and **randn** through translation and scaling, it is possible to produce random sequences with specified means and variances. For example,

```
x = 10 + 5*rand(n,1);
```

generates a sequence of uniformly distributed numbers from the interval $(10, 15)$. Likewise,

```
x = 10 + 5*randn(n,1);
```

produces a sequence of normally distributed random numbers with mean 10 and standard deviation 5.

It is possible to generate random integers using **rand** (or **randn**) and the **floor** function. The command

```
x = rand(n,1);
z = floor(6*x+1);
```

computes a random vector of integers selected from $\{1, 2, 3, 4, 5, 6\}$ and assigns them to **z**. This is because **floor** rounds to $-\infty$. The command

```
z = ceil(6*x)
```

is equivalent because **ceil** rounds toward $+\infty$. In either case, the vector **z** looks like a recording of n dice throws. Notice that **floor** and **ceil** accept vector arguments and return vectors of the

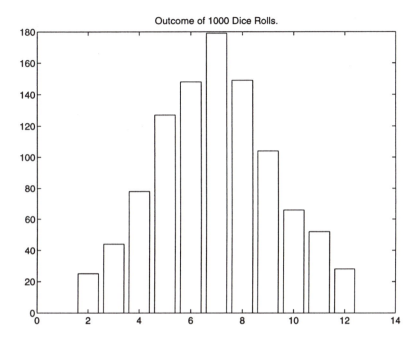

FIGURE 1.11 Rolling a pair of dice

same size. (See also `fix` and `round`.) Here is a script that simulates 1000 rolls of a pair of dice, displaying the outcome in histogram form:

```
% Script File: Dice
%
% Simulates 1000 rollings of a pair of dice.

    close all
    clc

    First  = 1 + floor(6*rand(1000,1));
    Second = 1 + floor(6*rand(1000,1));
    Throws = First + Second;
    hist(Throws, linspace(2,12,11));
    title('Outcome of 1000 Dice Rolls.')
```

(See Fig 1.11.)

Random simulations can be used to answer "nonrandom" questions. Suppose we throw n darts at the circle-in-square target depicted in Fig 1.12. Assume that the darts land anywhere on the square with equal probability and that the square has side 2 and center $(0,0)$. After a large number of throws, the fraction of the darts that land inside the circle should be approximately

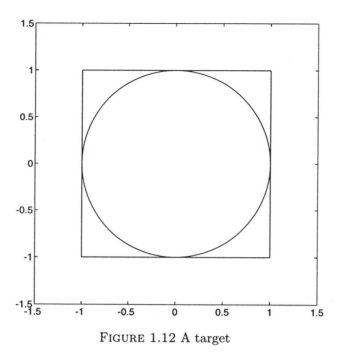

FIGURE 1.12 A target

equal to $\pi/4$, the ratio of the circle area to the square's area. Thus

$$\pi \approx \frac{\text{Number of Throws Inside the Circle}}{\text{Total Number of Throws}} \cdot 4$$

By simulating the throwing of a large number of darts, we can produce an estimate of π. Here is a script file that does just that:

```
% Script File Darts
%
% Displays random dart throws.
%
    close all; clc
    rand('seed',.123456)
    NumberInside = 0;
    PiEstimate = zeros(500,1);
    for k=1:500
        x = -1+2*rand(100,1); y = -1+2*rand(100,1);
        NumberInside = NumberInside + sum(x.^2 + y.^2 <= 1);
        PiEstimate(k) = (NumberInside/(k*100))*4;
    end
    plot(PiEstimate)
    title(sprintf('Monte Carlo Estimate of Pi = %5.3f',PiEstimate(500)));
    xlabel('Hundreds of Trials')
```

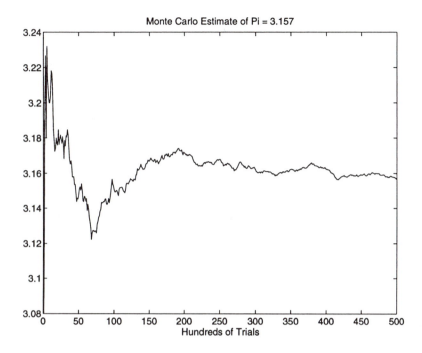

FIGURE 1.13 A Monte Carlo estimate of π

(See Fig 1.13.) Notice that the estimated values are gradually improving with n, but that the "progress" towards 3.14159... is by no means steady or fast. Simulation in this spirit is called *Monte Carlo*.

The `any` and `all` functions indicate whether any or all of the components of a vector are nonzero. Thus, if x and y are vectors of the same length, then

```
a = any( x.^2 + y.^2 <= 1);
```

assigns to a the value "1" if there is at least one (x_i, y_i) in the unit circle and "0" otherwise. Similarly,

```
b = all( x.^2 + y.^2 <= 1);
```

assigns "1" to a if all the (x_i, y_i) are in the unit circle and assigns "0" otherwise.

1.3.3 Polygon Smoothing

If x and y are column $n + 1$-vectors with the property that $x_1 = x_{n+1}$ and $y_1 = y_{n+1}$, then

```
plot(x,y,x,y,'*')
```

displays the polygon obtained by connecting the points $(x_1, y_1), \ldots, (x_{n+1}, y_{n+1})$ in order. If we compute

```
xnew = [(x(1:n)+x(2:n+1))/2;(x(1)+x(2))/2];
ynew = [(y(1:n)+y(2:n+1))/2;(y(1)+y(2))/2];
plot(xnew,ynew)
```

then a new polygon is displayed that is obtained by connecting the side midpoints of the original polygon. The following script enables us to explore what happens when the original polygon is repeatedly smoothed:

```
% Script File: Smooth
%
% Solicits n, draws an n-gon, and then smooths it.

    close all
    n = input('Enter the number of edges:');
    plot([],[])
    title(sprintf('Click in %2.0f points.',n))
    axis([0 1 0 1])
    axis('square','equal')
    hold on
    x = zeros(n,1);
    y = zeros(n,1);
    for k=1:n
       [x(k) y(k)] = ginput(1);
      plot(x(1:k),y(1:k), x(1:k),y(1:k),'*')
    end
    x = [x;x(1)];
    y = [y;y(1)];
    plot(x,y,x,y,'*')
    title('The Original Polygon')
    hold off
    k=0;
    xlabel('Click anywhere in window to smooth.')
    [a,b] = ginput(1);
    while (0<=a) & (a<=1) & (0<=b) & (b<=1)
       k = k+1;
       x = [(x(1:n)+x(2:n+1))/2;(x(1)+x(2))/2];
       y = [(y(1:n)+y(2:n+1))/2;(y(1)+y(2))/2];
       figure
       plot(x,y,x,y,'*')
       axis([0 1 0 1])
       axis('square','equal')
       title(sprintf('Number of Smoothings = %1.0f',k))
       xlabel('Click anywhere in window to smooth.')
       [a,b] = ginput(1);
    end
```

The script features two new ways that we can interact with the figure window. The `ginput` command permits mouseclick input. It returns the x-y-coordinates of the click with respect to the current axis. In `Smooth`, the vertices of the original polygon are obtained using `ginput`. The assignment `[x(k),y(k)] = ginput(1)` places the coordinates of the kth vertex in `x(k)` and `y(k)`. The `for`-loop displays the sides of the polygon as it is "built up." If we did not care about this kind of graphical feedback as we click in the vertices, then the command `[x,y] = ginput(n)` could be used. This just stores the coordinates of the next n mouseclicks in `x` and `y`. Notice how we set up an "empty" figure with a prescribed axis in advance of the `for`-loop that acquires vertices.

The `while`-loop oversees the repeated smoothing and this brings us to the second new feature that is introduced in `smooth`, the function `text`. To bring about termination of the loop with mouseclick input, the coordinates of the the mouseclick are checked to see if they fall within the plot window. (See Fig 1.14.) If they do not, then the smoothing iteration is brought to a close.

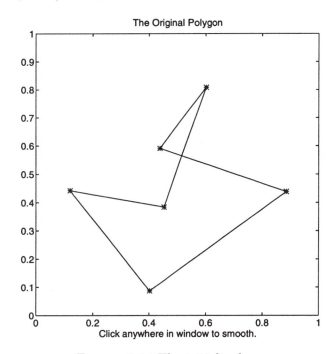

FIGURE 1.14 The initial polygon

Once the execution of the script is completed, you can review the evolution of the smoothed polygons by using `figure`. For example, for the initial polygon shown in Fig 1.14, if we enter the command `figure(2)`, then the polygon obtained after two smoothings is displayed. (See FIG 1.15.) Finally, we mention that the axes in `Smooth` are forced to be equally scaled in both the x- and y-directions. This ensures that the displayed polygons are not distorted. To guarantee the x- and y-values in the window range from 0 to 1, we force the window to be square. These autoscaling "overrides" are brought about by the command `axis('square','equal')`. The following command pair is equivalent:

```
axis('square')
axis('equal')
```

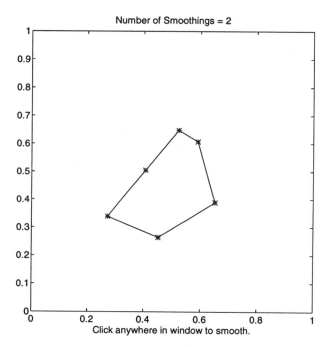

FIGURE 1.15 A smoothed polygon

Problems

P1.3.1 Let $x_{g(m)}$ be the first 1 in the up/down sequence with starting value m. Plot the function $g(m)$ for $m = 1{:}200$.

P1.3.2 What is the probability that the quadratic function $f(x) = ax^2 + bx + c$ has complex roots given that the three coefficients a, b, and c are random variables with uniform$(0, 1)$ distribution? What if they have normal$(0, 1)$ distribution?

P1.3.3 Write a simulation that estimates the volume of $\{(x_1, x_2, x_3, x_4) : x_1^2 + x_2^2 + x_3^2 + x_4^2 \leq 1\}$, the unit sphere in $\mathbf{R}^4$.

P1.3.4 Let S_0 be the set of points that are closer to the point $(.2, .4)$ than to an edge of the square with vertices $(1, 1)$, $(-1, 1)$, $(-1, -1)$, and $(1, -1)$. Write a MATLAB script that estimates the area of S_0.

1.4 Error

Errors of all kinds abound in scientific computation. Rounding errors attend floating point arithmetic, terminal screens are granular, analytic derivatives are approximated with divided differences, a polynomial is used in lieu of the sine function, the data acquired in a lab are correct to only three significant digits, etc. Life in computational science is like this, and we have to build up a facility for dealing with it. In this section we focus on the mathematical errors that arise through discretization and the rounding errors that arise due to finite precision arithmetic.

1.4.1   Absolute and Relative Error

If $\tilde{x}$ approximates a scalar x, then the *absolute error* in $\tilde{x}$ is given by $|\tilde{x} - x|$ and while the *relative error* is given by $|\tilde{x} - x|/|x|$. If the relative error is about 10^{-d}, then $\tilde{x}$ has approximately d correct significant digits in that there exists a number τ having the form

$$\tau = \pm(\underbrace{.00\ldots0}_{d \text{ times}} n_{d+1} n_{d+2} \ldots) \times 10^e$$

so that $\tilde{x} = x + \tau$.

As an exercise in relative and absolute error, let's examine the quality of the Stirling approximation

$$S_n = \sqrt{2\pi n}\left(\frac{n}{e}\right)^n \approx n! = 1 \cdot 2 \cdots n$$

where $e = \exp(1)$. Here is a script file that produces a table of errors:

```
% Script File: Stirling
%
% Prints a table showing error in Stirling's formula for n!
%

   close all; clc
   disp( ' ')
   disp('                           Stirling        Absolute    Relative')
   disp('    n          n!      Approximation         Error       Error')
   disp('-------------------------------------------------------------------')

   e=exp(1);
   nfact=1;
   for n=1:13
      nfact = n*nfact;
      s = sqrt(2*pi*n)*((n/e)^n);
      abserror = abs(nfact - s);
      relerror = abserror/nfact;
      s1 = sprintf('  %2.0f   %10.0f    %13.2f',n,nfact,s);
      s2 = sprintf('   %13.2f     %5.2e',abserror,relerror);
      disp([s1 s2])
   end
```

Notice how the strings s1 and s2 are concatenated before they are displayed. In general, you should think of a string as a row vector of characters. Concatenation is then just a way of obtaining a new row vector from two smaller ones. This is the logic behind the required square bracket.

The command clc clears the command window and home moves the cursor to the top. This ensures that the table produced is profiled nicely in the command window. Here it is:

n	n!	Stirling Approximation	Absolute Error	Relative Error
1	1	0.92	0.08	7.79e-02
2	2	1.92	0.08	4.05e-02
3	6	5.84	0.16	2.73e-02
4	24	23.51	0.49	2.06e-02
5	120	118.02	1.98	1.65e-02
6	720	710.08	9.92	1.38e-02
7	5040	4980.40	59.60	1.18e-02
8	40320	39902.40	417.60	1.04e-02
9	362880	359536.87	3343.13	9.21e-03
10	3628800	3598695.62	30104.38	8.30e-03
11	39916800	39615625.05	301174.95	7.55e-03
12	479001600	475687486.47	3314113.53	6.92e-03
13	6227020800	6187239475.19	39781324.81	6.39e-03

1.4.2   Taylor Approximation

The partial sums of the exponential satisfy

$$e^x = \sum_{k=0}^{n} \frac{x^k}{k!} + \frac{e^\eta}{(n+1)!} x^{n+1}$$

for some η in between 0 and x. The mathematics says that if we take enough terms, then the partial sums converge. The script ExpTaylor explores this by plotting the partial sum relative error as a function of n.

```
% Script File ExpTaylor
%
% Plots, as a function of n, the relative error in the
% Taylor approximation 1 + x + x^2/2! +...+ x^n/n! to exp(x).

close all
nTerms = 50;
for x=[10 5 1 -1 -5 -10]
    figure
    error = exp(x)*ones(nTerms,1);
    s = 1;
    term = 1;
    for k=1:50
        term = x.*term/k;
        s = s+ term;
        error(k) = abs(error(k) - s);
    end
```

```
        relerr = error/exp(x);
        semilogy(1:nTerms,relerr)
        ylabel('Relative Error in Partial Sum.')
        xlabel('Order of Partial Sum.')
        title(sprintf('x = %5.2f',x))
        pause(3)
    end
```

When plotting numbers that vary tremendously in range, it is useful to use `semilog`. It works just like `plot`, only the base-10 log of the y-vector is displayed. `ExpTaylor` produces six figure windows, one each for the six x-values. For example, the $x = 10$ plot is in figure 1. By entering the command `figure(1)`, this plot is "brought up" by making the Figure 1 window the active window. It could then (for example) be printed. (See FIGS. 1.16 and 1.17.)

1.4.3 Rounding Errors

The plots produced by `ExpTaylor` reveal that the mathematical convergence theory does not quite apply. The errors do *not* go to zero as the number of terms in the series increases. In each case, they seem to "bottom out" at some small value. Once that happens, the incorporation of more terms into the partial sum does not make a difference. Moreover, by comparing the plots in FIGS. 1.16 and 1.17, we observe that where the relative error bottoms out depends on x. The relative error for $x = -10$ is much worse than for $x = 10$.

An explanation of this phenomenon requires an understanding of floating point arithmetic. Like it or not, numerical computation involves working with an inexact computer arithmetic system. This will force us to rethink the connections between mathematics and the development of algorithms. Nothing will be simple ever again.

To dramatize this point, consider the plot of a rather harmless looking function: $p(x) = (x-1)^6$. The script `Zoom` graphs this polynomial over increasingly smaller neighborhoods around $x = 1$, but it uses the formula $p(x) = x^6 - 6x^5 + 15x^4 - 20x^3 + 15x_2 - 6x + 1$.

```
% Script File Zoom
%
% Plots (x-1)^6 near x=1 with increasingly refined scale.
% Evaluation via x^6 - 6x^5 + 15x^4 - 20x^3 + 15x^2 - 6x +1
% leads to severe cancellation.

    close all
    k=0;
    for delta = [.1 .01 .008 .007 .005  .003 ]
        x = linspace(1-delta,1+delta,100)';
        y = x.^6 - 6*x.^5 + 15*x.^4 - 20*x.^3 + 15*x.^2  - 6*x + ones(100,1);
        k=k+1;
        subplot(2,3,k)
        plot(x,y,x,zeros(1,100))
        axis([1-delta 1+delta -max(abs(y)) max(abs(y))])
    end
```

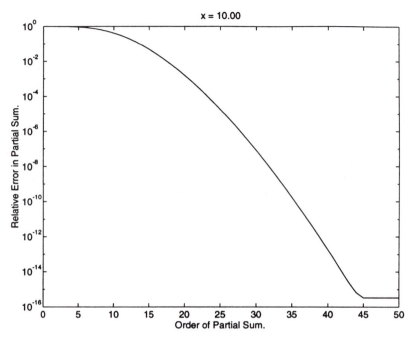

FIGURE 1.16 Taylor approximations to $e^x, x = 10$

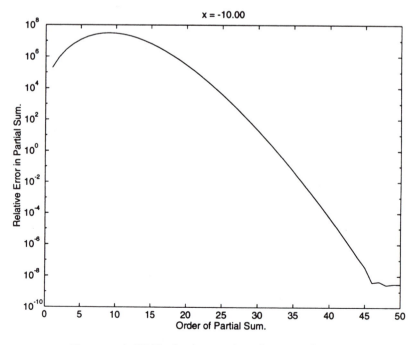

FIGURE 1.17 Taylor approximations to $e^x, x = -10$

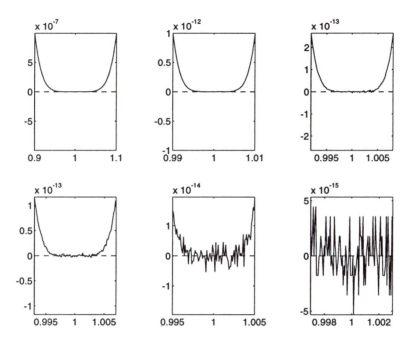

FIGURE 1.18 Plots of $(x-1)^6 = x^6 - 6x^5 + 15x^4 - 20x^3 + 15x^2 - 6x + 1$ near $x = 1$

Notice how the x-axis is plotted and how it is forced to appear across the middle of the window. (See Fig 1.18 for a display of the plots.) As we increase the "magnification," a very chaotic behavior unfolds. It seems that $p(x)$ has thousands of roots!

It turns out that if the plot is based on the formula $(x-1)^6$ instead of its expansion, then the expected graph is displayed and this gets right to the heart of the example. *Algorithms that are equivalent mathematically may behave very different numerically.* The time has come to look at floating point arithmetic.

1.4.4 The Floating Point Numbers

The numbers in a floating point system are defined by a base β, a mantissa length t, and an exponent range $[L, U]$. A nonzero floating point number x has the form

$$x = \pm .b_1 b_2 \ldots b_t \times \beta^e$$

Here, $.b_1 b_2 \ldots b_t$ is the *mantissa* and e is the exponent. The exponent satisfies $L \le e \le U$. The b_i are base-β digits and satisfy $0 \le b_i \le \beta - 1$. Nonzero floating point numbers are normalized if $b_1 \ne 0$. Zero is also a floating point number and we assume that in its representation both the mantissa and exponent are set to zero.

The model floating point system $(\beta, t, L, U) = (10, 2, -1, 2)$ can be used to clarify some of the key features associated with the floating point representation. There are 90 possible normalized positive mantissas (.10, .11, .12, ..., .98, and .99) and four possible exponents (-1, 0, 1, and 2). Accounting for negative numbers and zero, we see that there are $2 \cdot 90 \cdot 4 + 1 = 721$ normalized floating point numbers in total. *The set of floating point numbers is finite.* The smallest positive floating point number m is $.10 \times 10^{-1} = .01$ and the largest M is $.99 \times 10^2 = 99$. The spacing between the floating point numbers varies.

If x is a real number, then we let $\mathrm{fl}(x)$ be the nearest floating point number to x. Think of $\mathrm{fl}(x)$ as the stored version of x. The relative error in $\mathrm{fl}(x)$ depends on the mantissa length.

Theorem 1 *Suppose we are given the set of floating point numbers $F(\beta, t, L, U)$ and that $x \in \mathbb{R}$ satisfies $\beta^L < |x| < \beta^U$. It follows that*

$$\frac{|\mathrm{fl}(x) - x|}{|x|} \leq \frac{1}{2}\beta^{1-t}$$

Proof. Without loss of generality, assume that x is positive and that

$$x = .b_1 b_2 \ldots b_t b_{t+1} \ldots \times \beta^e$$

is its base β expansion with $b_1 \neq 0$. The spacing of the floating point numbers at x is β^{e-t}. Since $\mathrm{fl}(x)$ is the closest floating number to x, we have

$$|\mathrm{fl}(x) - x| \leq \frac{1}{2}\beta^{e-t}$$

From the inequality $\beta^{e-1} < x$ it follows that

$$\frac{|\mathrm{fl}(x) - x|}{|x|} \leq \frac{1}{2}\beta^{1-t} \quad \square$$

The quantity $\beta^{1-t}/2$ is the machine precision and is available in MATLAB through the built-in constant eps. Note that on a base-2 machine eps$= 2^{-t}$.

A reasonable way to model the addition, subtraction, multiplication, or division of two floating point numbers is as follows:

- Perform the operation exactly.

- Put the answer in normalized scientific form.

- Round the mantissa to the allotted number of mantissa digits.

Assume that $(\beta, t, L, U) = (10, 3, -9, 9)$ and consider the addition of x and y where

$$x \;=\; 12.3 \qquad\qquad \texttt{x} = \boxed{+\;|\;1\;|\;2\;|\;3\;\|\;+\;|\;2}$$

$$y \;=\; 5.27 \qquad\qquad \texttt{y} = \boxed{+\;|\;5\;|\;2\;|\;7\;\|\;+\;|\;1}$$

Proceed as follows:

- $12.3 + 5.27 = 17.57$

- $17.57 = .1757 \times 10^2$

- $\texttt{z = x+y} \qquad \texttt{z} = \boxed{+\;|\;1\;|\;7\;|\;6\;\|\;+\;|\;2}$

A consequence of this model of floating point arithmetic is that *rounding errors* usually attend every floating point arithmetic operation.

Some simple while-loop computations can be used to glean information about the underlying floating system. Here is a script that assigns to p the smallest positive integer so $1 + 1/2^p = 1$ in floating point arithmetic:

```
x=1; p=0; y=1; z=x+y;
while x~=z
    y=y/2; p=p+1; z=x+y;
end
```

The finiteness of the exponent range has ramifications too. When a floating point operation renders a nonzero result that is too small to represent, then an *underflow* results. Sometimes these are just set to zero. Sometimes they result in program termination. Here is a script that assigns to q the smallest positive integer so that $1/2^q = 0$ in floating point arithmetic:

```
x=1; q=0;
while x>0
    x=x/2; q=q+1;
end
```

At the other end of the scale, a floating point operation can result in an answer that is too big to represent. When this happens, it is called *floating point overflow* and a special value called inf is produced. Here is a script that assigns to r the smallest positive integer so $2^r = inf$ in floating point arithmetic:

```
x=1; r=0;
while x~=inf
    x=2*x;  r=r+1;
end
```

These computations and others are available in the script file FpFacts. The output depends on the underlying computer.

Problems

P1.4.1 The binomial coefficient n-choose-k is defined by

$$\left(\begin{array}{c} n \\ k \end{array} \right) = \frac{n!}{k!(n-k)!}$$

Let $B_{n,k} = S_n/(S_k S_{n-k})$. Write a script analogous to Stirling that explores the error in $B_{n,k}$ for the cases $(n,k) = (52,2), (52,3), \ldots, (52,13)$. There are no set rules on output except that it should look nice and clearly present the results.

P1.4.2 The sine function has the power series definition

$$\sin(x) = \sum_{k=0}^{\infty} (-1)^k \frac{x^{2k+1}}{(2k+1)!}$$

Write a script SinTaylor analogous to ExpTaylor that explores the relative error in the partial sums.

P1.4.3 Write a script that solicits n and plots both $\sin(x)$ and

$$S_n(x) = \sum_{k=0}^{n} (-1)^k \frac{x^{2k+1}}{(2k+1)!}$$

across the interval $[0, 2\pi]$.

P1.4.4 To affirm your understanding of the floating point representation, what is the largest value of n so that $n!$ can be exactly represented in a floating point number system where $(b, t, L, U) = (2, 24, -100, 100)$. Show your work.

P1.4.5 On a base-2 machine, the distance between 7 and the next largest floating point number is 2^{-12}. What is the distance between 70 and the next largest floating point number?

P1.4.6 Assume that x and y are normalized positive floating point numbers in a base-2 computer with t-bit mantissas. How small can $y - x$ be if $x < 8 < y$?

P1.4.7 Let F be the set of base-2 floating point numbers having t-bit mantissas and exponent range $[-L, L]$. Write a MATLAB function `k = Max10(t,L)` that returns the smallest positive integer k such that 10^k *cannot* be exactly represented as a floating point number from F.

1.5   Designing Functions

An ability to write good MATLAB functions is crucial. Three examples are used to clarify the essential ideas: Taylor series, numerical differentiation, and three-digit floating point simulation.

1.5.1   Four Ways to Compute the Exponential of a Vector

Consider once again the Taylor approximation

$$T_n(x) = \sum_{k=0}^{n} \frac{x^k}{k!}$$

to the exponential e^x. It is possible to write functions in MATLAB , and here is one that encapsulates this approximation:

```
    function y = MyExpF(x,n)
%
% Pre:
%    x = scalar
%    n = positive integer
%
% Post:
%    y = n-th order Taylor approximation to exp(x).
%
    y = 1;
    term = 1;
    for k = 1:n
        term = x*term/k;
        y = y + term;
    end
```

The function itself must be placed in a separate .m file having the same name as the function (e.g., `MyExpF.m`). Once that is done, it can be referenced like any of the built-in functions. Thus the script

```
m = 50;
x = linspace(-1,1,m);
y = zeros(1,m);
exact = exp(x);
k=0;
for n = [4 8 16 20]
   for i=1:m
      y(i) = MyExpF(x(i),n);
   end
   RelErr = abs(exact - y)./exact;
   k=k+1;
   subplot(2,2,k)
   plot(x,RelErr)
  Title(sprintf('n = %2.0f',n))
end
```

plots the relative error in $T_n(x)$ for $n = 4, 8, 16$, and 20 across the interval $[-1, 1]$. (See Fig 1.19.)

When writing a MATLAB function you must adhere to the following rules and guidelines:

- From the example we infer the following general structure for a MATLAB function:

  ```
  function ⟨Output Parameter⟩ = ⟨Name of Function⟩(⟨Input Parameters⟩)
  %
  % ⟨Comments that completely specify the function.⟩
  %
       ⟨function body⟩
  ```

- Somewhere in the function body the desired value must be assigned to the output variable.

- Comments that completely specify the function should be given immediately after the `function` statement. The specification should detail all input value assumptions (the *preconditions*) and what may be assumed about the output value (the *postconditions*).

- The lead block of comments after the `function` statement are displayed when the function is probed using `help` (e.g., `help MyExpF`).

- The input and output parameters are formal parameters. At the time of the call they are replaced by the actual parameters.

- All variables inside the function are local and are not part of the MATLAB workspace.

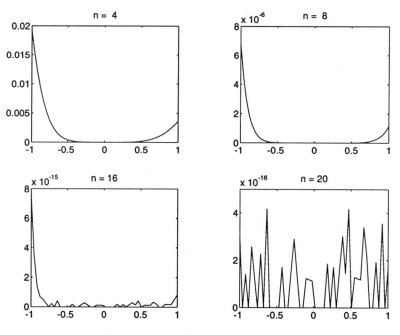

FIGURE 1.19 Relative error in $T_n(x)$.

- If the function file is not in the current directory, then it cannot be referenced unless the appropriate path is established. Type `help path`.

Further experimentation with `MyExpF` shows that if $n = 17$, then full machine precision exponentials are computed for all $x \in [-1, 1]$. With this understanding about the Taylor approximation across $[-1, 1]$, we are ready to develop a "vector version":

```
function y = MyExp1(x)
%
% Pre:
%    x: a column vector.
%
% Post:
%    y: a column vector with the property that
%        y(i) is a Taylor approximation to exp(x(i)).
%
   n = 17;
   p = length(x);
   y = ones(p,1);
   for i=1:p
      y(i) = MyExpF(x(i),n);
   end
```

This example shows several things: (1) A MATLAB function can have vector arguments and can return a vector, (2) the `length` function can be used to determine the size of an input vector, (3) one function can reference another. Here is a script that references `MyExp1`:

```
x = linspace(-1,1,50);
exact = exp(x);
RelErr = abs(exact - MyExp1(x')')./exact;
```

Notice the transpose that is required to ensure that the vector passed to `MyExp1` is a column vector. The other transpose is required to make `MyExp1(x')` a row vector so that it can be combined with `exact`.

Here is an alternative implementation that is not sensitive to the shape of `x`:

```
  function y = MyExp2(x)
%
% Pre:
%    x: a vector.
%
% Post:
%    y: a vector the same shape as x with the property that
%         y(i) is a Taylor approximation to exp(x(i)).
%
  y = ones(size(x));
  n = 17;
  term = ones(size(x));
  for k=1:n
     term = x.*term/k;
     y = y + term;
  end
```

The expression `ones(size(x))` creates a vector of ones that is exactly the same shape as `x`. In general, the command `[p,q] = size(A)` returns the number of rows and columns in `A` in `p` and `q`, respectively. If such a 2-vector is passed to `ones`, then the appropriate matrix of ones is established. (The same comment applies to `zeros`.) The new implementation "doesn't care" whether `x` is a row or column vector. The script

```
x = linspace(-1,1,50);
exact = exp(x);
RelErr = abs(exact - MyExp2(x))./exact;
```

produces a vector of relative error exactly the same size as `x`.

Notice the use of pointwise multiplication. In contrast to `MyExp1` which computes the component-level exponentials one at a time, `MyExp2` computes them "at the same time." In general, MATLAB runs faster in *vector mode*. Here is a script that quantifies this statement by *benchmarking* these two functions:

```
disp(' Length(x)        MyExp1(x)       MyExp2(x) ')
disp('                    Time            Time')
disp('--------------------------------------------------')
for L = 50:50:300
   xL = linspace(-1,1,L);
   tstart = clock;
   y = MyExp1(xL);
   tfinish = clock;
   t1 = etime(tfinish,tstart);
   tstart = clock;
   y = MyExp2(xL);
   tfinish = clock;
   t2 = etime(tfinish,tstart);
   disp(sprintf('%6.0f  %13.2f  %13.2f  %13.2f  ',L,t1,t2))
end
```

The script makes use of `clock` and `etime`. The function `clock` returns a 6-vector that specifies the time. The function `etime` takes two such time "snapshots" and returns the elapsed time in seconds. Here are some sample results:

Length(x)	MyExp1(x) Time	MyExp2(x) Time
50	2.67	0.10
100	5.67	0.13
150	7.92	0.17
200	10.58	0.20
250	13.23	0.20
300	15.90	0.28

It is important to stress that these are *sample* results. Different timings would result on different computers. Moreover, because the computer clocks are discrete, different timings would result if the benchmarking script is rerun.

The `for`-loop implementations in `MyExp1` and `MyExp2` are flawed in two ways. First, the value of n chosen is machine dependent. A different n would be required on a computer with a different machine precision. Second, the number of terms required for an x value near the origin may be considerably less than 17. To rectify this, we can use a `while`-loop that keeps adding in terms until the next term is less than or equal to `eps` times the size of the current partial sum:

```
   function y = MyExpW(x)
%
% Pre:
%    x = scalar
% Post:
%    y = n-th order Taylor approximation to exp(x).
```

```
      y = 0;
      term = 1;
      k=0;
      while abs(term) > eps*abs(y)
         k = k + 1;
         y = y + term;
         term = x*term/k;
      end
```

To produce a vector version, we can proceed as in MyExp1 and simply call MyExpW for each component:

```
      function y = MyExp3(x)
   %
   % Pre:
   %    x: a column vector.
   % Post:
   %    y: a column vector with the property that
   %        y(i) is a Taylor approximation to exp(x(i)).
   %
      p = length(x);
      y = ones(p,1);
      for i=1:p
         y(i) = MyExpW(x(i));
      end
```

Alternatively, we can follow the MyExp2 idea and vectorize as follows:

```
      function y = MyExp4(x)
   %
   % Pre:
   %    x: a vector.
   % Post:
   %    y: a vector the same shape as x with the property that
   %        y(i) is a Taylor approximation to exp(x(i)).
   %
      y = zeros(size(x));
      term = ones(size(x));
      k = 0;
      while any(abs(term) > eps*abs(y))
         y = y + term;
         k = k+1;
         term = x.*term/k;
      end
```

Observe the use of the **any** function. It returns a "1" as long as there is at least one component in abs(term) that is larger than eps times the corresponding term in abs(y). If **any** returns a zero, then this means that `term` is small relative to `y`. In fact, it is so small that the floating point sum of `y` and `term` is `y`. The `while`-loop terminates as this happens.

The fully vectorized `MyExp4` executes much faster than `MyExp3`. However, from the work point of view, it actually involves a little more arithmetic. This is because updates on the whole vector are performed even if more iterations are required for just a few of the components. The MATLAB function `flops` can be used to determine (approximately) the number off floating point operations required to execute a given fragment. The following script illustrates its use:

```
disp(' Length(x)       MyExp3(x)        MyExp4(x) ')
disp('                   Flops            Flops')
disp('------------------------------------------------')
for L = 50:50:300
   xL = linspace(-1,1,L);
   flops(0);
   y = MyExp3(xL);
   f3 = flops;
   flops(0);
   y = MyExp4(xL);
   f4 = flops;
   disp(sprintf('%6.0f  %13.0f  %13.0f  ',L,f3,f4))
end
```

It prints the following table:

Length(x)	MyExp3(x) Flops	MyExp4(x) Flops
50	3655	3869
100	7280	7719
150	10905	11569
200	14545	15419
250	18190	19269
300	21815	23119

To use `flops`, sandwich the fragment of interest in between a `flops(0)` reference and a command of the form

⟨*Variable*⟩ = flops

The former sets the "flop counter" to zero while the latter assigns the current state of the flop counter to the named variable.

The script file `testMyExp` produces all of the results discussed in this subsection.

1.5.2 Numerical Differentiation

Suppose $f(x)$ is a function whose derivative we wish to approximate at $x = a$. A Taylor series expansion about this point says that

$$f(a + h) = f(a) + f'(a)h + \frac{f''(\eta)}{2}h^2$$

for some $\eta \in [a, a+h]$. Thus

$$D_h = \frac{f(a+h) - f(a)}{h}$$

provides increasingly good approximations as h gets small since

$$D_h = f'(a) + f''(\eta)\frac{h}{2}$$

Here is a script that enables us to explore the quality of this approach when $f(x) = \sin(x)$:

```
a = input('Enter a:   ')
h = 10^(-1:-1:-16)
Dh = (sin(a+h) - sin(a))/h;
err = abs(Dh - cos(a))
```

Using this to find the derivative of sin at $a = 1$, we see the following:

h	Absolute Error
1.0e-01	0.0429385533327507
1.0e-02	0.0042163248562708
1.0e-03	0.0004208255078129
1.0e-04	0.0000420744495186
1.0e-05	0.0000042073622750
1.0e-06	0.0000004207468094
1.0e-07	0.0000000418276911
1.0e-08	0.0000000029698852
1.0e-09	0.0000000525412660
1.0e-10	0.0000000584810365
1.0e-11	0.0000011687040611
1.0e-12	0.0000432402169239
1.0e-13	0.0007339159003137
1.0e-14	0.0037069761981869
1.0e-15	0.0148092064444385
1.0e-16	0.5403023058681398

The loss of accuracy may be explained as follows. Any error in the computation of the numerator of D_h is magnified by $1/h$. Let us assume that the values returned by sin are within eps of their true values. Thus, instead of a precise calculus bound

$$|D_h - f'(a)| \leq \frac{h}{2}$$

as predicted earlier, we have a heuristic bound

$$|D_h - f'(a)| \approx \frac{h}{2} + \frac{2\text{eps}}{h}$$

The right-hand side incorporates the "truncation error" due to calculus and the computation error due to roundoff. This quantity is minimized when $h = 2\sqrt{\text{eps}}$.

Let's package these observations and write a function that does numerical differentiation. The input parameters will include

- The name of the function f that is to be differentiated

- The point of differentiation a

- Information about the second derivative of f so that we can estimate the truncation error

- Information about the accuracy of the computed f-evaluations

The key is the intelligent choice of h. If we have an upper bound on the second derivative of the form

$$|f''(x)| \le M_2$$

then the truncation error may be bounded as follows:

$$|D_h - f'(a)| \le \frac{M_2}{2}h. \tag{1.5.1}$$

If the absolute error in a computed function evaluation is bounded by δ, then

$$errD(h) = M_2\frac{h}{2} + \frac{2\delta}{h}$$

is a reasonable model for the total error. This quantity is minimized if

$$h_{opt} = 2\sqrt{\frac{\delta}{M_2}}$$

giving

$$errD(h_{opt}) = 2\sqrt{\delta M_2}.$$

Thus we obtain

```
    function d = Derivative(fname,a,delta,M2);
%
% Pre:
%    fname  a string that names an available function.
%    a:     the x-value at which an estimate of f'(x) is required.
%    delta: absolute error in function evaluation.
%    M2:    an estimate of the second derivative magnitude near a.
%
% Post:
%    d:     an approximation of f'(a).
%    err:   an estimate of the error in d.

  hopt = 2*sqrt(delta/M2);
  d = (feval(fname,a+hopt) - feval(fname,a))/hopt;
```

It follows that

```
d = derivative('sin10',1,eps,100)
```

assigns to d an approximate derivative of $\sin(10x)$ at $a = 1$. Since the second derivative of $\sin(10x)$ is $-100\sin(x)$, setting $M_2 = 100$ is appropriate.

We need to explain how the function name is extracted and used. The key is the built-in function feval. The feval function assumes that the first argument is a string that names the function. The remaining arguments are the arguments of the function. Thus feval('f',x1,x2,...,xn) only makes sense if f is a function of n variables. The value returned is $f(x1,...,xn)$.

A major shortcoming of this approach to numerical differentiation is that it requires second derivative information. To address this concern, it is handy to make use of the **nargin** function, which permits calls to a function with less than a full complement of input parameters.

```
    function d = Derivative(fname,a,delta,M2);
%
% Pre:
%    fname  a string that names an available function.
%    a:     the x-value at which an estimate of f'(x) is required.
%    delta: absolute error in function evaluation.
%    M2:    an estimate of the second derivative magnitude near a.
%
% Post:
%    d:     an approximation of f'(a).
%    err:   an estimate of the error in d.
%
% Usage:
%    d =  Derivative(fname,a)
%    d =  Derivative(fname,a,delta)
%    d =  Derivative(fname,a,delta,M2)

  if nargin <= 3
     % No derivative bound supplied, so assume the
     % second derivative bound is 1.
     M2 = 1;
  end

  if nargin == 2
     % No function evaluation error supplied, so
     % set delta to eps.
     delta = eps;
  end

% Compute optimum h and divided difference

  hopt = 2*sqrt(delta/M2);
  d = (feval(fname,a+hopt) - feval(fname,a))/hopt;
```

During the evaluation of a function, the value of **nargin** is the number of actual parameters that were involved in the call. With this feature we can call **derivative** with two, three, or four input parameters. If no second derivative information is provided, the function somewhat arbitrarily chooses $M_2 = 10$:

Form of Call	What It Does
da = derivative(fname,a,delta,M2)	User supplies δ and M_2.
da = derivative(fname,a,delta)	User supplies δ, M_2 set to 10.
da = derivative(fname,a)	δ set to **eps**, M_2 set to 10.

1.5.3   Three-digit Arithmetic

MATLAB's string processing capabilities are nicely presented by developing a three-digit, base-10 floating point arithmetic simulation package. Let's assume that the exponent range is $[-9, 9]$ and that we use a 4-vector to represent each floating point number with the convention that if $v \in \mathbb{R}^4$, then it represents the floating point number

$$.v_1 v_2 v_3 \times 10^{v_4}$$

The signs of the mantissa and exponent are the signs of v_1 and v_4, respectively. We will call this the *v-representation*. There are two special cases. We represent zero with $[0\ 0\ 0\ 0]$ and inf with inf.

Assume that we have the function

```
   function v = represent(x)
%
% Pre:
%     x is a real scalar.
%
% Post:
%     v is the 3-digit floating point representation of x.
%     In particular, x = .v(1)v(2)v(3) x 10^v(4)
```

The inverse of this is the following function, which can take a three-digit representation and compute its value:

```
   function x = convert(v)
%
% Pre:
%     v represents a 3-digit floating point number.
%
% Post:
%     x is the value of v.
%
   if abs(v) == inf
      x = v;
   else
      x = sign(v(1))*(v(3) + 10*v(2) + 100*abs(v(1)))*10^(v(4)-3);
   end
```

To simulate three-digit floating point arithmetic, we have

```
    function z = float(x,y,op)
%
% Pre:
%     x and y are in 3-digit floating point form.
%     op is one of the strings '+', '-', '*', or '/'.
%
% Post:
%     z is the 3-digit floating point representation of x op y.
%
    sx = num2str(convert(x));
    sy = num2str(convert(y));
    z = represent(eval(['(' sx ')' op '(' sy ')' ]));
```

Strings are enclosed in quotes. The conversion of a number to a string is handled by **num2str**. Strings are concatenated by assembling them in square brackets. The **eval** function takes a string for input and returns the value produced when that string is executed.

To "pretty print" the value of a floating point representation, we use

```
    function s = pretty(v)
%
% Pre:
%     v is a 3-digit floating point number
%
% Post:
%     s is a string so that disp(s)
%     "pretty prints" the value of v.

    if v == inf
        s = 'inf';
    elseif v==-inf
        s = '-inf';
    else
        % Convert the mantissa.
        m = ['.' num2str(abs(v(1))) num2str(v(2)) num2str(v(3))];
        if v(1) < 0
            m = ['-' m];
        end
        % Convert the exponent.
        e = num2str(v(4));
        s = [m 'x10^' e];
    end
```

As an illustration of how these functions can be used, the script file **Euler** generates the partial sums

$$s_n = 1 + \frac{1}{2} + \cdots + \frac{1}{n}$$

In exact arithmetic the s_n tend toward ∞, but when we run

```
% Script File: Euler
%
% Sums the series 1 + 1/2 + 1/3 + .. in 3-digit floating point arithmetic.
% Terminates when the addition of the next term does not change
% the value of the running sum.
```

```
    clc
    oldsum = represent(0);
    one = represent(1);
    sum = one;
    k = 1;
    while convert(sum) ~= convert(oldsum)
        k = k+1;
        kay  = represent(k);
        term = float(one,kay,'/');
        oldsum = sum;
        sum  = float(sum,term,'+');
        home
        disp(['Sum through ' num2str(k) ' terms = ' pretty(sum)])
    end
```

the loop terminates after 200 terms.

Problems

P1.5.1 Produce a plot that shows the error in the pause function.

P1.5.2 A function need not have any output parameters. Complete the following:

```
function DrawTriangle(x,y,n)
% draws a shaded triangle defined by the points
% (x(1),y(1)), (x(2),y(2)), (x(3),y(3))
```

Use `fill`.

P1.5.3 It can be shown that

$$C_h = \frac{f(a+h) - f(a-h)}{2h}$$

satisfies

$$|C_h - f'(a)| \leq \frac{M_3}{6}h^2$$

if

$$|f^{(3)}(x)| \leq M_3$$

for all x. Model the error in the evaluation of C_h by

$$errC(h) = \frac{M_3}{6} + 2\frac{\delta}{h}$$

Generalize `Derivative` so that it has a 5th optional argument M3 being an estimate of the 3rd derivative. It should compute $f'(a)$ using the better of the two approximations D_h and C_h.

P1.5.4 Consider the ellipse

$$
\begin{aligned}
x(t) &= a\cos(t) \\
y(t) &= b\sin(t)
\end{aligned}
$$

and assume that $0 = t_1 < t_2 < \ldots < t_n = \pi/2$. Define the points $Q_1, \ldots, Q_n$ by

$$Q_i = (x(t_i), y(t_i))$$

Let L_i be the tangent line to the ellipse at Q_i. This line is defined by the parametric equations

$$
\begin{aligned}
x(t) &= a\cos(t_i) - a\sin(t_i)t \\
y(t) &= b\sin(t_i) + b\cos(t_i)t
\end{aligned}
$$

define L_i. Next, define the points $P_0, \ldots, P_n$ by

$$
P_i = \begin{cases}
(a, 0) & i = 0 \\
\text{intersection of } L_i \text{ and } L_{i+1} & i = 1 \ldots n - 1 \\
(0, b) & i = n
\end{cases}
$$

For your information, if the lines defined by

$$
\begin{aligned}
x_1(t) &= \alpha_1 + \beta_1 t \\
y_1(t) &= \gamma_1 + \delta_1 t
\end{aligned}
$$

$$
\begin{aligned}
x_2(t) &= \alpha_2 + \beta_2 t \\
y_2(t) &= \gamma_2 + \delta_2 t
\end{aligned}
$$

intersect, then the point of their intersection (x_*, y_*) is given by

$$
x_* = \frac{\beta_2(\alpha_1\delta_1 - \beta_1\gamma_1) - \beta_1(\alpha_2\delta_2 - \beta_2\gamma_2)}{\delta_1\beta_2 - \beta_1\delta_2}
$$

$$
y_* = \frac{\delta_2(\alpha_1\delta_1 - \beta_1\gamma_1) - \delta_1(\alpha_2\delta_2 - \beta_2\gamma_2)}{\delta_1\beta_2 - \beta_1\delta_2}
$$

Complete the following function:

```
    function [P,Q] = Points(a,b,t)
% Pre: a and b are positive, n = length(t)>=2, and
%       0 = t(1) < t(2) <... < t(n) = pi/2
%
% Post:(Q(i,1),Q(i,2)) the ith Q-point, i=1..n
%       (P(i,1),P(i,2)) the ith P point, i=1..n-1
```

Write a script file that calls `Points` with $a = 5$, $b = 2$, and $t = linspace(0, pi/2, 4)$. The script should then plot in one window the first quadrant portion of the ellipse, the polygonal line that connects the Q points, and the polygonal line that connects the P points. Use `title` to display PL and QL, the lengths of these two polygonal lines. For example,

```
        title(sprintf(' QL = %10.6f PL = %10.6f ',QL,PL )).
```

P1.5.5 Write a MATLAB function `Ellipse(P,A,theta)` that plots the "tilted" ellipse defined by

$$
\begin{aligned}
x(t) &= \cos(\theta)\left[\frac{P-A}{2} + \frac{P+A}{2}\cos(t)\right] - \sin(\theta)\left[\sqrt{A \cdot P}\sin(t)\right] \\
y(t) &= \sin(\theta)\left[\frac{P-A}{2} + \frac{P+A}{2}\cos(t)\right] + \cos(\theta)\left[\sqrt{A \cdot P}\sin(t)\right]
\end{aligned}
$$

for $0 \le t \le 2\pi$. The function should also plot in the same window the smallest rectangle (with sides parallel to the coordinate axes) that encloses the ellipse. For example,

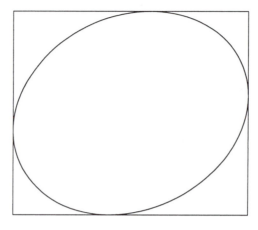

Your implementation should not have any loops.

M-Files and References

Script Files

SineTable	Prints a short table of sine evaluations.
SinePlot	Displays a sequence of sin(x) plots.
ExpPlot	Plots exp(x) and an approximation to exp(x).
TangentPlot	Plots tan(x).
SineAndCosPlot	Superimposes plots of sin(x) and cos(x).
Polygons	Displays nine regular polygons, one per window.
SumOfSines	Displays the sum of four sine functions.
SumOfSines2	Displays a pair of sum-of-sine functions.
UpDown	Sample core exploratory environment.
RunUpDown	Framework for running UpDown.
Histograms	Displays the distribution of rand and randn.
Clouds	Displays 2-dimensional rand and randn.
Dice	Histogram of 1000 dice rolls.
Darts	Monte Carlo computation of pi.
Smooth	Polygon smoothing.
Stirling	Relative and absolute error in Stirling formula.
ExpTaylor	Plots relative error in Taylor approximation to exp(x).
Zoom	Roundoff in the expansion of (x-1)^6.
FpFacts	Examines precision, overflow, and underflow.
TestMyExp	Examines MyExp1, MyExp2, MyExp3, and MyExp4.
TestDerivative	Applies Derivative to Sin10.
Euler	Three-digit arithmetic sum of $1 + 1/2 + ... + 1/n$.

Function Files

`MyExpF`	For-loop Taylor approximation to exp(x).
`MyExp1`	Vectorized version of MyExpF.
`MyExp2`	Better vectorized version of MyExpF.
`MyExpW`	While-loop Taylor approximation to exp(x).
`MyExp3`	Vectorized version of MyExpW.
`MyExp4`	Better vectorized version of MyExpW.
`Derivative`	Numerical differentiation.
`Sin10`	sin(10x).
`Represent`	Sets up 3-digit arithmetic representation.
`Convert`	Converts 3-digit representation to float.
`Float`	Simulates 3-digit arithmetic.
`Pretty`	Pretty prints a 3-digit representation.

References

The Student Edition of MATLAB, *Users Guide*, The MathWorks Inc., Natick, Massachusetts.

S.D. Conte and C. de Boor (1980). *Elementary Numerical Analysis: An Algorithmic Approach, Third Edition*, McGraw-Hill, New York.

D. Kahaner, C.B. Moler, and S. Nash (1989). *Numerical Methods and Software*, Prentice Hall, Englewood Cliffs, NJ.

M. Marcus (1993). *Matrices and* MATLAB: *A Tutorial*, Prentice Hall, Upper Saddle River, NJ.

R. Pratap (1995). *Getting Started with* MATLAB, Saunders College Publishing, Fort Worth, TX.

Chapter 2

Polynomial Interpolation

§**2.1** The Vandermonde Approach

§**2.2** The Newton Approach

§**2.3** Properties

§**2.4** Special Topics

In the problem of *data approximation*, we are given some points $(x_1, y_1), \ldots, (x_n, y_n)$ and are asked to find a function $\phi(x)$ that "captures the trend" of the data. If the trend is one of decay, then we may seek a ϕ of the form $a_1 e^{-\lambda_1 x} + a_2 e^{-\lambda_2 x}$. If the trend of the data is oscillatory, then a trigonometric approximant might be appropriate. Other settings may require a low-degree polynomial. Regardless of the type of function used, there are many different metrics for success. Least squares approximation is considered in Chapter 7.

A special form of the approximation problem ensues if we insist that ϕ actually "goes through" the data, as shown in Fig 2.1. This means that $\phi(x_i) = y_i$, $i = 1{:}n$ and we say that ϕ *interpolates* the data. The polynomial interpolation problem is particularly important:

> *Given $x_1, \ldots, x_n$ (distinct) and $y_1, \ldots, y_n$, find a polynomial $p_{n-1}(x)$ of degree $n - 1$ such that $p_{n-1}(x_i) = y_i$ for $i = 1{:}n$.*

Thus $p_2(x) = 1 + 4x - 2x^2$ interpolates the points $(-2, -15)$, $(3, -5)$, and $(1, 3)$.

Each (x_i, y_i) pair can be regarded as a snapshot of some function $f(x)$: $y_i = f(x_i)$. The function f may be explicitly available, as when we want to interpolate $\sin(x)$ at $x = 0, \pi/2$, and π with a quadratic. On other occasions, f is implicitly defined, as when we want to interpolate the solution to a differential equation at a discrete number of points.

The discussion of polynomial interpolation revolves around how it can be represented, computed, and evaluated:

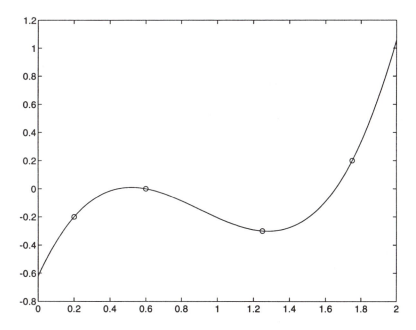

FIGURE 2.1 The interpolation of data

- How do we *represent* the interpolant $p_{n-1}(x)$? Instead of choosing to express the interpolant in terms of the "usual" basis polynomials 1, x, and x^2, we could use the alternative basis 1, $(x + 2)$, and $(x + 2)(x - 3)$. Thus

$$p_2(x) = -15 + 2(x + 2) - 2(x + 2)(x - 3)$$

  is another way to express the quadratic interpolant of the data $(-2, -15)$, $(3, -5)$, and $(1, 3)$. Different bases have different computational virtues.

- Once we have settled on a representation for the polynomial interpolant, how do we determine the associated coefficients? It turns out that this aspect of the problem involves the solution of a linear system of equations with highly structured coefficient matrices.

- After we have computed the coefficients, how can the interpolant be evaluated with efficiency? The act of plotting a polynomial interpolant sheds light on the evaluation problem and is a nice setting for discussing the tension between vectorization and arithmetic complexity. Moreover, visualization of the polynomial interpolant motivates some of its mathematical properties.

Other topics discussed briefly include divided differences, inverse interpolation, and the problem of interpolation in the plane. The problem of interpolating a function and its derivative is discussed in §3.2.1.

2.1   The Vandermonde Approach

In the Vandermonde approach, the interpolant is expressed as a linear combination of 1, x, x^2, etc. Although monomials are not the best choice for a basis, our familiarity with this way of "doing business" with polynomials makes it a good choice to initiate the discussion.

2.1.1   A Four-point Interpolation Problem

Let us find a cubic polynomial

$$p_3(x) = a_1 + a_2 x + a_3 x^2 + a_4 x^3$$

that interpolates the data $(-2, 10)$, $(-1, 4)$, $(1, 6)$, and $(2, 3)$. Each point of interpolation leads to a linear equation that relates the four unknowns a_1, a_2, a_3, and a_4:

$$
\begin{array}{rcllllllllll}
p_3(-2) &=& 10 & \Rightarrow & a_1 & - & 2a_2 & + & 4a_3 & - & 8a_4 & = & 10 \\
p_3(-1) &=& 4 & \Rightarrow & a_1 & - & a_2 & + & a_3 & - & a_4 & = & 4 \\
p_3(1) &=& 6 & \Rightarrow & a_1 & + & a_2 & + & a_3 & + & a_4 & = & 6 \\
p_3(2) &=& 3 & \Rightarrow & a_1 & + & 2a_2 & + & 4a_3 & + & 8a_4 & = & 3
\end{array}
$$

Expressing these four equations in matrix/vector terms gives

$$
\begin{bmatrix}
1 & -2 & 4 & -8 \\
1 & -1 & 1 & -1 \\
1 & 1 & 1 & 1 \\
1 & 2 & 4 & 8
\end{bmatrix}
\begin{bmatrix}
a_1 \\ a_2 \\ a_3 \\ a_4
\end{bmatrix}
=
\begin{bmatrix}
10 \\ 4 \\ 6 \\ 3
\end{bmatrix}
$$

The solution $a = [4.5000\ 1.9167\ 0.5000\ -0.9167]^T$ to this 4-by-4 system can be found as follows:

```
y = [10; 4; 6; 3];
V = [1 -2 4 -8; 1 -1 1 -1; 1 1 1 1; 1 2 4 8];
a = V\y;
```

2.1.2   The General n Case

From this example, it looks like the polynomial interpolation problem reduces to a linear equation problem. For general n, the goal is to determine $a_1, \ldots, a_n$ so that if

$$p_{n-1}(x) = a_1 + a_2 x + a_3 x^2 + \cdots + a_n x^{n-1}$$

then

$$p_{n-1}(x_i) = a_1 + a_2 x_i + a_3 x_i^2 + \cdots + a_n x_i^{n-1} = y_i$$

for $i = 1{:}n$. By writing these equations in matrix-vector form, we obtain

$$
\begin{bmatrix}
1 & x_1 & x_1^2 & \cdots & x_1^{n-1} \\
1 & x_2 & x_2^2 & \cdots & x_2^{n-1} \\
1 & x_3 & x_3^2 & \cdots & x_3^{n-1} \\
\vdots & \vdots & \vdots & \ddots & \vdots \\
1 & x_n & x_n^2 & \cdots & x_n^{n-1}
\end{bmatrix}
\begin{bmatrix}
a_1 \\ a_2 \\ a_3 \\ \vdots \\ a_n
\end{bmatrix}
=
\begin{bmatrix}
y_1 \\ y_2 \\ y_3 \\ \vdots \\ y_n
\end{bmatrix}
$$

Designate the matrix of coefficients by V. The solvability of the interpolation problem hinges on the nonsingularity of V. Suppose there is a vector c such that $Vc = 0$. It follows that the polynomial

$$q(x) = c_1 + c_2 x + \cdots + c_n x^{n-1}$$

is zero at $x = x_1, \ldots, x = x_n$. This says that we have a degree $n-1$ polynomial with n roots. The only way that this can happen is if q is the zero polynomial (i.e., $c = 0$). Thus V is nonsingular because the only vector that it zeros is the zero vector.

2.1.3 Setting Up and Solving the System

Let us examine the construction of the Vandermonde matrix V. Our first method is based on the observation that the ith row of V involves powers of x_i and that the powers increase from 0 to $n-1$ as the row is traversed from left to right. A conventional double-loop approach gives

```
n = length(x); V = zeros(n,n);
for i=1:n
    % Set up row i.
    for j=1:n
        V(i,j) = x(i)^(j-1);
    end
end
```

Algorithms that operate on a two-dimensional array in row-by-row fashion are *row oriented*.

The inner-loop in the preceding script can be *vectorized* because MATLAB supports point-wise exponentiation. For example, u = [1 2 3 4] .^[3 5 2 3] assigns to u the row vector [1 32 9 64]. The i-th row of V requires exponentiating the scalar x_i to each of the values in the row vector $0:n-1 = (0, 1, \ldots, n-1)$. Thus row = (x(i)*ones(1,n)).^(0:n-1) assigns the vector $(1, x_i, x_i^2, \ldots, x_i^{n-1})$ to row, precisely the values that make up the ith row of V. The ith row of a matrix V may be referenced by V(i,:), and so we obtain

```
n = length(x); V = zeros(n,n);
for i=1:n
    % Set up the i-th row of V.
    V(i,:)  = (x(i)*ones(1,n)).^(0:n-1);
end
```

By reversing the order of the loops in the original set-up script, we obtain a *column oriented* algorithm:

```
n = length(x); V = zeros(n,n);
for j=1:n
    % Set up column j.
    for i=1:n
        V(i,j) = x(i)^(j-1);
    end
end
```

If $j > 1$, then $V(i, j)$ is the product of $x(i)$ and $V(i, j - 1)$, the matrix entry to its left. This suggests that the required exponentiations can be obtained through repeated multiplication:

```
n = length(x); V = zeros(n,n);
V = ones(n,n);
for j=2:n
    % Set up column j.
    for i=1:n
        V(i,j) = x(i)*V(i,j-1)
    end
end
```

The generation of the jth column involves *vector multiplication*:

$$
\begin{bmatrix} x_1 \\ \vdots \\ x_n \end{bmatrix} .* \begin{bmatrix} v_{1,j-1} \\ \vdots \\ v_{n,j-1} \end{bmatrix} = \begin{bmatrix} v_{1,j} \\ \vdots \\ v_{n,j} \end{bmatrix}
$$

and this may be implemented by $V(:,j) = x .* V(:,j-1)$. Basing our final implementation on this, we obtain

```
    function a = InterpV(x,y)
%
% Pre:
%   x:   column n-vector with distinct components.
%   y:   column n-vector.
%
% Post:
%   a:   column n-vector with the property that
%        if p(x) = a(1) + a(2)x + ... a(n)x^(n-1) then
%        p(x(i)) = y(i), i=1:n
%
    n = length(x);
    V = ones(n,n);
    for j=2:n
        % Set up column j.
        V(:,j) = x.*V(:,j-1);
    end
    a = V\y;
```

Column-oriented, matrix-vector implementations will generally be favored in this text. One reason for doing this is simply to harmonize with the traditions of linear algebra, which is usually taught with a column-oriented perspective.

2.1.4 Nested Multiplication

We now consider the evaluation of $p_{n-1}(x) = a_1 + \cdots + a_n x^{n-1}$ at $x = z$, assuming that z and $a(1:n)$ are available. The routine approach

```
n = length(a);
zpower = 1;
pval = a(1);
for i=2:n
    zpower = z*zpower;
    pval = pval + a(i)*zpower;
end
```

assigns the value of $p_{n-1}(z)$ to pval.

A more efficient algorithm is based on a nested organization of the polynomial, which we illustrate for the case $n = 4$:

$$p_3(x) = a_1 + a_2 x + a_3 x^2 + a_4 x^3 = ((a_4 x + a_3)x + a_2)x + a_1$$

Note that the fragment

```
pval = a(4);
pval = z*pval + a(3);
pval = z*pval + a(2);
pval = z*pval + a(1);
```

assigns the value of $p_3(z)$ to pval. For general n, this nested multiplication idea takes on the following form:

```
n = length(a);
pval = a(n);
for i=n-1:-1:1
    pval = z*pval + a(i);
end
```

This is widely known as *Horner's rule*.

Before we encapsulate the Horner idea in a MATLAB function, let us examine the case when the interpolant is to be evaluated at many different points. To be precise, suppose $z(1:m)$ is initialized and that for $i = 1:m$, we want to assign the value of $p_{n-1}(z(i))$ to pval(i). One obvious approach is merely to repeat the preceding Horner iteration at each point:

```
m = length(z);
n = length(a);
pval = zeros(m,1);
for j=1:m
    %Evaluate p(z(j)).
    pval(j) = a(n);
    for i=n-1:-1:1
        pval(j) = z(j)*pval(j) + a(i);
    end
end
```

A vectorized version of this fragment can be obtained if we think about the "simultaneous" evaluation of the interpolants at each z_i. Suppose $m = 5$ and $n = 4$ (i.e, the case when a cubic interpolant is to be evaluated at five different points). The first step in the five applications of the Horner idea may be summarized as follows:

$$
\begin{bmatrix} \text{pval(1)} \\ \text{pval(2)} \\ \text{pval(3)} \\ \text{pval(4)} \\ \text{pval(5)} \end{bmatrix} = \begin{bmatrix} \text{a(4)} \\ \text{a(4)} \\ \text{a(4)} \\ \text{a(4)} \\ \text{a(4)} \end{bmatrix}
$$

In vector terms pval = a(n)*ones(m,1). The next step requires a multiply-add of the following form:

$$
\begin{bmatrix} \text{pval(1)} \\ \text{pval(2)} \\ \text{pval(3)} \\ \text{pval(4)} \\ \text{pval(5)} \end{bmatrix} = \begin{bmatrix} \text{z(1)*pval(1)} \\ \text{z(2)*pval(2)} \\ \text{z(3)*pval(3)} \\ \text{z(4)*pval(4)} \\ \text{z(5)*pval(5)} \end{bmatrix} + \begin{bmatrix} \text{a(3)} \\ \text{a(3)} \\ \text{a(3)} \\ \text{a(3)} \\ \text{a(3)} \end{bmatrix}
$$

That is,

$$
\text{pval} = \text{z.*pval + a(3)}
$$

The pattern is clear for the cubic case:

```
pval = a(4)*ones(m,1);
pval = z .* pval + a(3);
pval = z .* pval + a(2);
pval = z .* pval + a(1);
```

From this we generalize to the following:

```
    function pval = HornerV(a,z)
%
% Pre:
%     a:    a column n-vector.
%     z:    a column m-vector.
%
% Post:
%     pval: a column m-vector with the property that if
%           p(x) = a(1) + .. +a(n)x^(n-1), then
%           pval(i) = p(z(i)) for i=1:m.
%
      n = length(a);
      m = length(z);
      pval = a(n)*ones(m,1);
      for k=n-1:-1:1
          pval = z.*pval + a(k);
      end
```

Each update of `pval` requires $2m$ flops so approximately $2mn$ flops are required in total.

As an application, here is a script that displays cubic interpolants of $\sin(x)$ on $[0, 2\pi]$. The abscissas are chosen randomly.

```
% Script File: ShowV
%
% Plots 4 random cubic interpolants of sin(x) on [0,2pi].
% Uses the Vandermonde method.

    close all
    x0 = linspace(0,2*pi,100)';
    y0 = sin(x0);
    for eg=1:4
        x = 2*pi*sort(rand(4,1));
        y = sin(x);
        a = InterpV(x,y);
        pvals = HornerV(a,x0);
        subplot(2,2,eg)
        plot(x0,y0,x0,pvals,x,y,'*')
        axis([0 2*pi -2 2])
    end
```

See Fig 2.2 for sample output.

Problems

P2.1.1 Instead of expressing the polynomial interpolant in terms of the basis functions $1, x, \ldots, x^{n-1}$, we can work with the alternative representation

$$p_{n-1}(x) = \sum_{k=1}^{n} a_k \left(\frac{x - u}{v}\right)^{k-1}$$

Here u and v are scalars that serve to shift and scale the x-range. Generalize `InterpV` so that it can be called with either two, three, or four arguments. A call of the form `a = InterpV(x,y)` should assume that $u = 0$ and $v = 1$. A call of the form `a = InterpV(x,y,u)` should assume that $v = 1$ and that u houses the shift factor. A call of the form `a = InterpV(x,y,u,v)` should assume that u and v house the shift and scale factors, respectively.

P2.1.2 A polynomial of the form
$$p(x) = a_1 + a_2 x^2 + \cdots + a_m x^{2m-2}$$
is said to be *even*, while a polynomial of the form

$$p(x) = a_1 x + a_3 x^3 + \cdots + a_m x^{2m-1}$$

is said to be *odd*. Generalize `HornerV(a,z)` so that it has an optional third argument `type` that indicates whether or not the underlying polynomial is even or odd. In particular, a call of the form `HornerV(a,z,'even')` should assume that a_k is the coefficient of x^{2k-2}. A call of the form `HornerV(a,z,'odd')` should assume that a_k is the coefficient of x^{2k-1}.

P2.1.3 Assume that `z` and `a(1:n)` are initialized and define

$$p(x) = a_1 + a_2 x + \cdots + a_n x^{n-1}$$

Write a MATLAB fragment that evaluates (1) $p(z)/p(-z)$, (2) $p(z) + p(-z)$, (3) $p'(z)$, (4) $\int_0^1 p(x)dx$, and (e) $\int_{-z}^{z} p(x)dx$. Make effective use of `HornerV`.

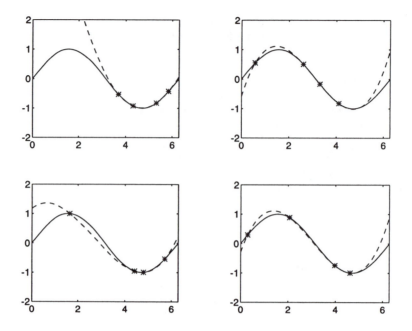

FIGURE 2.2 Random cubic interpolants of $\sin(x)$ on $[0, \pi]$

2.2   The Newton Representation

We now look at a form of the polynomial interpolant that is generally more useful than the Vandermonde representation.

2.2.1   A Four-Point Example

To motivate the idea, consider once again the problem of interpolating the four points (x_1, y_1), (x_2, y_2), (x_3, y_3), and (x_4, y_4) with a cubic polynomial $p_3(x)$. However, instead of expressing the interpolant in terms of the "canonical" basis 1, x, x^2, and x^3, we use the basis 1, $(x - x_1)$, $(x - x_1)(x - x_2)$, and $(x - x_1)(x - x_2)(x - x_3)$. This means that we are looking for coefficients c_1, c_2, c_3, and c_4 so that if

$$p_3(x) = c_1 + c_2(x - x_1) + c_3(x - x_1)(x - x_2) + c_4(x - x_1)(x - x_2)(x - x_3) \qquad (2.2.1)$$

then $y_i = p_3(x_i) = y_i$ for $i = 1{:}4$. In expanded form, these four equations state that

$$y_1 = c_1$$

$$y_2 = c_1 + c_2(x_2 - x_1)$$

$$y_3 = c_1 + c_2(x_3 - x_1) + c_3(x_3 - x_1)(x_3 - x_2)$$

$$y_4 = c_1 + c_2(x_4 - x_1) + c_3(x_4 - x_1)(x_4 - x_2) + c_4(x_4 - x_1)(x_4 - x_2)(x_4 - x_3)$$

By rearranging these equations, we obtain the following four-step solution process:

$$c_1 = y_1$$

$$c_2 = \frac{y_2 - c_1}{x_2 - x_1}$$

$$c_3 = \frac{y_3 - (c_1 + c_2(x_3 - x_1))}{(x_3 - x_1)(x_3 - x_2)}$$

$$c_4 = \frac{y_4 - (c_1 + c_2(x_4 - x_1) + c_3(x_4 - x_1)(x_4 - x_2))}{(x_4 - x_1)(x_4 - x_2)(x_4 - x_3)}$$

This sequential solution process is made possible by the clever choice of the basis polynomials and the result is the *Newton representation* of the interpolating polynomial.

To set the stage for the general-n algorithm, we redo the $n = 4$ case using matrix-vector notation to discover a number of simplifications. The starting point is the system of equations that we obtained previously which can be expressed in the following form:

$$\begin{bmatrix} 1 & 0 & 0 & 0 \\ 1 & (x_2 - x_1) & 0 & 0 \\ 1 & (x_3 - x_1) & (x_3 - x_1)(x_3 - x_2) & 0 \\ 1 & (x_4 - x_1) & (x_4 - x_1)(x_4 - x_2) & (x_4 - x_1)(x_4 - x_2)(x_4 - x_3) \end{bmatrix} \begin{bmatrix} c_1 \\ c_2 \\ c_3 \\ c_4 \end{bmatrix} = \begin{bmatrix} y_1 \\ y_2 \\ y_3 \\ y_4 \end{bmatrix}$$

From this we see immediately that $c_1 = y_1$. We can eliminate c_1 from equations 2, 3, and 4 by subtracting equation 1 from equations 2, 3, and 4:

$$\begin{bmatrix} 1 & 0 & 0 & 0 \\ 0 & (x_2 - x_1) & 0 & 0 \\ 0 & (x_3 - x_1) & (x_3 - x_1)(x_3 - x_2) & 0 \\ 0 & (x_4 - x_1) & (x_4 - x_1)(x_4 - x_2) & (x_4 - x_1)(x_4 - x_2)(x_4 - x_3) \end{bmatrix} \begin{bmatrix} c_1 \\ c_2 \\ c_3 \\ c_4 \end{bmatrix} = \begin{bmatrix} y_1 \\ y_2 - y_1 \\ y_3 - y_1 \\ y_4 - y_1 \end{bmatrix}$$

If we divide equations 2, 3, and 4 by $(x_2 - x_1)$, $(x_3 - x_2)$, and $(x_4 - x_3)$, respectively, then the system transforms to

$$\begin{bmatrix} 1 & 0 & 0 & 0 \\ 0 & 1 & 0 & 0 \\ 0 & 1 & (x_3 - x_2) & 0 \\ 0 & 1 & (x_4 - x_2) & (x_4 - x_2)(x_4 - x_3) \end{bmatrix} \begin{bmatrix} c_1 \\ c_2 \\ c_3 \\ c_4 \end{bmatrix} = \begin{bmatrix} y_1 \\ y_{21} \\ y_{31} \\ y_{41} \end{bmatrix}$$

where y_{21}, y_{31}, and y_{41} are defined by

$$y_{21} = \frac{y_2 - y_1}{x_2 - x_1} \qquad y_{31} = \frac{y_3 - y_1}{x_3 - x_1} \qquad y_{41} = \frac{y_4 - y_1}{x_4 - x_1}$$

Notice that

$$\begin{bmatrix} y_{21} \\ y_{31} \\ y_{41} \end{bmatrix} = \left(\begin{bmatrix} y_2 \\ y_3 \\ y_4 \end{bmatrix} - \begin{bmatrix} y_1 \\ y_1 \\ y_1 \end{bmatrix} \right) ./ \left(\begin{bmatrix} x_2 \\ x_3 \\ x_4 \end{bmatrix} - \begin{bmatrix} x_1 \\ x_1 \\ x_1 \end{bmatrix} \right) = (y(2{:}4) - y(1{:}3))./(x(2{:}4) - x(1))$$

The key point is that we have reduced the size of problem by one. The remaining unknowns satisfy a 3-by-3 system:

$$\begin{bmatrix} 1 & 0 & 0 \\ 1 & (x_3 - x_2) & 0 \\ 1 & (x_4 - x_2) & (x_4 - x_2)(x_4 - x_3) \end{bmatrix} \begin{bmatrix} c_2 \\ c_3 \\ c_4 \end{bmatrix} = \begin{bmatrix} y_{21} \\ y_{31} \\ y_{41} \end{bmatrix}$$

This is exactly the system obtained were we to seek the coefficients of the quadratic

$$q(x) = c_2 + c_3(x - x_2) + c_4(x - x_2)(x - x_3)$$

that interpolates the data (x_2, y_{21}), (x_3, y_{31}), and (x_4, y_{41}).

2.2.2   The General n Case

For general n, we see that if $c_1 = y_1$ and

$$q(x) = c_2 + c_3(x - x_2) + \cdots + c_n(x - x_2) \cdots (x - x_{n-1})$$

interpolates the data

$$\left(x_i, \frac{y_i - y_1}{x_i - x_1} \right) \qquad i = 2{:}n$$

then

$$p(x) = c_1 + (x - x_1)q(x)$$

interpolates $(x_1, y_1), \ldots, (x_n, y_n)$. This is easy to verify. Indeed, for $j = 1{:}n$

$$p(x_j) = c_1 + (x_j - x_1)q(x_j) = y_1 + (x_j - x_1)\frac{y_j - y_1}{x_j - x_1} = y_j$$

This sets the stage for a recursive formulation of the whole process:

```
function c = InterpNRecur(x,y)
%
% Pre:
%   x:   column n-vector with distinct components.
%   y:   column n-vector.
%
% Post:
%   c:   a column n-vector with the property that if
%           p(x) = c(1) + c(2)(x-x(1))+...+ c(n)(x-x(1))...(x-x(n-1)),
%        then p(x(i)) = y(i), i=1:n.
%
    n = length(x)
    c = zeros(n,1);
    c(1) = y(1);
    if n>1
        c(2:n) =InterpNRecur(x(2:n),(y(2:n)-y(1))./(x(2:n)-x(1)));
    end
```

If $n = 1$, then the constant interpolant $p(x) \equiv y_1$ is returned (i.e., $c_1 = y_1$.) Otherwise, the final c-vector is a "stacking" of y_1 and the solution to the reduced problem. The recursive call obtains the coefficients of the interpolant $q(x)$ mentioned earlier.

To develop a nonrecursive implementation, we return to our four-point example and the equation

$$
\begin{bmatrix}
1 & 0 & 0 & 0 \\
0 & 1 & 0 & 0 \\
0 & 1 & (x_3 - x_2) & 0 \\
0 & 1 & (x_4 - x_2) & (x_4 - x_2)(x_4 - x_3)
\end{bmatrix}
\begin{bmatrix}
c_1 \\
c_2 \\
c_3 \\
c_4
\end{bmatrix}
=
\begin{bmatrix}
y_1 \\
y_{21} \\
y_{31} \\
y_{41}
\end{bmatrix}
$$

From this we see that $c_2 = y_{21}$. Now subtract equation 2 from equation 3 and divide by $(x_3 - x_2)$. Next, subtract equation 2 from equation 4 and divide by $(x_4 - x_2)$. With these operations we obtain

$$
\begin{bmatrix}
1 & 0 & 0 & 0 \\
0 & 1 & 0 & 0 \\
0 & 0 & 1 & 0 \\
0 & 0 & 1 & (x_4 - x_3)
\end{bmatrix}
\begin{bmatrix}
c_1 \\
c_2 \\
c_3 \\
c_4
\end{bmatrix}
=
\begin{bmatrix}
y_1 \\
y_{21} \\
y_{321} \\
y_{421}
\end{bmatrix}
$$

where

$$
y_{321} = \frac{y_{31} - y_{21}}{x_3 - x_2} \qquad y_{421} = \frac{y_{41} - y_{21}}{x_4 - x_2}
$$

At this point we see that $c_3 = y_{321}$. Finally, by subtracting the third equation from the fourth equation and dividing by $(x_4 - x_3)$, we obtain

$$
\begin{bmatrix}
1 & 0 & 0 & 0 \\
0 & 1 & 0 & 0 \\
0 & 0 & 1 & 0 \\
0 & 0 & 0 & 1
\end{bmatrix}
\begin{bmatrix}
c_1 \\
c_2 \\
c_3 \\
c_4
\end{bmatrix}
=
\begin{bmatrix}
y_1 \\
y_{21} \\
y_{321} \\
y_{4321}
\end{bmatrix}
$$

where
$$y_{4321} = \frac{y_{421} - y_{321}}{x_4 - x_3}$$

Clearly, $c_4 = y_{4321}$. The pattern for the general n case should be apparent:

```
for k=1:n-1
    c(k) = y(k)
    for j=k+1:n
        Subtract equation k from equation j.
        Divide equation j by (x_j - x_k).
    end
end
c(n) = y(n)
```

However, when updating the equations *we need only keep track of the changes in the y-vector.* For example,

$$y(k+1{:}n) \;\; \leftarrow \;\; \left(\begin{bmatrix} y_{k+1} \\ \vdots \\ y_n \end{bmatrix} - \begin{bmatrix} y_k \\ \vdots \\ y_k \end{bmatrix} \right) ./ \left(\begin{bmatrix} x_{k+1} \\ \vdots \\ x_n \end{bmatrix} - \begin{bmatrix} x_k \\ \vdots \\ x_k \end{bmatrix} \right)$$

$$= \;\; (y(k+1{:}n) - y(k)) \; ./ \; (x(k+1) - x(k))$$

This leads to

```
    function c = InterpN(x,y)
%
% Pre:
%    x:   column n-vector with distinct components.
%    y:   column n-vector.
%
% Post:
%    c:   a column n-vector with the property that if
%         p(x) = c(1) + c(2)(x-x(1))+...+ c(n)(x-x(1))...(x-x(n-1))
%         p(x(i)) = y(i), i=1:n.
%
    n = length(x);
    for k = 1:n-1
        y(k+1:n) = (y(k+1:n)-y(k)) ./ (x(k+1:n) - x(k));
    end
    c = y;
```

2.2.3   Nested Multiplication

As with the Vandermonde representation, the Newton representation permits an efficient nested multiplication scheme. For example, to evaluate $p_3(x)$ at $x = z$, we have the nesting

$$p_3(x) = ((c_4(x - x_3) + c_3)(x - x_2) + c_2)(x - x_1) + c_1$$

The fragment

```
pval = c(4);
pval = (z-x(3))*pval + c(3);
pval = (z-x(2))*pval + c(2);
pval = (z-x(1))*pval + c(1);
```

assigns the value of $p_3(z)$ to pval. If z is a vector, then this becomes

```
pval = c(4)*ones(size(z));
pval = (z-x(3)).*pval + c(3);
pval = (z-x(2)).*pval + c(2);
pval = (z-x(1)).*pval + c(1);
```

In general, we have

```
    function pval = HornerN(c,x,z)
%
% Pre:
%    c:    a vector.
%    x:    a vector with at least length(c)-1 components
%    z:    a vector.
%
% Post:
%   pval:   a vector the same size as z with the property that if
%            p(x) = c(1) +  c(2)(x-x(1))+ ... + c(n)(x-x(1))...(x-x(n-1))
%            then pval(i) = p(z(i)) for i=1:m.

  n = length(c);
  pval = c(n)*ones(size(z));
  for k=n-1:-1:1
     pval = (z-x(k)).*pval + c(k);
  end
```

The script ShowN illustrates the use of HornerN and InterpN.

Problems

P2.2.1 Complete the following function:

```
    function a = N2V(c,x);
% Pre:
%     c:  column n-vector
%     x:  column (n-1)-vector
% Post:
%     a:  column n-vector so that if
%          p(x) = c(1) + c(2)(x-x(1)) + ... + c(n)(x-x(1))(x-x(2))...(x-x(n-1))
%          then p(x) = a(1) + a(2)x + ... + a(n)x^(n-1)
%
```

P2.2.2 Suppose we are given the data (x_i, y_i), $i = 1{:}n$. Assume that the x_i are distinct and that $n \geq 2$. Let $p_L(x)$ and $p_R(x)$ be degree $n - 2$ polynomials that satisfy

$$p_L(x_i) \;\; = \;\; y_i \qquad i = 1{:}n - 1$$

$$p_R(x_i) \;\; = \;\; y_i \qquad i = 2{:}n$$

Note that if

$$p(x) = \frac{(x - x_n)p_L(x) - (x - x_1)p_R(x)}{x_1 - x_n}$$

then $p(x_i) = y_i$, $i = 1{:}n$. In other words, $p(x)$ is the unique degree $n - 1$ interpolant of (x_i, y_i), $i = 1{:}n$. Using this result, complete the following function:

```
function pval = RecurEval(x,y,z);
%
% Pre:
%   x:   column n-vector with distinct entries.
%   y:   column n-vector
%   z:   column m-vector
%
% Post:
%   pval: column m-vector with the property that
%               pval(i) = p(z(i))
%          where p(x) is the degree n-1 polynomial
%          interpolant of (x(i),y(i)), i=1:n.
%
```

The implementation should be recursive and vectorized. No loops are necessary! Use `RecurEval` to produce an interpolant of $\sin(2\pi x)$ at $x = 0{:}.25{:}1$.

P2.2.3 Write a MATLAB script that solicits the name of a built-in function (as a string), the left and right limits of an interval $[L, R]$, and a positive integer n and then displays both the function and the $n - 1$ degree interpolant of it at `linspace(L,R,n)`.

2.3 Properties

With two approaches to the polynomial interpolation problem, we have an occasion to assess their relative merits. Speed and accuracy are the main concerns.

2.3.1 Efficiency

We start with a flop-counting script

```
% Script File: InterpEff
%
% Compares the Vandermonde and Newton Approaches

   clc home
   disp('Flop Counts:')
   disp(' ')
   disp('  n    InterpV    InterpN    InterpNRecur')
   disp('------------------------------------------')
```

```
for n = [4 8 16]
    x = linspace(0,1,n)'; y = sin(2*pi*x);
    flops(0); a = InterpV(x,y);      f1 = flops;
    flops(0); c = InterpN(x,y);      f2 = flops;
    flops(0); c = InterpNRecur(x,y); f3 = flops;
    disp(sprintf('%3.0f %7.0f    %7.0f     %7.0f',n,f1,f2,f3));
end
```

and discover that the Vandermonde approach is decidedly less efficient. (See Fig 2.3.) This is because that method involves the solution of an n-by-n linear system. In Chapter 6 we learn that "backslash" requires about $2n^3/3$ flops when applied to an n-by-n problem. This means that if we double n, the number of flops required goes up by a factor of 8. This is evident from the table. Algorithms with this property are said to be *cubic*, and the notation $O(n^3)$ is used to designate this fact.

n	InterpV	InterpN	InterpNRecur
4	171	28	18
8	863	106	84
16	4831	406	360

FIGURE 2.3 Flops required to compute a degree $n - 1$ interpolation polynomial

On the other hand, the Newton approach is quadratic (i.e., $O(n^2)$. Work goes up by a factor of 4 (approximately) if n is doubled. This can be confirmed analytically, because the kth pass through the loop in InterpN involves three length-$(n - k)$ vector operations and

$$\sum_{k=1}^{n-1} 3(n - k) = 3\sum_{k=1}^{n-1} k = 3\frac{(k - 1)k}{2} \approx \frac{3}{2}n^2$$

The Newton approach is therefore much more efficient even though it too involves the solution of a linear system. The key is that the structure of the linear system is exploited unlike in the Vandermonde case.[1]

So far we have just discussed execution time efficiency. Memory efficiency is also important. InterpV requires an n-by-n array, while InterpN needs just a few n-vectors. On the other hand, InterpNRecur requires a couple of n-vectors with each level of the recursion. It turns out that n cannot be very big before this reality begins to take a serious toll.

2.3.2 Accuracy

We know that the polynomial interpolant exists and is unique, but how well does it approximate? The answer to the question depends on the derivatives of the function that is being interpolated.

---

[1]It turns out that $O(n^2)$ Vandermonde system solvers exist, but we shall not go into the details.

Theorem 2 *Suppose $p_{n-1}(x)$ interpolates the function $f(x)$ at the distinct points $x_1, \ldots, x_n$. If f is n times continuously differentiable on an interval I containing the x_i, then for any $x \in I$*

$$f(x) = p_{n-1}(x) + \frac{f^{(n)}(\eta)}{n!}(x - x_1) \cdots (x - x_n)$$

where $a \leq \eta \leq b$.

Proof. For clarity and not a tremendous loss of generality, we prove the theorem for the $n = 4$ case. Consider the function

$$F(t) = f(t) - p_3(t) - cL(t)$$

where

$$c = \frac{f(x) - p_3(x)}{(x - x_1)(x - x_2)(x - x_3)(x - x_4)}$$

and $L(t) = (t - x_1)(t - x_2)(t - x_3)(t - x_4)$. Note that $F(x) = 0$ and $F(x_i) = 0$ for $i = 1:4$. Thus, F has at least five zeros in I. In between these zeros $F'(t)$ has a zero. Thus $F'(t)$ has at least four zeros in I. Continuing in this way, we conclude that

$$F^{(4)}(t) = f^{(4)}(t) - p_3^{(4)}(t) - cL^{(4)}(t)$$

has at least one root in I which we designate by η_x. Since p_3 has degree 3, $p_3^{(4)}(t) \equiv 0$. Since L is a monic polynomial with degree 4, $L_3^{(4)}(t) = 4!$. Thus

$$0 = F^{(4)}(\eta_x) = f^{(4)}(\eta_x) - p_3^{(4)}(\eta_x) - cL^{(4)}(\eta_x) = f^{(4)}(\eta_x) - c \cdot 4! \qquad \square$$

This result shows that the quality of $p_{n-1}(x)$ depends on the size of the nth derivative. If we have a bound on this derivative, then we can compute a bound on the error. To illustrate this point in a practical way, suppose $|f^{(n)}(x)| \leq M_n$ for all $x \in [a, b]$. It follows that for any $z \in [a, b]$ we have

$$|f(z) - p_{n-1}(z)| \leq \frac{M_n}{n!} \max_{a \leq x \leq b} |(x - x_1)(x - x_2) \cdots (x - x_n)|$$

If we base the interpolant on the equally spaced points

$$x_i = a + \left(\frac{b - a}{n - 1}\right)(i - 1) \qquad i = 1:n$$

then, by a simple change of variable,

$$|f(z) - p_{n-1}(z)| \leq M_n \left(\frac{b - a}{n - 1}\right)^n \max_{0 \leq s \leq n-1} \left| \frac{s(s - 1) \cdots (s - n + 1)}{n!} \right|$$

It can be shown that the max is no bigger than $1/n$, from which we conclude that

$$|f(z) - p_{n-1}(z)| \leq \frac{M_n}{n} \left(\frac{b - a}{n - 1}\right)^n \tag{2.3.1}$$

Thus, if a function has ill-behaved higher derivatives, then the quality of the polynomial interpolants may actually decrease as the degree increases.

A classic example of this is the problem of interpolating the function $f(x) = 1/(1 + 25x^2)$ across the interval $[-1, 1]$. The script **RungeEg** explores this issue:

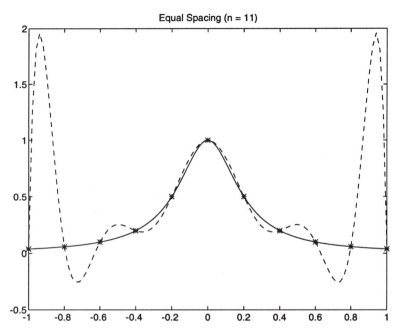

FIGURE 2.4 The Runge phenomena.

```
% Script File: RungeEg
%
% For n=10:13, interpolants of f(x) = 1/(1+25x^2) on [-1,1]'
% are of plotted.
%
  close all
  x = linspace(-1,1,100)';
  y = ones(100,1)./(1 + 25*x.^2);
  for n=10:13
     figure
     xEqual = linspace(-1,1,n)';
     yEqual = ones(size(xEqual))./(1+25*xEqual.^2);;
     cEqual=InterpN(xEqual,yEqual);
     pvalsEqual = HornerN(cEqual,xEqual,x);
     plot(x,y,x,pvalsEqual,xEqual,yEqual,'*')
     title(sprintf('Equal Spacing (n = %2.0f)',n))
  end
```

(See Fig 2.4 for sample output.) While the interpolant "captures" the trend of the function in the middle part of the interval, it blows up near the endpoints.

Problems

P2.3.1 Write a MATLAB script that compares HornerN and HornerV from the flop point of view.

P2.3.2 Write a MATLAB script that repeatedly solicits an integer n and produces a reasonable plot of the function $e(s) = |s(s-1)\cdots(s-n+1)/n!|$ on the interval $[0, n-1]$. Verify experimentally that this function is never bigger than 1, a fact that we used to establish (2.3.1).

P2.3.3 Write a MATLAB function nBest(L,R,a,delta) that returns an integer n such that if $p_{n-1}(x)$ interpolates e^{ax} at $L + (i-1)(R-L)/(n-1)$, $i = 1{:}n$, then $|p_{n-1}(z) - e^{az}| \leq \delta$ for all $z \in [L, R]$. Try to make the value of n as small as you can.

2.4 Special Topics

As a follow-up to the preceding developments, we briefly discuss properties and algorithms associated with divided differences, the notion of inverse interpolation, and the basics of two-dimensional linear interpolation. We also review MATLAB's polynomial interpolation tools.

2.4.1 Divided Differences

Returning to the $n = 4$ example used in the previous section, we can express c_1, c_2, c_3, and c_4 in terms of the x_i and f:

$$c_1 = f(x_1)$$

$$c_2 = \frac{f(x_2) - f(x_1)}{x_2 - x_1}$$

$$c_3 = \frac{\dfrac{f(x_3) - f(x_1)}{x_3 - x_1} - \dfrac{f(x_2) - f(x_1)}{x_2 - x_1}}{x_3 - x_2}$$

$$c_4 = \frac{\dfrac{\dfrac{f(x_4) - f(x_1)}{x_4 - x_1} - \dfrac{f(x_2) - f(x_1)}{x_2 - x_1}}{x_4 - x_2} - \dfrac{\dfrac{f(x_3) - f(x_1)}{x_3 - x_1} - \dfrac{f(x_2) - f(x_1)}{x_2 - x_1}}{x_3 - x_2}}{x_4 - x_3}.$$

The coefficients are called *divided differences*. To stress the dependence of c_k on f and $x_1, \ldots, x_k$, we write

$$c_k = f[x_1, \ldots, x_k]$$

and refer to this quantity as the *kth order divided difference*. Thus

$$p_{n-1}(x) = \sum_{k=1}^{n} f[x_1, \ldots, x_k] \left(\prod_{j=1}^{k-1} (x - x_j) \right)$$

is the degree $n - 1$ polynomial interpolant of f at $x_1, \ldots, x_n$.

We now establish another recursive property that relates the divided differences of f on designated subsets of $\{x_1, \ldots, x_n\}$. Suppose $p_L(x)$ and $p_R(x)$ are the interpolants of f on $\{x_1, \ldots, x_{k-1}\}$ and $\{x_2, \ldots, x_k\}$, respectively. It is easy to confirm that if

$$p(x) = \frac{(x - x_k)p_L(x) - (x - x_1)p_R(x)}{x_1 - x_k} \tag{2.4.1}$$

then $p(x_i) = f(x_i)$, $i = 1{:}n$. Thus $p(x)$ is the interpolant of f on $\{x_1, \ldots, x_k\}$ and so

$$p(x) = f[x_1] + f[x_1, x_2](x - x_1) + \cdots + f[x_1, \ldots, x_n](x - x_1) \cdots (x - x_{k-1}) \tag{2.4.2}$$

Note that since

$$p_L(x) = f[x_1] + f[x_1, x_2](x - x_1) + \cdots + f[x_1, \ldots, x_{k-1}](x - x_1) \cdots (x - x_{k-2})$$

the coefficient of x^{k-2} is given by $f[x_1, \ldots, x_{k-1}]$. Likewise, since

$$p_R(x) = f[x_2] + f[x_1, x_2](x - x_1) + \cdots + f[x_2, \ldots, x_k](x - x_2) \cdots (x - x_{k-1})$$

the coefficient of x^{k-2} is given by $f[x_2, \ldots, x_k]$. Comparing the coefficients of x^{k-1} in (2.4.1) and (2.4.2), we conclude that

$$f[x_1, \ldots, x_k] = \frac{f[x_2, \ldots, x_k] - f[x_1, \ldots, x_{k-1}]}{x_k - x_1} \tag{2.4.3}$$

The development of higher-order divided differences from lower order divided differences is illustrated in Fig 2.5. Observe that the sought-after divided differences are along the left edge of the tree. Pruning the excess, we see that the required divided differences can be built up as shown in Fig 2.6. This enables us to rewrite InterpN as follows:

```
function c = InterpN2(x,y)
%
% Pre:
%   x:   column n-vector with distinct components.
%   y:   column n-vector.
%
% Post:
%   c:   a column n-vector with the property that if
%        p(x) = c(1) + c(2)(x-x(1))+...+ c(n)(x-x(1))...(x-x(n-1))
%        p(x(i)) = y(i), i=1:n.
%
%
   n = length(x);
   for k = 1:n-1
      y(k+1:n) = (y(k+1:n)-y(k:n-1)) ./ (x(k+1:n) - x(1:n-k));
   end
   c = y;
```

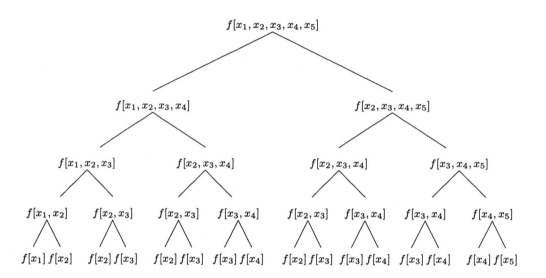

FIGURE 2.5 Divided differences

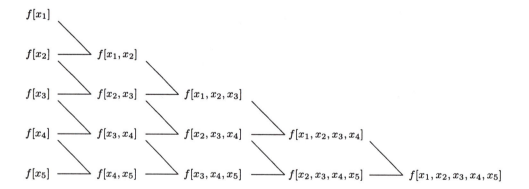

FIGURE 2.6 Efficient computation of divided differences

A number of simplifications result if the x_i are equally spaced. Assume that

$$x_i = x_1 + (i-1)h$$

where $h > 0$ is the spacing. From (2.4.3) we see that

$$f[x_1, \ldots, x_k] = \frac{f[x_2, \ldots, x_k] - f[x_1, \ldots, x_{k-1}]}{h(k-1)}$$

This makes divided difference a scaling of the the *differences* $\Delta f[x_1, \ldots, x_k]$ which we define by

$$\Delta f[x_1, \ldots, x_k] = \begin{cases} f(x_1) & \text{if } k = 1 \\ \Delta f[x_2, \ldots, x_k] - \Delta f[x_1, \ldots, x_{k-1}] & \text{if } k > 1 \end{cases}$$

For example,

0th Order	1st Order	2nd Order	3rd Order	4th Order
f_1				
f_2	$f_2 - f_1$			
f_3	$f_3 - f_2$	$f_3 - 2f_2 + f_1$		
f_4	$f_4 - f_3$	$f_4 - 2f_3 + f_2$	$f_4 - 3f_3 + 3f_2 - f_1$	
f_5	$f_5 - f_4$	$f_5 - 2f_4 + f_3$	$f_5 - 3f_4 + 3f_3 - f_2$	$f_5 - 4f_4 + 6f_3 - 4f_2 + f_1$

It is not hard to show that

$$f[x_1, \ldots, x_k] = \frac{\Delta f[x_1, \ldots, x_k]}{h^{k-1}(k-1)!}$$

The built-in function `diff` can be used to compute differences. In particular, if `y` is an n-vector, then

    d = diff(y)

and

    d = y(2:n) - y(1:n-1)

are equivalent. A second argument can be used to compute higher-order differences. For example,

    d = diff(y,2)

computes the second-order differences:

    d = y(3:n) - 2*y(2:n-1) + y(1:n-2)

Problems

P2.4.1 Compare the computed c_i produced by `InterpN` and `InterpN2`.

P2.4.2 Complete the following MATLAB function:

    function [c,x,y] = InterpNEqual(fname,L,R,n)

Make effective use of the `diff` function.

2.4.2   Inverse Interpolation

Suppose the function $f(x)$ has an inverse on $[a, b]$. This means that there is a function g so that $g(f(x)) = x$ for all $x \in [a, b]$. Thus $g(x) = \sqrt{x}$ is the inverse of $f(x) = x^2$ on $[0, 1]$. If

$$a = x_1 < x_2 < \cdots < x_n = b$$

and $y_i = f(x_i)$, then the polynomial that interpolates the data (y_i, x_i), $i = 1{:}n$ is an interpolate of f's inverse. Thus the script

```
x = linspace(0,1,6);
y = x.*x;
a = InterpV(y,x);
yvals = linspace(y(1),y(6));
xvals = HornerV(a,yvals);
plot(yvals,xvals);
```

plots a quintic interpolant of the square root function. This is called *inverse* interpolation, and it has an important application in zero finding. Suppose $f(x)$ is continuous and either monotone increasing or decreasing on $[a, b]$. If $f(a)f(b) < 0$, then f has a zero in $[a, b]$. If $q(y)$ is an inverse interpolant, then $q(0)$ can be thought of as an approximation to this root.

Problems

P2.4.3 Suppose we have three data points (x_1, y_1), (x_2, y_2), and (x_3, y_3) with the property that $x_1 < x_2 < x_3$ and that y_1 and y_3 are opposite in sign. Write a function `root = InverseQ(x,y)` that returns the value of the inverse quadratic interpolant at 0.

2.4.3   Interpolation in Two Dimensions

Suppose $(\tilde{x}, \tilde{y})$ is inside the rectangle

$$R = \{(x, y) : a \le x \le b, \qquad c \le y \le d\}$$

Suppose $f(x, y)$ is defined on R and that we have its values on the four corners $f_{ac} = f(a, c)$, $f_{bc} = f(b, c)$, $f_{ad} = f(a, d)$, $f_{bd} = f(b, d)$. Our goal is to use linear interpolation to obtain an estimate of $f(\tilde{x}, \tilde{y})$. Suppose $\lambda \in [0, 1]$ with the property that $\tilde{x} = (1 - \lambda)a + \lambda b$. It follows that

$$
\begin{aligned}
f_{xc} &= (1 - \lambda)f_{ac} + \lambda f_{bc} \\
f_{xd} &= (1 - \lambda)f_{ad} + \lambda f_{bd}
\end{aligned}
$$

are linearly interpolated estimates of $f(\tilde{x}, c)$ and $f(\tilde{x}, d)$, respectively. Consequently, if $\mu \in [0, 1]$ with $\tilde{y} = (1 - \mu)c + \mu d$, then a second interpolation between f_1 and f_2 gives an estimate of $f(\tilde{x}, \tilde{y})$:

$$z = (1 - \mu)f_{xc} + \mu f_{xd} \approx f(\tilde{x}, \tilde{y})$$

Putting it all together, we see that

$$
\begin{aligned}
z &= (1 - \mu)((1 - \lambda)f_{ac} + \lambda f_{bc}) + \mu((1 - \lambda)f_{ad} + \lambda f_{bd}) \\
&\approx f((1 - \lambda)a + \lambda b, (1 - \mu)c + \mu d)
\end{aligned}
$$

Figure 2.7 depicts the interpolation points. To try out these ideas, we write a function that stores in a matrix the values of a specified function of two variables:

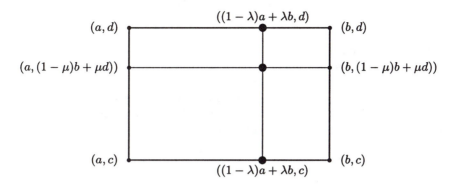

FIGURE 2.7 Linear interpolation in two dimensions

```
        function fA = SetUp(fname,a,b,n,c,d,m)
%
% Pre:
%   fname: string that names a valid function.
%   a,b: scalars that satisfy a<=b
%   c,d: scalars that satisfy c<=d
%   n,m:   integers >=2.
%
% Post:
%   fA:     an n-by-m matrix with the property that
%           A(i,j) = f(a+(i-1)(b-a)/(n-1),c+(j-1)(d-c)/(m-1))
%           where f is the function named by fname..
%
%
    x = linspace(a,b,n);
    y = linspace(c,d,m);
    fA = zeros(n,m);
    for i=1:n
       for j=1:m
          fA(i,j) = feval(fname,x(i),y(j));
       end
    end
```

In general, the function passed to feval can have an arbitrary number of arguments. If n arguments are required, then they must also be specified in the call to feval.

To interpolate the values in a matrix computed by SetUp, it is necessary to "locate" the point at which the interpolation is required. The four relevant values from the array must then be combined as described previously.

```
   function z = LinInterp2D(x,y,a,b,n,c,d,m,fA)
%
% Pre:
%    n,m:        integers >=2
%    x,a,b:      scalars that satisfy a<=x<=b
%    y,c,d:      scalars that satisfy c<=y<=d
%    fA:         an n-by-m matrix with the property that
%                fA(i,j) = f(a+(i-1)(b-a)/(n-1),c+(j-1)(d-c)/(m-1))
%                for some function f(.,.).
% Post:
%    z:          linearly interpolated value for f(x,y).
%
  hx = (b-a)/(n-1);
  i  = max([1 ceil((x-a)/hx)]);
  dx = (x - (a+(i-1)*hx))/hx;
% x  = a+(i-1+dx)*hx      0<=dx<=1

  hy = (d-c)/(m-1);
  j  = max([1 ceil((y-c)/hy)]);
  dy = (y - (c+(j-1)*hy))/hy;
% y  = c+(j-1+dy)*hy      0<=dy<=1

  z = (1-dy)*((1-dx)*fA(i,j)+dx*fA(i+1,j)) + dy*((1-dx)*fA(i,j+1)+dx*fA(i+1,j+1));
```

The script file TableLookUp2D illustrates the use of these functions.

Problems

P2.4.4 Write a function CubicInterp2D. Determine i and j as in LinearInterp2D. Let $c_r(x)$ be the cubic interpolant of f at (x_{i-1}, y_r), (x_i, y_r), (x_{i+1}, y_r), and (x_{i+2}, y_r), $r = j-1:j+2$. Evaluate these four cubics at x. With the four values so obtained, construct another cubic interpolant $d(y)$ which then can be evaluated at y.

2.4.4 MATLAB's Polynomial Tools

MATLAB has a number of functions that support computation with polynomials. The script

```
x = [-2 3 1];
y = [-15 -5 3];
a = polyfit(x,y,3)
xvals = linspace(-3,2,100);
pvals = polyval(a,xvals);
plot(xvals,pvals)
```

plots the polynomial interpolant of the data $(-2, -15)$, $(3, -5)$, and $(1, 3)$. The interpolant is given by $p(x) = 1 + 4x - 2x^2$ and the call to polyfit computes this polynomial. However, MATLAB represents polynomials in the reverse of our Vandermonde approach. In particular, the vector a will be assigned the vector [-2 4 1]. In general, if x and y are n-vectors, then a = polyfit(x,y,n) assigns a length-n vector to a with the property that the polynomial

$$p(x) = a_n + a_{n-1}x + a_{n-2}x^2 + \cdots + a_1 x^{n-1}$$

interpolates the data $(x_1, y_1), \ldots, (x_n, y_n)$. The function `polyval` is used to evaluate polynomials in the MATLAB representation. In our example, `polyval(a,xvals)` is a vector of interpolant evaluations.

Problems

P2.4.5 Write a function `PlotDerPoly(x,y)` that plots the derivative of the polynomial interpolant of the data (x_i, y_i), $i = 1{:}n$. Assume that $x_1 < \cdots < x_n$ and the plot should be across the interval $[x_1, x_n]$. Use `polyfit` and `polyval`.

M-Files and References

Script Files

ShowV	Illustrates InterpV and HornerV
ShowN	Illustrates InterpN and HornerN
InterpEff	Compares Vandermonde and Newton approaches.
RungeEg	Examines accuracy of interpolating polynomial.
TableLookUp2D	Illustrates SetUp and LinInterp2D.

Function Files

InterpV	Construction of Vandermonde interpolating polynomial.
HornerV	Evaluates the Vandermonde interpolating polynomial.
InterpNRecur	Recursive construction of the Newton interpolating polynomial.
InterpN	Nonrecursive construction of the Newton interpolating polynomial.
HornerN	Evaluates the Newton interpolating polynomial.
SetUp	Sets up matrix of f(x,y) evaluation.
LinInterp2D	2-Dimensional Linear Interpolation.

References

S.D. Conte and C. de Boor (1980). *Elementary Numerical Analysis: An Algorithmic Approach, Third Edition*, McGraw-Hill, New York.

P. Davis (1963). *Interpolation and Approximation*, Blaisdell, New York.

Chapter 3

Piecewise Polynomial Interpolation

§**3.1** Piecewise Linear Interpolation

§**3.2** Piecewise Cubic Hermite Interpolation

§**3.3** Cubic Splines

An important lesson from the previous chapter is that high-degree polynomial interpolants should generally be avoided. This can pose a problem if we are to produce an accurate interpolant across a wide interval $[\alpha, \beta]$. One way around this difficulty is to partition $[\alpha, \beta]$,

$$\alpha = x_1 < x_2 < \cdots < x_n = \beta$$

and then interpolate the given function on each subinterval $[x_i, x_{i+1}]$ with a polynomial of low degree. This is the *piecewise polynomial* interpolation idea. The x_i are called *breakpoints*.

We begin with piecewise linear interpolation working with both fixed and adaptively determined breakpoints. The latter requires a classical divide-and-conquer approach that we shall use again in later chapters.

Piecewise linear functions do not have a continuous first derivative, and this creates problems in certain applications. Piecewise cubic hermite interpolants address this issue. In this setting, the value of the interpolant and its derivative is specified at each breakpoint. The local cubics join in a way that forces first derivative continuity.

Second derivative continuity can be achieved by carefully choosing the first derivative values at the breakpoints. This leads to the topic of splines, a very important idea in the area of approximation and interpolation. It turns out that cubic splines produce the smoothest solution to the interpolation problem.

3.1 Piecewise Linear Interpolation

Assume that $x(1{:}n)$ and $y(1{:}n)$ are given where $\alpha = x_1 < \cdots < x_n = \beta$ and $y_i = f(x_i)$, $i = 1{:}n$. If you connect the dots $(x_1, y_1), \ldots, (x_n, y_n)$ with straight lines, as in Fig 3.1, then the graph

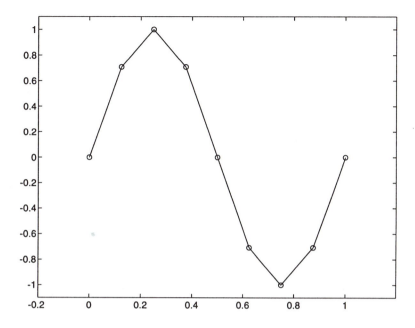

FIGURE 3.1 A piecewise linear function

of a *piecewise linear* function is displayed. We already have considerable experience with such functions, for this is what `plot(x,y)` displays.

3.1.1   Set-Up

The piecewise linear interpolant is built upon the local linear interpolants

$$L_i(z) = a_i + b_i(x - x_i)$$

where for $i = 1:n-1$ the coefficients are defined by

$$a_i = y_i \qquad b_i = \frac{y_{i+1} - y_i}{x_{i+1} - x_i}$$

Note that $L_i(z)$ is just the linear interpolant of f at the points $x = x_i$ and $x = x_{i+1}$. We then define

$$L(z) = \begin{cases} L_1(z) & \text{if} & x_1 \le z < x_2 \\ L_2(z) & \text{if} & x_2 \le z < x_3 \\ \;\;\vdots & & \;\;\vdots \\ L_{n-1}(z) & \text{if} & x_{n-1} \le z \le x_n \end{cases}$$

The act of setting up L is the act of solving each of the local linear interpolation problems. The $n - 1$ divided differences $b_1, \ldots, b_{n-1}$ can obviously be computed by a loop,

```
for i=1:n-1
    b(i) = (y(i+1)-y(i))/(x(i+1)-x(i));
end
```

or by using pointwise division,

```
b = (y(2:n)-y(1:n-1)) ./ (x(2:n)-x(1:n-1))
```

or by using the built-in function `diff`:

```
b = diff(y) ./ diff(x)
```

Packaging these operations we obtain

```
function [a,b] = pwL(x,y)
%
% Pre:
%   x    column n-vector with x(1) < x(2) < ... < x(n)
%   y    column n-vector.
%
% Post:
%   a,b  column (n-1)-vectors with the property that the
%        line L(z) = a(i) + b(i)z passes though the points
%        (x(i),y(i)) and (x(i+1),y(i+1)).
%
  n = length(x);
  a = y(1:n-1);
  b = diff(y) ./ diff(x);
```

Thus

```
z = linspace(0,1,9);
[a,b] = pwL(z,sin(2*pi*z));
```

sets up a piecewise linear interpolant of $\sin(2\pi z)$ on a uniform, nine-point partition of $[0,1]$.

3.1.2   Evaluation

To evaluate L at a point $z \in [\alpha, \beta]$, it is necessary to determine the subinterval that contains z. In our problem x has the property that $x_1 < \cdots < x_n$ and so `sum(x<=z)` is the number of x_i that are to the right of z or equal to z. It follows that

```
if z == x(n);
    i = n-1;
else
    i = sum(x<z);
end
```

determines the index i so that $x_i \leq z \leq x_{i+1}$. Notice the special handling of the case when z equals x_n. (Why?) A total of n comparisons are made because every component in x is compared to z.

A better approach is to exploit the monotonicity of the x_i and to use binary search. Here is the main idea. Suppose we have indices Left and Right so that $x_{\text{Left}} \leq z \leq x_{\text{Right}}$. If mid = floor((Left+Right)/2), then by checking z's relation to x_{mid} we can halve the search space by redefining Left or Right accordingly:

```
mid = floor((Left+Right)/2);
if z < x(mid)
    Right = mid;
else
    Left = mid;
end
```

Repeated application of this process eventually identifies the subinterval that houses z:

```
if z == x(n)
    i = n-1;
else
    Left = 1;
    Right = n;
    while Right > Left+1
        % z is in [x(Left),x(Right)].
        mid = floor((Left+Right)/2);
        if z < x(mid)
            Right = mid;
        else
            Left = mid;
        end
    end
end
i = Left;
```

Upon completion, i contains the index of the subinterval that contains z. If $n = 10$ and $z \in [x_6, x_7]$, then here is the succession of Left and Right values produced by the binary search method:

Left	Right	mid
1	10	5
5	10	7
5	7	6
6	7	-

Roughly $\log_2(n)$ comparisons are required to locate the appropriate subinterval. If n is large, then this is much more efficient than the sum(x<z) method, which requires n comparisons.

For "random" z, we can do no better than binary search. However, if L is to be evaluated at an ordered succession of points, then we can improve the subinterval location process. For example, suppose we want to plot L on $[\alpha, \beta]$. This requires the assembly of the values $L(z_1), \ldots, L(z_m)$ in a vector where m is a typically large integer and

$$\alpha \leq z_1 \leq \cdots \leq z_m \leq \beta$$

Rather than locate each z_i via binary search, it is more efficient to exploit the systematic "migration" of the evaluation point as it moves left to right across the subintervals. Chances are that if i is the subinterval index associated with the current z-value, then i will be the correct index for the next z-value. This "guess" at the correct subinterval can be checked before we launch the binary search process.

```
    function i = Locate(x,z,g)
%
% Pre:
%   x    column n-vector with x(1) < x(2) <...<x(n).
%   z    scalar with x(1) <= z <= x(n).
%   g    optional 3rd argument that satisfies 1 <= g <= n-1.
% Post:
%   i    integer such that x(i) <= z <= x(i+1).
%        Before the general search for i begins,
%        the value i=g is tried.
%
  if nargin==3
    % Try the initial guess.
    if (x(g)<=z) & (z<=x(g+1))
       i = g;
       return
    end
  end
  n = length(x);
  if z==x(n)
     i = n-1;
  else
     % Binary Search
     Left = 1;
     Right = n;
     while Right > Left+1
        % x(Left) <= z <= x(Right)
        mid = floor((Left+Right)/2);
        if z < x(mid)
           Right = mid;
        else
           Left = mid;
        end
     end
     i = Left;
  end
```

Locate makes use of the **return** command. This terminates the execution of the function. It is possible to restructure Locate to avoid the **return**, but the resulting logic would be cumbersome.

As an application of locate, here is a function that produces a vector of *L*-values:

```
function LVals = pwLEval(a,b,x,zvals)
%
% Pre:
%       x    column n-vector with x(1) < ... < x(n).
%    zvals   column m-vector with each component in [x(1),x(n)].
%     a,b    column (n-1)-vectors
%
% Post:
%    LVals   column m-vector with the property that
%            LVals(j) = L(zvals(j)) for j=1:m where
%            L(z) = a(i) + b(i)(z-x(i)) for x(i)<=z<=x(i+1).
%
  m = length(zvals);
  LVals = zeros(m,1);
  g = 1;
  for j=1:m
     i = Locate(x,zvals(j),g);
     LVals(j) = a(i) + b(i)*(zvals(j)-x(i));
     g = i;
  end
```

The following script illustrates the use of this function, producing a sequence of piecewise linear approximations to the built-in function

$$humps(x) = \frac{1}{(x-.3)^2 + .01} + \frac{1}{(x-.9)^2 + .04} - 6$$

```
% Script File: ShowPWL1
%
% Convergence of the piecewise linear interpolant to
% humps(x) on [0,2*pi].

  z = linspace(0,3,200)';
  fvals = humps(z);
  for n = [5 10 25 50]
     x = linspace(0,3,n)';
     y = humps(x);
     [a,b] = pwL(x,y);
     Lvals = pwLEval(a,b,x,z);
     plot(z,Lvals,z,fvals,x,y,'*');
     title(sprintf('Interpolation of humps(x) with pwL, n = %2.0f',n))
     pause
  end
```

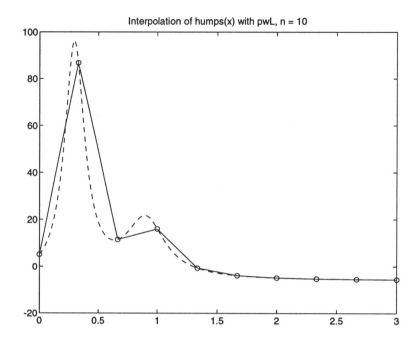

FIGURE 3.2 Piecewise linear approximation

(See Fig 3.2 for the 10-point case.) Observe that more interpolation points are required in regions where *humps* is particularly nonlinear.

3.1.3   Á Priori Determination of the Breakpoints

Let us consider how many breakpoints we need to obtain a satisfactory piecewise linear interpolant. If $z \in [x_i, x_{i+1}]$, then from Theorem 2

$$f(z) = L(z) + \frac{f^{(2)}(\eta)}{2}(z - x_i)(z - x_{i+1})$$

where $\eta \in [x_i, x_{i+1}]$. If the second derivative of f on $[\alpha, \beta]$ is bounded by M_2 and if $\bar{h}$ is the length of the longest subinterval in the partition, then it is not hard to show that

$$|f(z) - L(z)| \ \leq \ \frac{M_2 \bar{h}^2}{8}$$

for all $z \in [\alpha, \beta]$.

A typical situation where this error bound can be put to good use is in the design of the underlying partition upon which L is based. Assume that $L(x)$ is based on the uniform partition

$$\alpha = x_1 < x_2 < \cdots < x_n = \beta$$

where

$$x_i = \alpha + \frac{i-1}{n-1}(\beta - \alpha)$$

To ensure that the error between L and f is less than or equal to a given positive tolerance δ, we insist that

$$|f(z) - L(z)| \leq \frac{M_2 \bar{h}^2}{8} = \frac{M_2}{8}\left(\frac{\beta - \alpha}{n-1}\right)^2 \leq \delta.$$

From this we conclude that n must satisfy

$$n \geq 1 + (\beta - \alpha)\sqrt{M_2/8\delta}.$$

For the sake of efficiency, it makes sense to let n be the smallest integer that satisfies this inequality:

```
    function [x,y] = pwLStatic(fname,M2,alpha,beta,delta)
%
% Pre:
%   fname        string that names an available function f(x).
%                Assume that f can take vector arguments.
%   M2           an upper bound for|f"(x)| on [alpha,beta].
%   alpha,beta   scalars with alpha<beta
%   delta        positive scalar
%
% Post:
%   x,y          column n-vectors with the property that y(i) = f(x(i)),
%                i=1:n. The piecewise linear interpolant L(x) of this
%                data satisfies |L(x) - f(x)| <= delta, where
%                x = linspace(alpha,beta,n).

    n = max(2,ceil(1+(beta-alpha)*sqrt(M2/(8*delta))));
    x = linspace(alpha,beta,n)';
    y = feval(fname,x);
```

The partition produced by pwLStatic *does not* taken into account the sampled values of f. As a result, the uniform partition produced may be much too refined in regions where f'' is much smaller than the bound M_2.

3.1.4 Adaptive Piecewise Linear Interpolation

Suppose f is very nonlinear over just a small portion of $[\alpha, \beta]$ and very smooth elsewhere. (See Fig 3.2.) This means that if we use pwLStatic to generate the partition, then we are compelled to use a large M2. Lots of subintervals and (perhaps costly) f-evaluations will be required. Over regions where $f(x)$ is smooth, the partition will be overly refined.

To address this problem, we develop a recursive partitioning algorithm that "discovers" where f is "extra nonlinear" and that clusters the breakpoints accordingly. A definition simplifies the discussion. We say that the subinterval $[xL, xR]$ is *acceptable* if

$$\left| f\left(\frac{xL + xR}{2} \right) - \frac{f(xL) + f(xR)}{2} \right| \leq \delta$$

or if

$$xR - xL \leq h_{min}$$

where $\delta > 0$ and $h_{min} > 0$ are (typically small) refinement parameters. The first condition measures the discrepancy between the line that connects $(xL, f(xL))$ and $(xR, f(xR))$ and the function $f(x)$ at the interval midpoint $m = (xL + xR)/2$. The second condition says that sufficiently short subintervals are also acceptable where "sufficiently short" means less than h_{min} in length.

One more definition is required before we can describe the complete partitioning process. A partition $x_1 < \cdots < x_n$ is *acceptable* if each subinterval is acceptable. Note that if

$$xL = x_1^{(L)} < \cdots < x_n^{(L)} = m$$

is an acceptable partition of $[xL, m]$ and if

$$m = x_1^{(R)} < \cdots < x_n^{(R)} = xR$$

is an acceptable partition of $[m, xR]$, then

$$xL = x_1^{(L)} < \cdots < x_n^{(L)} < x_2^{(R)} < \cdots < x_n^{(R)} = xR$$

is an acceptable partition of $[xL, xR]$. This sets the stage for a recursive determination of an acceptable partition:

```
    function [x,y] = pwLAdapt(fname,xL,fL,xR,fR,delta,hmin)
%
%   Pre:
%   fname        string that names an available function of the form
%                y = f(u).
%   xL,fL        real, fL = f(xL)
%   xR,fR        real, fR = f(xR)
%   delta        positive real
%   hmin         positive real
%
%   Post:
%        x       column n-vector with the property that
%                xL = x(1) < ... < x(n) = xR. Each subinterval
%                is either <= hmin in length or has the property
%                that at its midpoint m, |f(m) - L(m)| <= delta
%                where L(x) is the line that connects (xR,fR).
%
%        y       column n-vector with the property  that
%                        y(i) = f(x(i)).
```

```
    if (xR-xL) <= hmin
       % Subinterval is acceptable
       x = [xL;xR];
       y = [fL;fR];
    else
       mid  = (xL+xR)/2;
       fmid = feval(fname,mid);
       if (abs(((fL+fR)/2) - fmid) <= delta )
         % Subinterval accepted.
          x = [xL;xR];
          y = [fL;fR];
       else
         % Produce left and right partitions, then synthesize.
          [xLeft,yLeft]   = pwLAdapt(fname,xL,fL,mid,fmid,delta,hmin);
          [xRight,yRight] = pwLAdapt(fname,mid,fmid,xR,fR,delta,hmin);
          x = [ xLeft;xRight(2:length(xRight))];
          y = [ yLeft;yRight(2:length(yRight))];
       end
    end
 end
```

The idea behind the function is to check and see if input interval is acceptable. If it is not, then acceptable partitions are obtained for the left and right half intervals. These are then "glued" together to obtain the final, acceptable partition.

The distinction between static and adaptive piecewise linear interpolation is revealed by running the following script:

```
% Script File: ShowPWL2
%
% Compares pwLStatic and pwLAdapt on [0,1] using the function
%   humps(x) = 1/((x-.3)^2 + .01)  +  1/((x-.9)^2+.04).
%
   close all
   % Second derivative estimate based on divided differences
   z = linspace(0,1,101);
   humpvals = humps(z);
   M2 = max(abs(diff(humpvals,2)/(.01)^2));
   for delta = [1 .5 .1 .05 .01]
      figure
      [x,y] = pwlStatic('humps',M2,0,1,delta);
      subplot(1,2,1)
      plot(x,y,'.');
      title(sprintf('delta = %8.4f Static  n= %2.0f',delta,length(x)))
      [x,y] = pwlAdapt('humps',0,humps(0),1,humps(1),delta,.001);
      subplot(1,2,2)
      plot(x,y,'.');
      title(sprintf('delta = %8.4f  Adapt n= %2.0f',delta,length(x)))
   end
```

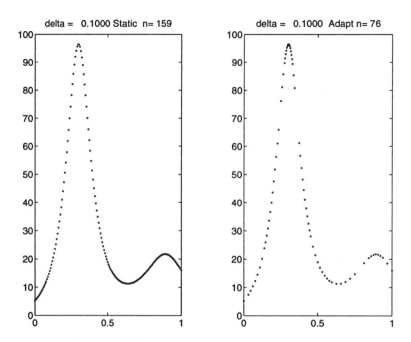

FIGURE 3.3 Static versus adaptive approximation

(See Fig 3.3.) The humps function is very nonlinear in the vicinity of $x = .3$. A second derivative bound is approximated with differences and used in pwLStatic. In the example approximately twice as many function evaluations are required when the static approach is taken.

Problems

P3.1.1 Generalize Locate so that it tries $i = g + 1$ and $i = g - 1$ before resorting to binary search. (Take care to guard against subscript out-of-range.) Implement pwLEval with this modified subinterval locator and document the speed-up.

P3.1.2 Write a function i = LocateUniform(alpha,beta,n,z) that returns the index of the interval that houses z given that $[\alpha, \beta]$ is uniformly partitioned into $n - 1$ subintervals.

P3.1.3 What happens if pwLAdapt is applied to $\sin(x)$ with $[\alpha, \beta] = [0, 2\pi]$?

P3.1.4 Describe what would happen if pwLAdapt is called with delta = 0.

P3.1.5 Describe why the number of recursive calls in pwLAdapt is bounded if $|f''(x)|$ is bounded on $[\alpha, \beta]$.

P3.1.6 Modify pwLAdapt so that a subinterval is accepted if $|f(p) - \lambda(p)|$ *and* $|f(q) - \lambda(q)|$ are less than or equal to delta, where $p = (2xL + xR)/3$, $q = (xL + 2xR)/3$, and $\lambda(x)$ is the line that connects (xL, fL) and (xR, fR). Avoid redundant function evaluations.

P3.1.7 If pwLAdapt is applied to the function $f(x) = \sqrt{x}$ on the interval $[0,1]$, then a partition $x(1:n)$ is produced that satisfies $x_2 - x_1 \le x_3 - x_2 \le \cdots \le x_n - x_{n-1}$. Why?

P3.1.8 Generalize pwLAdapt(fname,xL,fL,xR,fR,delta,hmin) to

    function [x,y,eTaken] = pwLAdapt(fname,xL,fL,xR,fR,delta,hmin,eMax)

so that no more than eMax function evaluations are taken. The value of eTaken should be the actual number of function evaluations spent. Let $n = length(x)$. In a "successful" call, x(n) should equal xR, meaning that a satisfactory piecewise linear approximation was found extending across the entire interval $[xL, xR]$. If this is not the case, then the evaluation limit was encountered before xR was reached and x(n) will be less than xR. In this situation vectors returned define a satisfactory piecewise linear approximation across $[x(1), x(n)]$.

P3.1.9 Notice that in pwLAdapt the vector y does not include all the computed function evaluations. So that these evaluations are not lost, generalize pwLAdapt to

    [x,y,xUnused,yUnused] = pwLAdaptGen(fname,xL,fL,xR,fR,delta,hmin,...)

where (1) the x and y vectors are identical to what pwLAdapt computes and (2) xUnused and yUnused are column vectors that contain the x-values and function values that were computed, but not included in x and y. Thus the xUnused and yUnused vectors should have the property that yUnused(i) = feval(fname,xUnused(i)), i = 1:length(xUnused). You are allowed to extend the calling sequence if convenient. In that case, indicate the values that should be passed through these new parameters at the top-level call. xUnused and yUnused should be assigned the empty vector [] if xR-xL<hmin. The order of the values in xUnused is not important.

3.2 Piecewise Cubic Hermite Interpolation

Now let's graduate to piecewise cubic functions. With the increase in degree we can obtain a smoother fit to a given set of n points. The idea is to interpolate both f and its derivative with a cubic on each of the subintervals.

3.2.1 Cubic Hermite Interpolation

So far we have only considered the interpolation of function values at distinct points. In the *Hermite* interpolation problem, both the function and its derivative are interpolated. To illustrate the idea, we consider the interpolation of the function $f(z) = \cos(z)$ at the points $x_1 = 0$, $x_2 = \delta$, $x_3 = 3\pi/2 - \delta$, and $x_4 = 3\pi/2$ by a cubic $p_3(z)$. For small δ we notice that $p_3(z)$ seems to interpolate both f and f' at $z = 0$ and $z = 3\pi/2$. The interpolation shown in Fig 3.4 was obtained by running the following script:

```
% Script File: ShowHermite
%
% Plots the cubic interpolant to cos(x)
% at x = [ 0 delta 3*pi/2-delta 3*pi/2]
% where delta gets increasingly small.
%
  close all

  z = linspace(-pi/2,2*pi,100);
  CosValues = cos(z);
```

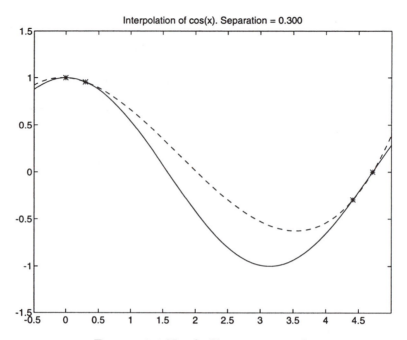

FIGURE 3.4 Nearly Hermite interpolation

```
for delta = [1 .5  .3  .1  .05 .001]
   figure
   xvals = [0;delta;(3*pi/2)-delta;3*pi/2];
   yvals = cos(xvals);
   c = InterpN(xvals,yvals);
   CubicValues = HornerN(c,xvals,z);
   plot(z,CosValues,z,CubicValues,xvals,yvals,'*')
   axis([-.5 5 -1.5 1.5])
   title(sprintf('Interpolation of cos(x). Separation = %5.3f',delta))
end
```

As the points coalesce, the cubic converges to a cubic interpolant of the cosine and its derivative at the points 0 and $3\pi/2$. This is called the *Hermite cubic interpolant*.

In the general cubic Hermite interpolation problem, we are given function values y_L and y_R and derivative values s_L and s_R and seek coefficients a, b, c, and d so that if

$$q(z) = a + b(z - x_L) + c(z - x_L)^2 + d(z - x_L)^2(z - x_R)$$

then

$$q(x_L) = y_L \qquad q(x_R) = y_R$$
$$q'(x_L) = s_L \qquad q'(x_R) = s_R$$

Each of these equations "says" something about the unknown coefficients. Noting that

$$q'(z) = b + 2c(z - x_L) + d(2(x - x_L)(z - x_R) + (z - x_L)^2)$$

we see that

$$a. = y_L \qquad a + b\Delta x + c(\Delta x)^2 = y_R$$
$$b = s_L \qquad b + 2c\Delta x + d(\Delta x)^2 = s_R$$

where $\Delta x = x_R - x_L$. Expressing this in matrix-vector we obtain

$$
\begin{bmatrix}
1 & 0 & 0 & 0 \\
0 & 1 & 0 & 0 \\
1 & \Delta x & (\Delta x)^2 & 0 \\
0 & 1 & 2\Delta x & (\Delta x)^2
\end{bmatrix}
\begin{bmatrix}
a \\ b \\ c \\ d
\end{bmatrix}
=
\begin{bmatrix}
y_L \\ s_L \\ y_R \\ s_R
\end{bmatrix}
$$

The solution to this triangular system is straightforward:

$$
\begin{aligned}
a &= y_L \\
b &= s_L \\
c &= \frac{y_L' - s_L}{\Delta x} \\
d &= \frac{s_R + s_L - 2y_L'}{(\Delta x)^2}
\end{aligned}
$$

where

$$
y_L' = \frac{y_R - y_L}{\Delta x} = \frac{y_R - y_L}{x_R - x_L}
$$

Thus we obtain

```
function [a,b,c,d] = HCubic(xL,yL,sL,xR,yR,sR)
%
% Pre:
%    xL,xR    distinct reals
%
% Post:
%  a,b,c,d    real numbers with the property that if
%             p(z) = a + b(z-xL) + c(z-xL)^2 + d(z-xL)^2(z-xR)
%             then p(xL)=yL, p'(xL)=sL, p(xR)=yR, p'(xR)=sR.
%
  a = yL;
  b = sL;
  delx = xR - xL;
  yp = (yR - yL)/delx;
  c = (yp - sL)/delx;
  d = (sL - 2*yp + sR)/(delx*delx);
```

An error expression for the cubic Hermite interpolant can be derived from Theorem 3.

Theorem 3 *Suppose $f(z)$ and its first four derivatives are continuous on $[x_L, x_R]$ and that the constant M_4 satisfies*

$$|f^{(4)}(z)| \leq M_4$$

for all $z \in [L, R]$. If q is the cubic Hermite interpolant of f at x_L and x_R, then

$$|f(z) - q(z)| \leq \frac{M_4}{384} h^4$$

where $h = x_R - x_L$.

Proof. If $q_\delta(z)$ is the cubic interpolant of f at x_L, $x_L + \delta$, $x_R - \delta$, and x_R, then from Theorem 2 we have

$$|f(z) - q_\delta(z)| \leq \frac{M_4}{24} |(z - x_L)(z - x_L - \delta)(z - x_R + \delta)(z - x_R)|$$

for all $z \in [x_L, x_R]$. As we have seen,

$$\lim_{\delta \to 0} q_\delta(z) = q(z)$$

and so

$$|f(z) - q(z)| \leq \frac{M_4}{24} |(z - x_L)(z - x_L)(z - x_R)(z - x_R)|$$

The maximum value of the quartic polynomial on the right occurs at the midpoint $z = x_L + h/2$ and so for all z in the interval $[x_L, x_R]$ we have

$$|f(z) - q(z)| \leq \frac{M_4}{24} \left(\frac{h}{2}\right)^4 = \frac{M_4}{384} h^4 \qquad \square$$

Theorem 3 says that if the interval length is divided by 10, then the error bound is reduced by a factor of 10^4.

3.2.2   Representation and Set-Up

We now show how to glue a sequence of Hermite cubic interpolants together so that the resulting piecewise cubic polynomial $C(z)$ interpolates the data $(x_1, y_1), \ldots, (x_n, y_n)$, with the prescribed slopes $s_1, \ldots, s_n$. To that end we assume $x_1 < x_2 < \cdots < x_n$ and define the ith *local cubic* by

$$q_i(z) = a_i + b_i(z - x_i) + c_i(z - x_i)^2 + d_i(z - x_i)^2(z - x_{i+1})$$

Define the piecewise cubic polynomial by

$$C(z) = \begin{cases} q_1(z) & \text{if} & x_1 \leq z < x_2 \\ q_2(z) & \text{if} & x_2 \leq z < x_3 \\ \vdots & & \vdots \\ q_{n-1}(z) & \text{if} & x_{n-1} \leq z \leq x_n \end{cases}$$

Our goal is to determine $a(1{:}n-1)$, $b(1{:}n-1)$, $c(1{:}n-1)$, and $d(1{:}n-1)$ so that

$$\begin{aligned} C(x_i) &= y_i \\ & \qquad\qquad i = 1{:}n \\ C'(x_i) &= s_i \end{aligned}$$

This will be the case if we simply solve the following $n - 1$ cubic Hermite problems:

$$
\begin{aligned}
q_i(x_i) &= y_i \\
q_i'(x_i) &= s_i \\
q_i(x_{i+1}) &= y_{i+1} \\
q_i'(x_{i+1}) &= s_{i+1}
\end{aligned}
$$

The results of §3.2.1 apply:

$$
a_i = y_i \qquad b_i = s_i \qquad c_i = \frac{y_i' - s_i}{\Delta x_i} \qquad d_i = \frac{s_{i+1} + s_i - 2y_i'}{(\Delta x_i)^2}
$$

where $\Delta x_i = x_{i+1} - x_i$ and

$$
y_i' = \frac{y_{i+1} - y_i}{\Delta x_i} = \frac{y_{i+1} - y_i}{x_{i+1} - x_i}
$$

We could use HCubic to resolve the coefficients:

```
for i=1:n-1
    [a(i), b(i), c(i), d(i)] = HCubic(x(i),y(i),s(i),x(i+1),y(i+1),s(i+1))
end
```

But a better solution is to vectorize the computation, and this gives

```
    function [a,b,c,d] = pwC(x,y,s)
%
% Pre:
%       x,y,s  column n-vectors with x(1) < ... < x(n)
%
% Post:
%       a,b,c,d  column (n-1)-vectors that define a
%                continuous, piecewise cubic polynomial q(z)
%                with the property that for i = 1:n,
%                q(x(i)) = y(i) and q'(x(i)) = s(i).
%                On the interval [x(i),x(i+1)], q(z) is
%                given by a(i) + b(i)(z-x(i)) + c(i)(z-x(i))^2
%                + d(i)(z-x(i))^2(z-x(i+1)).
%
    n  = length(x);
    a  = y(1:n-1);
    b  = s(1:n-1);
    Dx = diff(x);
    Dy = diff(y);
    yp = Dy ./ Dx;
    c  = (yp - s(1:n-1)) ./ Dx;
    d  = (s(2:n) + s(1:n-1) - 2*yp) ./ (Dx.* Dx);
```

If M_4 bounds $|f^{(4)}(x)|$ on the interval $[x_1, x_n]$, then Theorem 3 implies that

$$
|f(z) - C(z)| \le \frac{M_4}{384} \bar{h}^4
$$

for all $z \in [x_1, x_n]$, where $\bar{h}$ is the length of the longest subinterval (i.e., $\max_i |x_{i+1} - x_i|$.)

3.2.3   Evaluation

The evaluation of $C(z)$ has two parts. As with any piecewise polynomial that must be evaluated, the position of z in the partition must be ascertained. Once that is accomplished, the relevant local cubic must be evaluated. Here is a function that can be used to evaluate C at a vector of z values:

```
   function Cvals = pwCEval(a,b,c,d,x,z)
%
% Pre:
%     a,b,c,d   column (n-1)-vectors that represent a piecewise
%               cubic polynomial.
%     x         column n-vector with x(1) < ... < x(n).
%     z         column m-vector with each z(i) in [x(1),x(n)].
%
% Post:
%     Cvals     column m-vector with the property that
%               Cvals(j) = C(z(j)), j=1:m. If z(j) is in
%               the interval [x(i),x(i+1)], then
%               C(z(j)) = a(i) +
%                         b(i)(z(j)-x(i)) +
%                         c(i)(z(j)-x(i))^2 +
%                         d(i)(z(j)-x(i))^2(z(j)-x(i+1))
%
  m = length(z);
  Cvals = zeros(m,1);
  g=1;
  for j=1:m
     i = Locate(x,z(j),g);
     Cvals(j) = d(i)*(z(j)-x(i+1)) + c(i);
     Cvals(j) = Cvals(j)*(z(j)-x(i)) + b(i);
     Cvals(j) = Cvals(j)*(z(j)-x(i)) + a(i);
     g = i;
  end
```

Analogous to pwLeval, we use Locate to determine the subinterval that houses the jth evaluation point z_j. The cubic version of HornerN is then used to evaluate the appropriate local cubic.

The following script file illustrates the use of pwC and pwCEval:

```
% Script File: ShowpwCH
%
% Convergence of the piecewise cubic hermite  interpolant to
% exp(-2x)sin(10*pi*x) on [0,1].)

   close all
   z = linspace(0,1,200)';
   fvals = exp(-2*z).*sin(10*pi*z);
```

```
for n = [4 8 16 24]
    x = linspace(0,1,n)';
    y = exp(-2*x).*sin(10*pi*x);
    s = 10*pi*exp(-2*x).*cos(10*pi*x)-2*y;
    [a,b,c,d] = pwC(x,y,s);
    Cvals = pwCEval(a,b,c,d,x,z);
    figure
    plot(z,fvals,z,Cvals,x,y,'*');
    title(sprintf('Interpolation of exp(-2x)sin(10pi*x) with pwCH, n = %2.0f',n))
    pause
end
```

Sample output is displayed in Fig 3.5.

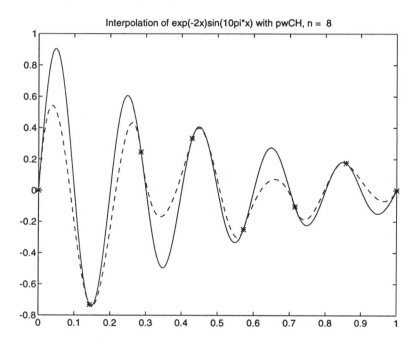

FIGURE 3.5 Piecewise cubic Hermite interpolant of $e^{-2x}\sin(10\pi x)$, $n = 8$

Problems

P3.2.1 Write a function [a,b,c,d] = pwCStatic(fname,fpname,M4,alpha,beta,delta) analogous to pwLStatic. It should produce a piecewise cubic Hermite approximation with uniform spacing. It should use the error result of Theorem 3 and the 4th derivative bound M4 to determine the partition. The parameters fname and fpname should house the name of the function and the derivative function.

P3.2.2 Write a recursive function

```
function [x,y,s] = pwCAdapt(fname,fpname,L,fL,DfL,R,fR,DfR,delta,hmin)
```

analogous to pwLAdapt. On each subinterval, the cubic Hermite interpolant should be the same kind of hmin/delta test used in pwLAdapt. The string fpname should name the derivative function.

Use both pwLAdapt and pwCAdapt to produce approximations to $f(x) = \sqrt{x}$ on the interval $[.001, 9]$. Fix $h_{min} = .001$. Print a table that shows the number of partition points computed by pwLRefine and pwCRefine for delta $= .1, .01, .001, .0001,$ and $.00001$.

P3.2.3 (1) Complete the following function:

```
    function [R,fR] = stretch(L,fL,tol);
%
% Pre:
%   L,fL       scalars that satisfy fL = exp(-L)
%   tol        positive real.
% Post:
%   R,fR       scalars that satisfy fR = exp(-R) with the
%              property that if  q(z) is the cubic hermite interpolant
%              of exp(-z) at z=L and z=R, then |q(z) - exp(-z)| <= tol on [L,R].
```

Make effective use of the error bound in Theorem 3 when choosing R. Hint: How big can you make R and still guarantee the required accuracy? (2) Assuming that stretch is available, complete the following:

```
    function [x,y] = pwCHExp(a,b,tol)
%
% Pre:
%   a,b        real scalars with a < b.
%   tol        positive real scalar.
% Post:
%   x,y        column n-vectors where a = x(1) < x(2) < ... < x(n) = b
%              and y(i) = exp(-x(i)), i=1:n.
%              The partition is chosen so that if C(z) is the piecewise cubic
%              hermite interpolant of exp(-z) on this partition,
%              then |C(z) - exp(-z)| <= tol for all z in [a,b]
```

Make effective use of stretch.

3.3 Cubic Splines

In the piecewise cubic Hermite interpolation problem, we are given n triplets

$$(x_1, y_1, s_1), \ldots, (x_n, y_n, s_n)$$

and determine a function $C(x)$ that is piecewise cubic on the partition $x_1 < \cdots < x_n$ with the property that $C(x_i) = y_i$ and $C'(x_i) = s_i$ for $i = 1:n$. This interpolation strategy is subject to a number of criticisms:

- Because the function $C(z)$ does not have a continuous second derivative, its display may be too crude in graphical applications because the human eye can detect discontinuities in the second derivative.

- In other applications where C and its derivatives are part of a larger mathematical landscape, there may be difficulties if $C''(x)$ is discontinuous. For example, trouble arises if C is a distance function.

- In experimental settings where the y_i are "instrument readings," we may not have the first derivative information required by the cubic Hermite process. Indeed, the underlying function f may not be known explicitly.

These reservations prompt us to pose the *cubic spline interpolation problem*:

> Given $(x_1, y_1), \ldots, (x_n, y_n)$ with $\alpha = x_1 < \cdots < x_n = \beta$, find a piecewise cubic interpolant $S(z)$ with the property that S, S', and S'' are continuous.

The function $S(z)$ that solves this problem is a *cubic spline interpolant*. This can be accomplished by *choosing* the appropriate slope values $s_1, \ldots, s_n$.

3.3.1 Continuity at the Interior Breakpoints

Assume that $S(z)$ is the cubic Hermite interpolant of the data (x_i, y_i, s_i) for $i = 1{:}n$. We ask the following question: Is it possible to choose $s_1, \ldots, s_n$ so that the second derivative of S is continuous? Let us look at what happens to S'' at each of the "interior" breakpoints $x_2, \ldots, x_{n-1}$. To the left of x_{i+1}, $S(z)$ is defined by the local cubic

$$q_i(z) = y_i + s_i(z - x_i) + \frac{y_i' - s_i}{\Delta x_i}(z - x_i)^2 + \frac{s_i + s_{i+1} - 2y_i'}{(\Delta x_i)^2}(z - x_i)^2(z - x_{i+1})$$

where $y_i' = (y_{i+1} - y_i)/(x_{i+1} - x_i)$ and $\Delta x_i = x_{i+1} - x_i$. The second derivative of this local cubic is given by

$$q_i''(z) = 2\frac{y_i' - s_i}{\Delta x_i} + \frac{s_i + s_{i+1} - 2y_i'}{(\Delta x_i)^2}[4(z - x_i) + 2(z - x_{i+1})] \qquad (3.3.1)$$

Likewise, to the right of x_{i+1} the piecewise cubic $C(z)$ is defined by

$$q_{i+1}(z) = y_{i+1} + s_{i+1}(x - x_{i+1}) + \frac{y_{i+1}' - s_{i+1}}{\Delta x_{i+1}}(z - x_{i+1})^2 + \frac{s_{i+1} + s_{i+2} - 2y_{i+1}'}{(\Delta x_{i+1})^2}(z - x_{i+1})^2(z - x_{i+2})$$

The second derivative of this local cubic is given by

$$q_{i+1}''(z) = 2\frac{y_{i+1}' - s_{i+1}}{\Delta x_{i+1}} + \frac{s_{i+1} + s_{i+2} - 2y_{i+1}'}{(\Delta x_{i+1})^2}[4(z - x_{i+1}) + 2(z - x_{i+2})] \qquad (3.3.2)$$

To force second derivative continuity at x_{i+1}, we insist that

$$q_i''(x_{i+1}) = \frac{2}{\Delta x_i}(2s_{i+1} + s_i - 3y_i')$$

and

$$q_{i+1}''(x_{i+1}) = \frac{2}{\Delta x_{i+1}}(3y_{i+1}' - 2s_{i+1} - s_{i+2})$$

be equal. That is,

$$\Delta x_{i+1}s_i + 2(\Delta x_i + \Delta x_{i+1})s_{i+1} + \Delta x_i s_{i+2} = 3(\Delta x_{i+1}y_i' + \Delta x_i y_{i+1}') \qquad (3.3.3)$$

for $i = 1{:}n - 2$. If we choose $s_1, \ldots, s_n$ to satisfy these equations, then $S''(z)$ is continuous.

Before we plunge into the resolution of these equations for general n, we acquire some intuition by examining the $n = 7$ case. The equations designated by (3.3.3) are as follows:

$$
\begin{aligned}
i = 1 &\Rightarrow \Delta x_2 s_1 + 2(\Delta x_1 + \Delta x_2)s_2 + \Delta x_1 s_3 = 3(\Delta x_2 y_1' + \Delta x_1 y_2') \\
i = 2 &\Rightarrow \Delta x_3 s_2 + 2(\Delta x_2 + \Delta x_3)s_3 + \Delta x_2 s_4 = 3(\Delta x_3 y_2' + \Delta x_2 y_3') \\
i = 3 &\Rightarrow \Delta x_4 s_3 + 2(\Delta x_3 + \Delta x_4)s_4 + \Delta x_3 s_5 = 3(\Delta x_4 y_3' + \Delta x_3 y_4') \\
i = 4 &\Rightarrow \Delta x_5 s_4 + 2(\Delta x_4 + \Delta x_5)s_5 + \Delta x_4 s_6 = 3(\Delta x_5 y_4' + \Delta x_4 y_5') \\
i = 5 &\Rightarrow \Delta x_6 s_5 + 2(\Delta x_5 + \Delta x_6)s_6 + \Delta x_5 s_7 = 3(\Delta x_6 y_5' + \Delta x_5 y_6')
\end{aligned}
$$

Notice that we have five constraints and seven parameters and therefore two "degrees of freedom." If we move two of the parameters (s_1 and s_7) to the right hand side and assemble the results in matrix-vector form, then we obtain a 5-by-5 linear system

$$
Ts(2{:}6) = T
\begin{bmatrix}
s_2 \\ s_3 \\ s_4 \\ s_5 \\ s_6
\end{bmatrix}
=
\begin{bmatrix}
3(\Delta x_2 y_1' + \Delta x_1 y_2') - \Delta x_2 s_1 \\
3(\Delta x_3 y_2' + \Delta x_2 y_3') \\
3(\Delta x_4 y_3' + \Delta x_3 y_4') \\
3(\Delta x_5 y_4' + \Delta x_4 y_5') \\
3(\Delta x_6 y_5' + \Delta x_5 y_6') - \Delta x_5 s_7
\end{bmatrix}
= r
$$

where

$$
T =
\begin{bmatrix}
2(\Delta x_1 + \Delta x_2) & \Delta x_1 & 0 & 0 & 0 \\
\Delta x_3 & 2(\Delta x_2 + \Delta x_3) & \Delta x_2 & 0 & 0 \\
0 & \Delta x_4 & 2(\Delta x_3 + \Delta x_4) & \Delta x_3 & 0 \\
0 & 0 & \Delta x_5 & 2(\Delta x_4 + \Delta x_5) & \Delta x_4 \\
0 & 0 & 0 & \Delta x_6 & 2(\Delta x_5 + \Delta x_6)
\end{bmatrix}
$$

Matrices like this that are zero everywhere except on the diagonal, subdiagonal, and superdiagonal are said to be *tridiagonal*.

Different choices for the end slopes s_1 and s_n yield different cubic spline interpolants. Having defined the end slopes, the interior slopes $s(2{:}n-1)$ are determined by solving an $(n-2)$-by-$(n-2)$ linear system. In each case that we consider here, the matrix of coefficients looks like

$$
T =
\begin{bmatrix}
t_{11} & t_{12} & 0 & \cdots & & 0 \\
\Delta x_3 & 2(\Delta x_2 + \Delta x_3) & \Delta x_2 & & & \vdots \\
\vdots & \ddots & \ddots & \ddots & & \vdots \\
0 & & & \Delta x_{n-2} & 2(\Delta x_{n-3} + \Delta x_{n-2}) & \Delta x_{n-3} \\
0 & \cdots & & 0 & t_{n-2,n-3} & t_{n-2,n-2}
\end{bmatrix}
$$

while the right-hand side r has the form

$$r = \begin{bmatrix} r_1 \\ 3(\Delta x_3 y_2' + \Delta x_2 y_3') \\ \vdots \\ 3(\Delta x_{n-2} y_{n-3}' + \Delta x_{n-3} y_{n-2}') \\ r_{n-2} \end{bmatrix}$$

As we show in the next subsection, the values of t_{11}, t_{12}, and r_1 depend on how s_1 is chosen. The values of $t_{n-2,n-3}$, $t_{n-2,n-2}$, and r_{n-2} depend on how s_n is defined. Moreover, the T matrices that emerge can be shown to be nonsingular.

The following fragment summarizes what we have established so far about the linear system $Ts(2{:}n-1) = r$:

```
n=length(x);
Dx = diff(x);
yp = diff(y) ./ Dx;
T = zeros(n-2,n-2);
r = zeros(n-2,1);
for i=2:n-3
   T(i,i) = 2(Dx(i)+Dx(i+1));
   T(i,i-1) = Dx(i+1);
   T(i,i+1) = Dx(i);
   r(i) = 3(Dx(i+1)*yp(i) + Dx(i)*yp(i+1));
end
```

This sets up all but the first and last rows of T and all but the first and last components of r. How T and r are completed depends on the end conditions that are imposed on the spline.

3.3.2   The Complete Spline

The *complete spline* is obtained by setting $s_1 = \mu_L$ and $s_n = \mu_R$, where μ_L and μ_R are given real values. With these constraints, setting $i = 1$ and $i = n - 2$ in (3.3.3) gives

$$\Delta x_2 \mu_L + 2(\Delta x_1 + \Delta x_2)s_2 + \Delta x_1 s_3 = 3(\Delta x_2 y_1' + \Delta x_1 y_2')$$

$$\Delta x_{n-1} s_{n-2} + 2(\Delta x_{n-2} + \Delta x_{n-1})s_{n-1} + \Delta x_{n-2}\mu_R = 3(\Delta x_{n-1} y_{n-2}' + \Delta x_{n-2} y_{n-1}')$$

and so the first and last equations are given by

$$2(\Delta x_1 + \Delta x_2)s_2 + \Delta x_1 s_3 = 3(\Delta x_2 y_1' + \Delta x_1 y_2') - \Delta x_2 \mu_L$$

$$\Delta x_{n-1} s_{n-2} + 2(\Delta x_{n-2} + \Delta x_{n-1})s_{n-1} = 3(\Delta x_{n-1} y_{n-2}' + \Delta x_{n-2} y_{n-1}') - \Delta x_{n-2}\mu_R$$

Thus the setting up of T and r and the resolution of s are completed with the fragment

```
T(1,1) = 2*(Dx(1) + Dx(2));
T(1,2) = Dx(1);
r(1) = 3*(Dx(2)*yp(1) + Dx(1)*yp(2)) - Dx(2)*muL;
T(n-2,n-2) = 2*(Dx(n-2) + Dx(n-1));
T(n-2,n-3) = Dx(n-1);
r(n-2) = 3*(Dx(n-1)*yp(n-2) + Dx(n-2)*yp(n-1)) - Dx(n-2)*muR;
s = [ muL; T \ r(1:n-2) ; muR];
```

assuming that muL and muR house μ_L and μ_R, respectively.

3.3.3   The Natural Spline

Instead of prescribing the slope of the spline at the endpoints, we can prescribe the value of its second derivative. In particular, if we insist that $\mu_L = q_1''(x_1)$, then from (3.3.2) it follows that

$$\mu_L = 2\frac{y_1' - s_1}{\Delta x_1} - 2\frac{s_1 + s_2 - 2y_1'}{\Delta x_1}$$

from which we conclude that

$$s_1 = \frac{1}{2}\left(3y_1' - s_2 - \frac{\mu_L}{2}\Delta x_1\right)$$

Substituting this result into the $i = 1$ case of (3.3.1) and rearranging, we obtain

$$(2\Delta x_1 + 1.5\Delta x_2)s_2 + \Delta x_1 s_3 = 1.5\Delta x_2 y_1' + 3\Delta x_1 y_2' + \frac{\mu_L}{4}\Delta x_1\Delta x_2$$

Likewise, by setting $\mu_R = q_{n-1}''(x_n)$, then (3.3.2) implies

$$\mu_R = 2\frac{y_{n-1}' - s_{n-1}}{\Delta x_{n-1}} + 4\frac{s_{n-1} + s_n - 2y_{n-1}'}{\Delta x_{n-1}}$$

from which we conclude that

$$s_n = \frac{1}{2}\left(3y_{n-1}' - s_{n-1} + \frac{\mu_R}{2}\Delta x_{n-1}\right)$$

Substituting this result into the $i = n - 2$ case of (3.3.1) and rearranging we obtain

$$\Delta x_{n-1}s_{n-2} + (1.5\Delta x_{n-2} + 2\Delta x_{n-1})s_{n-1} = 3\Delta x_{n-1}y_{n-2}' + 1.5\Delta x_{n-2}y_{n-1}' - \frac{\mu_R}{4}\Delta x_{n-2}\Delta x_{n-1}$$

Thus the setting up of T and r and the resolution of s are completed with the fragment

```
T(1,1) = 2*Dx(1) + 1.5*Dx(2);
T(1,2) = Dx(1);
r(1) = 1.5*Dx(2)*yp(1) + 3*Dx(1)*yp(2)) + Dx(1)*Dx(2)*muL/4;
T(n-2,n-2) = 1.5*Dx(n-2) + 2*Dx(n-1);
T(n-2,n-3) = Dx(n-1);
r(n-2) = 3*Dx(n-1)*yp(n-2) + 1.5*Dx(n-2)*yp(n-1) -Dx(n-2)*Dx(n-1)*muR;
stilde = T \ r;
s1 = (3*yp(1) - stilde(1) - muL*Dx(1)/2)/2;
sn = (3*yp(n-1) - stilde(n-2) + muR*Dx(n-1)/2)/2;
s = [s1; stilde; sn];
```

If $\mu_L = \mu_R = 0$, then the resulting spline is called *the natural spline*.

3.3.4   The Not-a-Knot Spline

This method for prescribing the end conditions is appropriate if no endpoint derivative information is available. It produces the *not-a-knot* spline. The idea is to ensure third derivative continuity at both x_2 and x_{n-1}. Note from (3.3.1) that

$$q_i'''(x) = 6\frac{s_i + s_{i+1} - 2y_i'}{(\Delta x_i)^2}$$

and so $q_1'''(x_2) = q_2'''(x_2)$ says that

$$\frac{s_1 + s_2 - 2y_1'}{(\Delta x_1)^2} = \frac{s_2 + s_3 - 2y_2'}{(\Delta x_2)^2}$$

It follows that this will be the case if we set

$$s_1 = -s_2 + 2y_1' + \left(\frac{\Delta x_1}{\Delta x_2}\right)^2 (s_2 + s_3 - 2y_2')$$

As a result of making the third derivative continuous at x_2, the cubics $q_1(x)$ and $q_2(x)$ are identical.

Likewise, $q_{n-2}'''(x_{n-1}) = q_{n-1}'''(x_{n-1})$ says that

$$\frac{s_{n-2} + s_{n-1} - 2y_{n-2}'}{(\Delta x_{n-2})^2} = \frac{s_{n-1} + s_n - 2y_{n-1}'}{(\Delta x_{n-1})^2}$$

It follows that this will be the case if we set

$$s_n = -s_{n-1} + 2y_{n-1}' + \left(\frac{\Delta x_{n-1}}{\Delta x_{n-2}}\right)^2 (s_{n-2} + s_{n-1} - 2y_{n-2}')$$

Thus the first and last equations for the not-a-knot spline are set up as follows:

```
q = Dx(1)*Dx(1)/Dx(2);
T(1,1)= 2*Dx(1) +Dx(2) + q;
T(1,2) = Dx(1) + q;
r(1) = Dx(2)*yp(1) + Dx(1)*yp(2)+2*yp(2)*(q+Dx(1));
q= Dx(n-1)*Dx(n-1)/Dx(n-2);
T(n-2,n-2) = 2*Dx(n-1) + Dx(n-2)+q;
T(n-2,n-3) = Dx(n-1)+q;
r(n-2) = Dx(n-1)*yp(n-2) + Dx(n-2)*yp(n-1) +2*yp(n-2)*(Dx(n-1)+q);
stilde = T\ r;
s1 = -stilde(1)+2*yp(1);
s1 = s1 + ((Dx(1)/Dx(2))^2)*(stilde(1)+stilde(2)-2*yp(2));
sn = -stilde(n-2) +2*yp(n-1);
sn=sn+((Dx(n-1)/Dx(n-2))^2)*(stilde(n-3)+stilde(n-2)-2*yp(n-2));
s=[s1;stilde;sn];
```

3.3.5   The Cubic Spline Interpolant

The function CubicSpline can be used to construct the cubic spline interpolant with any of the three aforementioned types of end conditions. Here is its specification:

```
    function [a,b,c,d] = CubicSpline(x,y,derivative,muL,muR)
%
% Pre:
%   x,y          column n-vectors. n >= 4 and x(1) < ... x(n)
%   derivative   the order of the spline's derivative that are
%                used in the end conditions (= 1 or 2)
%   muL,muR      the value of the specified derivative at the
%                left and right endpoint.
%
% Post:
%   a,b,c,d      column (n-1)-vectors that define the spline.
%                On [x(i),x(i+1)], the spline S(z) is specified by the cubic
%
%                a(i) + b(i)(z-x(i)) + c(i)(z-x(i))^2 + d(i)(z-x(i))^2(z-x(i+1)).
%
% Usage:
%   [a,b,c,d] = CubicSpline(x,y,1,muL,muR)   S'(x(1))  = muL, S'(x(n))  = muR
%   [a,b,c,d] = CubicSpline(x,y,2,muL,muR)   S''(x(1)) = muL, S''(x(n)) = muR
%   [a,b,c,d] = CubicSpline(x,y)             S'''(z) cont. at x(2) and x(n-1)
%
```

Notice that a two-argument call is all that is required to produce the not-a-knot spline. The script ShowSpline examines various CubicSpline interpolants to the sine function.

Error bounds for the cubic spline interpolant are complicated to derive. The bounds are *not* good if the end conditions are improperly chosen. Figure 3.6 shows what can happen if the complete spline is used with end conditions that are at variance with the behavior of the function being interpolated. However, if the end values are properly chosen or if the not-a-knot approach is used, then the error bound has the form $M_4 \bar{h}^4$ where $\bar{h}$ is the maximum subinterval length and M_4 bounds the 4th derivative of the function being interpolated. The script SplineErr confirms this for the case of an "easy" $f(x)$. It produces the plots shown in Fig 3.7. Notice that the error is reduced by a factor of 10^4 if the subinterval length is reduced by a factor of 10.

3.3.6   MATLAB Spline Tools

The MATLAB function spline can be used to compute not-a-knot spline interpolants. It can be called with either two or three arguments. The script

```
z = linspace(-5,5);
x = linspace(-5,5,9);
y = atan(x);
Svals = spline(x,y,z);
plot(z,Svals);
```

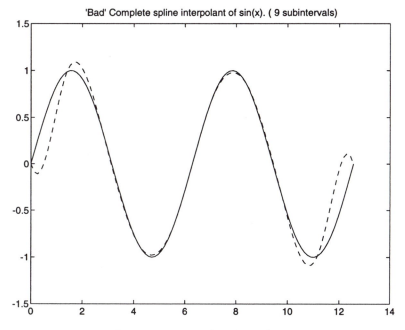

FIGURE 3.6 Bad end conditions.

illustrates a three-argument call. It plots the $n = 9$ not-a-knot spline interpolant of the function $f(x) = \arctan(x)$ across the interval $[-5, 5]$. The first two arguments in the call to spline specify the interpolation points that define the spline $S(z)$. The spline is then evaluated at z with the values returned in Svals. Thus $S(x_i) = y_i$ for $i = 1{:}length(x)$ and $S(z_i) = Svals_i$ for $i = 1{:}length(z)$.

A 2-argument call to spline returns what is called the *pp-representation* of the spline. This type of reference is required whenever one has to manipulate the local cubics that make up the spline. The *pp*-representation of a spline is different from the four-vector representation that we have been using for piecewise cubics. For one thing, it is more general because it can accommodate piecewise polynomials of arbitrary degree.

To gain a facility with MATLAB's piecewise polynomial tools, let's consider the problem of constructing the *pp*-representation of the derivative of a cubic spline $S(z)$. In particular, let's plot $S'(z)$ where $S(z)$ is nine-point, equally spaced, not-a-knot spline interpolant of the arctangent function across the interval $[-5, 5]$. We start by constructing the *pp*-representation of $S(z)$:

```
x = linspace(-5,5,9);
y = atan(x);
S = Spline(x,y)
```

A two-argument call to Spline such as this produces the *pp*-representation of the spline. The ppval function can be used to evaluate a piecewise polynomial in this representation:

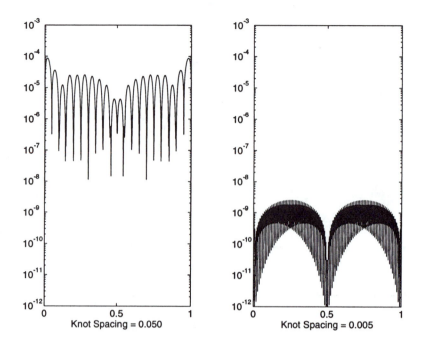

FIGURE 3.7 Not-a-knot spline error

```
z = linspace(-5,5);
Svals = ppval(S,z);
plot(z,Svals)
```

The call to ppval returns the value of the spline at z. The vector Svals contains the values of the spline on z. These values are then plotted.

The derivative of the spline is a piecewise *quadratic* polynomial, and by using the functions unmkpp and mkpp we can produce its *pp*-representation. A call to unmkpp unveils the four major components of the *pp*-representation:

```
[x,rho,L,k] = unmkpp(S)
```

The x-values are returned in x. The coefficients of the local polynomials are assembled in an L-by-k matrix rho. L is the number of local polynomials and k is their degree. So in our case, $x = linspace(-5, 5, 9)$, $L = 8$, and $k = 4$. The coefficients of the i-th local cubic are stored in ith row of the rho matrix. In particular, the spline is defined by

$$S(z) = \rho_{i,4} + \rho_{i,3}(z - x_i) + \rho_{i,2}(z - x_i)^2 + \rho_{i,1}(z - x_i)^3$$

on the interval $[x_i, x_{i+1}]$. Thus, rho(i,j) is the ith local polynomial coefficient of $(x - x_i)^{k-j+1}$.

The function mkpp takes the array of coefficients and the breakpoints and produces the *pp*-representation of the piecewise polynomial so defined. Thus to set up the *pp*-representation of the spline's derivative, we execute

```
drho = [3*rho(:,1) 2*rho(:,2) rho(:,3)];
dS = mkpp(x,drho);
```

The set-up of the three-column matrix drho follows from the observation that

$$S'(x) = \rho_{i,3} + 2\rho_{i,2}(x - x_i) + 3\rho_{i,1}(x - x_i)^2$$

on the interval $[x_i, x_{i+1}]$. Putting it all together, we obtain

```
% Script File: ShowSplineTools
%
% Illustrates the Matlab functions spline, ppval, mkpp, unmkpp

    close all

%   Set Up Data:
    n = 9;
    x = linspace(-5,5,n);
    y = atan(x);

%   Compute the spline interpolant and its derivative:
    S   = spline(x,y);
    [x,rho,L,k] = unmkpp(S);
    drho = [3*rho(:,1) 2*rho(:,2) rho(:,3)];
    dS = mkpp(x,drho);

%   Evaluate S and dS:
    z       = linspace(-5,5);
    Svals   = ppval(S,z);
    dSvals  = ppval(dS,z);

%   Plot:
    atanvals = atan(z);
    figure
    plot(z,atanvals,z,Svals,x,y,'*');
    Title(sprintf('n = %2.0f Spline Interpolant of atan(x)',n))
    datanvals = ones(size(z))./(1 + z.*z);
    figure
    plot(z,datanvals,z,dSvals)
    Title(sprintf('Deivative of n = %2.0f Spline Interpolant of atan(x)',n))
```

Problems

P3.3.1 What can you say about the $n = 4$ not-a-knot spline interpolant of $f(x) = x^3$?

P3.3.2 Suppose $S(z)$ is the not-a-knot spline interpolant of (x_1, y_1), (x_2, y_2), (x_3, y_3), and (x_4, y_4) where it is assumed that the x_i are distinct. Suppose $p(x)$ is the cubic interpolant at same four points. Explain why $S(z) = p(z)$ for all z.

P3.3.3 Let $S(z)$ be the natural spline interpolant of z^3 at $z = -3$, $z = -1$, $z = 1$, $z = 3$. What is $S(0)$?

P3.3.4 Given $\sigma > 0$, (x_i, y_i, s_i), and $(x_{i+1}, y_{i+1}, s_{i+1})$, show how to determine a_i, b_i, c_i, and d_i so that

$$g_i(x) = a_i + b_i(x - x_i) + c_i e^{\sigma(x - x_i)} + d_i e^{-\sigma(x - x_i)}$$

satisfies $g_i(x_i) = y_i$, $g_i'(x_i) = s_i$, $g_i(x_{i+1}) = y_{i+1}$, and $g_i'(x_{i+1}) = s_{i+1}$.

P3.3.5 Another approach that can be used to make up for a lack of endpoint derivative information is to glean that information from a four-point cubic interpolant. For example, if $q_L(x)$ is the cubic interpolant of (x_1, y_1), (x_2, y_2), (x_3, y_3), and (x_4, y_4), then either of the endpoint conditions

$$\begin{aligned} q_1'(x_1) &= q_L'(x_1) \\ q_1''(x_1) &= q_L''(x_1) \end{aligned}$$

is reasonable, where $q_1(x)$ is the leftmost local cubic. Likewise, if $q_R(x)$ is the cubic interpolant of (x_{n-3}, y_{n-3}), (x_{n-2}, y_{n-2}), (x_{n-1}, y_{n-1}), and (x_n, y_n), then either of the right endpoint conditions

$$\begin{aligned} q_{n-1}'(x_n) &= q_R'(x_n) \\ q_{n-1}''(x_n) &= q_R''(x_n) \end{aligned}$$

is reasonable, where $q_{n-1}(x)$ is the rightmost local cubic. Modify `CubicSpline` so that it invokes this strategy whenever the function call involves just three arguments, (i.e., `[a,b,c,d] = CubicSpline(x,y,derivative)`.) The value of `derivative` should determine which derivative is to be matched at the endpoints. (Its value should be 1 or 2.) Augment the script file `ShowSpline` so that it that graphically depicts the splines that are produced by this method.

P3.3.6 Complete the following function:

```
function [a,b,c,d] = SmallSpline(z,y)

% Pre:
%    z is a scalar and y is 3-vector.
%
% Post:
%    a,b,c,d are column 2-vectors with the property that if
%
%       S(x) = a(1) + b(1)(x - z) + c(1)(x - z)^2 + d(1)(x - z)^3   on [z-1,z]
%    and
%       S(x) = a(2) + b(2)(x - z) + c(2)(x - z)^2 + d(2)(x - z)^3   on [z,z+1]
%    then
%                        (a) S(z-1) = y(1),  S(z) = y(2),  S(z+1) = y(3),
%                        (b) S''(z-1) = S''(z+1) = 0
%                        (c) S, S', and S'' are continuous on [z-1,z+1]
%
```

P3.3.7 In computerized typography the problem arises of finding an interpolant to points that lie on a path in the plane (e.g., a printed capital S). Such a shape cannot be represented as a function of x because it is not single valued. One approach is to number the points $(x_1, y_1), \ldots, (x_n, y_n)$ as we traverse the curve. Let d_i be the straight-line distance between (x_i, y_i) and (x_{i+1}, y_{i+1}), $i = 1{:}n - 1$. Set $t_i = d_1 + \cdots + d_{i-1}$, $i = 1{:}n$. Suppose $S_x(t)$ is a spline interpolant of $(x_1, t_1), \ldots, (x_n, t_n)$ and that $S_y(t)$ is a spline interpolant of $(y_1, t_1), \ldots, (y_n, t_n)$.

It follows that the curve $\Lambda = \{(S_x(t), S_y(t)) : t_1 \le t \le t_n\}$ is smooth and passes through the n points. Write a MATLAB function [xi,yi] = SplineInPlane(x,y,m) that returns in xi(1:m) and yi(1:m) the x-y coordinates of m points on the curve Λ. Use the MATLAB Spline function to determine the splines $S_x(t)$ and $S_y(t)$.

P3.3.8 Explain how MATLAB's spline tools can be used to compute

$$\int_\alpha^\beta [S''(x)]^2 dx$$

where $S(x)$ is a cubic spline.

P3.3.9 Suppose $S(x)$ is a cubic spline interpolant of the data $(x_1, y_1), \ldots, (x_n, y_n)$ obtained using spline. Write a MATLAB function d3 = MaxJump(S) that returns the maximum jump in the third derivative of the spline S assumed to be in the *pp*-representation. Vectorize as much as possible. Use the **max** function.

P3.3.10 Write a MATLAB function S = Convert(a,b,c,d,x) that takes our piecewise cubic interpolant representation and converts it into *pp* form.

M-Files and References

Script Files

ShowPWL1	Illustrates pwL and pwLEval.
ShowPWL2	Compares pwLStatic and pwLAdapt.
ShowHermite	Illustrates the Hermite interpolation idea.
ShowpwCH	Illustrates pwC and pwCEval.
ShowSpline	Illustrates CubicSpline.
SplineErr	Explores the not-a-knot spline interpolant error.
ShowSplineTools	Illustrates MATLAB spline tools.

Function Files

pwL	Sets up a piecewise linear interpolant.
Locate	Determines the subinterval in a mesh that houses a given x-value.
pwLEval	Evaluates a piecewise linear function.
pwLStatic	A priori determination of a mesh for a pwL approximation.
pwLAdapt	Dynamic determination of a mesh for a pwL approximation.
HCubic	Constructs the cubic Hermite interpolant.
pwC	Sets up a piecewise cubic Hermite interpolant.
pwCEval	Evaluates a piecewise cubic function.
CubicSpline	Constructs complete, natural, or not-a-knot spline.

References

R. Bartels, J. Beatty, and B. Barsky (1987). *An Introduction to Splines for Use in Computer Graphics and Geometric Modeling*, Morgan Kaufmann, Los Altos, CA.

C. de Boor (1978). *A Practical Guide to Splines*, Springer, Berlin.

Chapter 4

Numerical Integration

§**4.1** The Newton-Cotes Rules

§**4.2** Composite Rules

§**4.3** Adaptive Quadrature

§**4.4** Special Topics

§**4.5** Shared Memory Adaptive Quadrature

An m-point *quadrature rule* Q for the definite integral

$$I = \int_a^b f(x)dx$$

is an approximation of the form

$$Q = (b - a)\sum_{k=1}^{m} w_k f(x_k)$$

The x_k are the *abscissas* and the w_k are the *weights*.[1] The abscissas and weights define the rule and are chosen so that $Q \approx I$. *Efficiency essentially depends upon the number of function evaluations.* This is because the time needed to evaluate f at the x_i is typically *much* greater than the time needed to form the required linear combination of function values. Thus, a six-point quadrature rule is twice as expensive as a three-point rule.

We start with the *Newton-Cotes* family of quadrature rules. These rules are derived by integrating a polynomial interpolant of the integrand $f(x)$. Composite rules based on a partition of $[a, b]$ into subintervals are then discussed. In a composite rule, a simple rule is applied to each subintegral and the result summed. The adaptive determination of the partition with error control is stressed and leads to an important class of quadrature procedures that adapt to the

---

[1]In the linear combination for Q, we "pulled out" the factor $(b - a)$ partly for the sake of tradition and partly because it simplifies the discussion. Recognize that the "real" weights are $(b - a)w_i$, $i = 1{:}m$.

behavior of the integrand. Important special topics discussed include *Gauss quadrature* and methods for multiple integrals.

The quadrature problem is ideal for introducing some of the key ideas associated with parallel computation. In the final section we discuss the shared memory model and outline the implementation of several numerical integration procedures.

4.1   The Newton-Cotes Rules

One way to derive a quadrature rule Q is to integrate a polynomial approximation $p(x)$ of the integrand $f(x)$. The philosophy is that $p(x) \approx f(x)$ implies

$$\int_a^b f(x)dx \approx \int_a^b p(x)dx$$

(See Fig 4.1.) The *Newton-Cotes* quadrature rules are obtained by integrating uniformly spaced

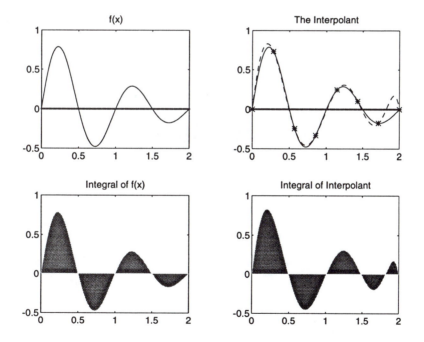

FIGURE 1.1 The Newton-Cotes idea

polynomial interpolants of the integrand. The m-point Newton-Cotes rule ($m \geq 2$) is defined by

$$Q_{\mathrm{NC}(m)} = \int_a^b p_{m-1}(x)dx \tag{4.1.1}$$

where $p_{m-1}(x)$ interpolates $f(x)$ at

$$x_i = a + \frac{i-1}{m-1}(b-a) \qquad i = 1{:}m$$

If $m = 2$, then we obtain the *trapezoidal rule*:

$$
\begin{aligned}
Q_{\mathrm{NC}(2)} &= \int_a^b \left(f(a) + \frac{f(b)-f(a)}{b-a}(x-a) \right) dx \\
&= (b-a)\left(\frac{1}{2}f(a) + \frac{1}{2}f(b) \right)
\end{aligned}
$$

If $m = 3$ and $c = (a+b)/2$, then we obtain the *Simpson rule*:

$$
\begin{aligned}
Q_{\mathrm{NC}(3)} &= \int_a^b \left(f(a) + \frac{f(c)-f(a)}{c-a}(x-a) + \frac{\frac{f(b)-f(c)}{b-c} - \frac{f(c)-f(a)}{c-a}}{b-a}(x-a)(x-c) \right) dx \\
&= \frac{b-a}{6}\left(f(a) + 4f\left(\frac{a+b}{2}\right) + f(b) \right)
\end{aligned}
$$

From these low-degree examples, it appears that a linear combination of f-evaluations is obtained upon expansion of the right-hand side in (4.1.1).

4.1.1 Derivation

For general m, we proceed by substituting the Newton representation

$$p_{m-1}(x) = \sum_{k=1}^m \left(c_k \prod_{i=1}^{k-1}(x-x_i) \right)$$

into (4.1.1):

$$Q_{\mathrm{NC}(m)} = \int_a^b p_{m-1}(x)dx = \sum_{k=1}^m c_k \int_a^b \left(\prod_{i=1}^{k-1}(x-x_i) \right) dx$$

If we set $x = a + sh$, where $h = (b-1)/(m-1)$, then this transforms to

$$Q_{\mathrm{NC}(m)} = \int_a^b p_{m-1}(x)dx = h\int_0^{m-1} p_{m-1}(a+sh)ds = \sum_{k=1}^m c_k h^k S_{mk}$$

where

$$S_{mk} = \int_0^{m-1} \left(\prod_{i=1}^{k-1}(s-i+1) \right) ds$$

The c_k are divided differences. Because of the equal spacing, they have a particularly simple form in terms of the f_i, as was shown in §2.4.3. For example,

$$
\begin{aligned}
c_1 &= f_1 \\
c_2 &= (f_2 - f_1)/h \\
c_3 &= (f_3 - 2f_2 + f_1)/(2h^2) \\
c_4 &= (f_4 - 3f_3 + 3f_2 - f_1)/(3!h^3)
\end{aligned}
$$

Recipes for the S_{mk} can also be derived. Here are a few examples:

$$
\begin{aligned}
S_{m1} &= \int_0^{m-1} 1 \cdot ds & &= (m-1) \\
S_{m2} &= \int_0^{m-1} s\,ds & &= (m-1)^2/2 \\
S_{m3} &= \int_0^{m-1} s(s-1)ds & &= (m-1)^2(m-5/2)/3 \\
S_{m4} &= \int_0^{m-1} s(s-1)(s-2)ds & &= (m-1)^2(m-3)^2/4
\end{aligned}
$$

Using these tabulations we can readily derive the weights for any particular m-point rule. For example, if $m = 4$, then $S_{41} = 3$, $S_{42} = 9/2$, $S_{43} = 9/2$, and $S_{44} = 9/4$. Thus

$$
\begin{aligned}
Q_{\mathrm{NC}(4)} &= S_{41}c_1h + S_{42}c_2h^2 + S_{43}c_3h^3 + S_{44}c_4h^4 \\
&= 3f_1h + \frac{9}{2}\frac{f_2-f_1}{h}h^2 + \frac{9}{2}\frac{f_3-2f_2+f_1}{2h^2}h^3 + \frac{9}{4}\frac{f_4-3f_3+3f_2-f_1}{6h^3}h^4 \\
&= \frac{3h}{8}(f_1 + 3f_2 + 3f_3 + f_4) \\
&= (b-a)(f_1 + 3f_2 + 3f_3 + f_4)/8
\end{aligned}
$$

showing that $[1\ 3\ 3\ 1]/8$ is the weight vector for $Q_{\mathrm{NC}(4)}$.

4.1.2 Implementation

For convenience in subsequent computations, we "package" the Newton-Cotes weight vectors in the following function:

```
    function w=WNC(m);
%
% Pre:
%   m      integer that satisfies 2 <= m <= 11
%
% Post:
%   w      column m-vector consisting of the weights for the
%          the m-point Newton-Cotes rule.
%
  if m==2
    w=[1 1]'/2;
  elseif m==3
    w=[1 4 1]'/6;
  elseif m==4
    w=[1 3 3 1]'/8;
  elseif m==5
    w=[7 32 12 32 7]'/90;
    :
    :
  end
```

Notice that the weight vectors are symmetric about their middle in that $w(1{:}m) = w(m{:}-1{:}1)$.

Turning now to the evaluation of $Q_{NC(m)}$ itself, we see from

$$Q_{NC(m)} = (b-a)\sum_{i=1}^{m} w_i f_i = (b-a)\begin{bmatrix} w_1 & \cdots & w_m \end{bmatrix} \begin{bmatrix} f(x_1) \\ \vdots \\ f(x_m) \end{bmatrix}$$

that it is a scaled inner product of the weight vector w and the vector of function values. Therefore, we obtain

```
function numI = QNC(fname,a,b,m)
%
% Pre:
%    fname      string that names an available function of the
%               form f(x) that is defined on [a,b]. f should
%               return a column vector if x is a column vector.
%      a,b      real scalars
%        m      integer that satisfies 2 <= m <=11
% Post:
%    numI       the m-point Newton-Cotes approximation of the
%               integral of f(x) from a to b.
%
   w = wNC(m);
   x = linspace(a,b,m)';
   f = feval(fname,x);
   numI = (b-a)*(w'*f);
```

The following script supports experimentation with this function:

```
% Script File: ShowQNC
%
% Examines the Newton-Cotes rules.

   while input('Another exammple? (1=yes, 0=no). ')
      fname = input('Enter within quotes the name of  integrand function:');
      a = input('Enter left endpoint: ');
      b = input('Enter right endpoint: ');
      s = ['QNC(' fname sprintf(',%6.3f,%6.3f,m )',a,b)];
      clc
      disp(['  m         ' s])
      disp(' ')
      for m=2:11
         numI = QNC(fname,a,b,m);
         disp(sprintf(' %2.0f          %20.16f',m,numI))
      end
   end
```

For the integral of the sine function from 0 to $\pi/2$, the script gives

m	QNC('sin',0,pi/2,m)
2	0.7853981633974483
3	1.0022798774922104
4	1.0010049233142790
5	0.9999915654729927
6	0.9999952613861667
7	1.0000000258372355
8	1.0000000158229039
9	0.9999999999408976
10	0.9999999999621675
11	1.0000000000001021

4.1.3 Newton-Cotes Error

How good are the Newton-Cotes rules? Since they are based on the integration of a polynomial interpolant, the answer clearly depends on the quality of the interpolant. Here is a result for Simpson's rule:

Theorem 4 *If $f(x)$ and its first four derivatives are continuous on $[a,b]$, then*

$$\left| \int_a^b f(x)dx - Q_{NC(3)} \right| \le \frac{(b-a)^5}{2880} M_4$$

where M_4 is an upper bound on $|f^{(4)}(x)|$ on $[a,b]$.

Proof. Suppose

$$p(x) = c_1 + c_2(x-a) + c_3(x-a)(x-b) + c_4(x-a)(x-b)(x-c)$$

is the Newton form of the cubic interpolant to $f(x)$ at the points a, b, c, and d. If c is the midpoint of the interval $[a,b]$, then

$$\int_a^b \left(c_1 + c_2(x-a) + c_3(x-a)(x-b) \right) dx = Q_{NC(3)}$$

because the first three terms in the expression for $p(x)$ specify the quadratic interpolant of $(a, f(a))$, $(c, f(c))$, and $(b, f(b))$, on which the three-point Newton-Cotes rule is based. By symmetry we have

$$\int_a^b (x-a)(x-b)(x-c)dx = 0$$

and so

$$\int_a^b p(x)dx = Q_{NC(3)}$$

The error in $p(x)$ is given by Theorem 2,

$$f(x) - p(x) = \frac{f^{(4)}(\eta_x)}{24}(x-a)(x-b)(x-c)(x-d)$$

and thus

$$\int_a^b f(x)dx - Q_{NC(3)} = \int_a^b \left(\frac{f^{(4)}(\eta_x)}{24}(x-a)(x-b)(x-c)(x-d) \right) dx$$

Taking absolute values, we obtain

$$\left| \int_a^b f(x)dx - Q_{NC(3)} \right| \le \frac{M_4}{24} \int_a^b |(x-a)(x-b)(x-c)(x-d)|\, dx$$

If we set $d = c$, then $(x-a)(x-b)(x-c)(x-d)$ is always negative and it is easy to verify that

$$\int_a^b |(x-a)(x-b)(x-c)(x-d)|\, dx = \frac{(b-a)^5}{120}$$

and so

$$\left| \int_a^b f(x)dx - Q_{NC(3)} \right| \le \frac{M_4}{24} \frac{(b-a)^5}{120} = \frac{M_4}{2880} \quad \square$$

Note that if $f(x)$ is a cubic polynomial, then $f^{(4)} = 0$ and so Simpson's rule is exact. This is somewhat surprising because the rule is based on the integration of a *quadratic* interpolant.

In general, it can be shown that

$$\int_a^b f(x)dx = Q_{NC(m)} + c_m f^{(d+1)}(\eta) \left(\frac{b-a}{m-1} \right)^{d+2} \tag{4.1.2}$$

where c_m is a small constant, η is in the interval $[a, b]$, and

$$d = \begin{cases} m-1 & \text{if } m \text{ is even} \\ m & \text{if } m \text{ is odd} \end{cases}$$

Notice that if m is odd, as in Simpson's rule, then an extra degree of accuracy results. See P4.1.5 for details.

From (4.1.3), we see that knowledge of $f^{(d+1)}$ is required in order to say something about the error in $Q_{NC(m)}$. For example, if $|f^{(d+1)}(x)| \le M_{d+1}$ on $[a, b]$, then

$$\left| Q_{NC(m)} - \int_a^b f(x)dx \right| \le |c_m| M_{d+1} \left(\frac{b-a}{m-1} \right)^{d+2} \tag{4.1.3}$$

The following function can be used to return this upper bound given the interval $[a, b]$, m, and the appropriate derivative bound:

```
     function error = NCErr(a,b,m,DerBound)
%
% Pre:
%          a,b    real scalars that satsify a<=b
%            m    integer that satisfies 2<=m<=11 and
```

```
%   DerBound    an upper bound for the d-th derivative of a
%               function f(x) defined on [a,b] where
%               d = m if m is odd, and d = m-1 if m is even.
%
% Post:
%      error    an upper bound for the absolute error of the
%               m-point Newton-Cotes rule when it is
%               applied to the integral of f(x) from a to b.
%
   if m==2,      d = 1;  c = -1/12;
   elseif m==3,  d = 3;  c = -1/90;
   elseif m==4,  d = 3;  c = -3/80;
   elseif m==5,  d = 5;  c = -8/945;
   elseif m==6,  d = 5;  c = -275/12096;
   elseif m==7,  d = 7;  c = -9/1400;
   elseif m==8,  d = 7;  c = -8183/518400;
   elseif m==9,  d = 9;  c = -2368/467775;
   elseif m==10, d = 9;  c = -173/14620;
   else          d =11;  c = -1346350/326918592;
   end;
   error = abs( c*DerBound*((b-a)/(m-1))^(d+2) );
```

From this we see that if you are contemplating an even m rule, then the $(m-1)$-point rule is probably just as good and requires one less function evaluation. The following script illustrates this point:

```
% Script File: ShowQNCError
%
% Examines the quality of the Newton-Cotes error bound.
   clc
   disp('Easy case: Integral from 0 to pi/2 of sin(x)')
   disp(' ')
   disp('Take DerBound = 1.')
   disp(' ')
   disp('   m             QNC(m)              Error       Error Bound')
   disp(' ')
   for m=2:11
      numI = QNC('sin',0,pi/2,m);
      err = abs(numI-1);
      errBound = ErrNC(0,pi/2,m,1);
      s = sprintf('%20.16f    %10.3e     %10.3e',numI,err,errBound);
      disp([ sprintf('  %2.0f    ',m) s])
   end
```

The results are summarized in the following table

m	$Q_{NC}(m)$	Actual Error	Error Bound
2	0.7853981633974483	2.146e-01	3.230e-01
3	1.0022798774922104	2.280e-03	3.321e-03
4	1.0010049233142790	1.005e-03	1.476e-03
5	0.9999915654729927	8.435e-06	1.219e-05
6	0.9999952613861667	4.739e-06	6.867e-06
7	1.0000000258372355	2.584e-08	3.714e-08
8	1.0000000158229039	1.582e-08	2.277e-08
9	0.9999999999408976	5.910e-11	8.466e-11
10	0.9999999999621675	3.783e-11	5.417e-11
11	1.0000000000001021	1.021e-13	1.460e-13

4.1.4   A Note on the Open Rules

The Newton-Cotes rules presented previously are actually the *closed* Newton-Cotes rules because $f(x)$ is evaluated at the left and right endpoints. The m-point "open" Newton-Cotes rule places the abscissas at $a + (i - (1/2))h$, where $h = (b - a)/m$ and $i = 1{:}m$. The abscissas for the open and closed four-point rules are depicted in Fig 4.2. The one-point open Newton-Cotes rule is

FIGURE 4.2 Four-point Newton-Cotes abscissa placement

called the *midpoint rule*. For $m = 3, 5, 6, 7, \ldots$, the open rules involve negative weights, a feature that can undermine the numerical stability of the rule. (Numerical cancellation can occur with positive integrands, and this can jeopardize the relative error.) The closed Newton formulas do not "go negative" until $m = 9$, making them a more attractive family of quadrature rules.

The weights and errors associated with the open rules are available through the functions WNCOpen and ErrNCOpen.

Problems

P4.1.1 Write a function QNCopen(fname,a,b,m) analogous to QNC that is based on the m-point open Newton-Cotes rule. Use

```
function w = WNCOpen(m)
% w is the weight vector for the open m-point Newton-Cotes rule.
% Assumes 1<=m<=11.
```

P4.1.2 Derive the four-point open Newton-Cotes rule.

P4.1.3 Let $C(x)$ be the cubic Hermite interpolant of $f(x)$ at $x = a$ and b. Show that

$$\int_a^b C(x)dx = \frac{h}{2}(f(a) + f(b)) + \frac{h^2}{12}(f'(a) - f'(b))$$

This is sometimes called the *corrected trapezoidal rule*. Write a function CorrTrap(fname,fpname,a,b) that computes this value. Here, fname and fpname are strings that respectively contain the name of the integrand function and its derivative. The error in this rule has the form $ch^4 f^{(4)}(\eta)$. Determine c (approximately) through experimentation.

P4.1.4 This problem is about the computation of the closed Newton Cotes weights by solving an appropriate linear system. Observe that m-point rule should compute the integral

$$\int_0^1 x^{i-1}dx = \frac{1}{i}$$

exactly for $i = 1{:}m$. For this calculation, the abscissas are given by $x_j = (j-1)/(m-1)$, $i = 1{:}m$. Thus the weights $w_1, \ldots, w_m$ satisfy

$$w_1 x_1^{i-1} + w_2 x_2^{i-1} + \cdots + w_m x_m^{i-1} = \frac{1}{i}$$

for $i = 1{:}m$. This defines a linear system whose solution is the weight vector for the m-point rule. Write a function MywNC(m) that computes the weights by setting up the preceding linear system and solving for w using the backslash operation. Compare the output of wNC and MywNC for $m = 2{:}11$.

P4.1.5 (1) Suppose m is odd and that $c = (a + b)/2$. Show that $Q_{\text{NC}(m)}$ is exact when applied to

$$I = \int_a^b (x - c)^k dx$$

when k is odd. (2) If $p(x)$ has degree m, then it can be written in the form $p(x) = q(x) + \alpha(x - c)^m$ where q has degree $m - 1$ and c is a scalar. Use this fact with $c = (a + b)/2$ to show that if m is odd, then $Q_{\text{NC}(m)}$ is exact when applied to

$$I = \int_a^b p(x)dx$$

P4.1.6 Augment ShowNCError so that it also prints a table of errors and error bounds for the integral

$$I = \int_0^1 \frac{dx}{1 + 10x}.$$

Explain clearly the derivative bounds that are used.

P4.1.7 Write a script analogous to ShowNCError that examines the quality of the open Newton-Cotes rules. Make use of wNCOpen and errNCOpen.

4.2 Composite Rules

We will not be happy with the error bound (4.1.3) unless $b - a$ is sufficiently small. Fortunately, there is an easy way to organize the computation of an integral so that small-interval quadratures prevail.

4.2.1   Derivation

If we have a partition

$$a = z_1 < z_2 < \cdots < z_{n+1} = b$$

then

$$\int_a^b f(x)dx \;=\; \sum_{i=1}^n \int_{z_i}^{z_{i+1}} f(x)dx$$

If we apply $Q_{\mathrm{NC}(m)}$ to each of the subintegrals, then a *composite quadrature rule* based on $Q_{\mathrm{NC}(m)}$ results. For example, if $\Delta_i = z_{i+1} - z_i$ and $z_{i+1/2} = (z_i + z_{i+1})/2$, $i = 1{:}n$, then

$$Q = \sum_{i=0}^{n-1} \frac{\Delta_i}{6} \left(f(z_i) + 4f(z_{i+1/2}) + f(z_{i+1}) \right) \qquad (4.2.1)$$

is a composite Simpson rule. In general, if `z` houses a partition of $[a, b]$ and `fname` is a string that names a function, then

```
numI=0
for i=1:length(z)-1
    numI = numI + QNC(fname,z(i),z(i+1),m);
end
```

assigns to `numI` the composite m-point Newton-Cotes estimate of the integral based on the partition housed in `z`.

In §4.3 we show how to automate the choice of a good partition. In the remainder of this section, we focus on composite rules that are based on uniform partitions. In these rules, $n \geq 1$,

$$z_i = a + (i-1)\Delta, \qquad \Delta = \frac{b-a}{n}$$

for $i = 1{:}n+1$, and the composite rule evaluation has the form

```
numI = 0;
Delta=(b-a)/n;
for i=1:n
    numI = numI + QNC('f',a+(i-1)*Delta,a+i*Delta,m);
end
```

We designate the estimate produced by this quadrature rule by $Q_{\mathrm{NC}(m)}^{(n)}$. The computation is a little inefficient because it involves a surplus of $n-1$ function evaluations. The rightmost f-evaluation in the ith call to QNC is the same as the leftmost f-evaluation in the $i+1$st call. Figure 4.3 depicts the situation in the four-subinterval, five-point rule case.

To avoid redundant f-evaluation, it is better not to apply QNC to each of the n subintegrals. Instead, we precompute *all* the required function evaluations and store them in a single column vector `fval(1:n(m-1)+1)`. The linear combination that defines the composite rule is then calculated. In the preceding $Q_{\mathrm{NC}(5)}^{(4)}$ example, the 17 required function evaluations are assembled in `fval(1:17)`. If w is the weight vector for $Q_{\mathrm{NC}(5)}$, then

$$Q_{\mathrm{NC}(5)}^{(4)} = \Delta \left(w^T fval(1{:}5) + w^T f(5{:}9) + w^T fval(9{:}13) + w^T fval(13{:}17) \right)$$

FIGURE 4.3 Function evaluations in $Q^{(4)}_{\mathrm{NC}(5)}$

From this we conclude that $Q^{(n)}_{\mathrm{NC}(m)}$ is a summation of n inner products, each of which involves the weight vector w of the underlying rule and a portion of the $fval$-vector. The following function is organized around this principle:

```
function numI = CompQNC(fname,a,b,m,n)
%
% Pre:
%    fname     string that names an available function of the
%              form f(x) that is defined on [a,b]. f should
%              return a column vector if x is a column vector.
%       a,b    real scalars
%        m     integer that satisfies 2 <= m <=11
%        n     positive integer
%
% Post:
%    numI      the composite m-point Newton-Cotes approximation of the
%              integral of f(x) from a to b. The rule is applied on
%              each of n equal subintervals of [a,b].
%
   Delta = (b-a)/n;
   w = WNC(m);
   x = linspace(a,b,n*(m-1)+1)';
   f = feval(fname,x);
   numI = 0;
   first = 1;
   last = m;
   for i=1:n
      %Add in the inner product for the i-th subintegral.
      numI = numI + w'*f(first:last);
      first = last;
      last = last+m-1;
   end
   numI = Delta*numI;
```

4.2.2   Error

Let us examine the error. Suppose Q_i is the m-point Newton-Cotes estimate of the ith subintegral. If this rule is exact for polynomials of degree d, then using (4.2.1) we obtain

$$\int_a^b f(x)dx = \sum_{i=1}^{n+1} \int_{z_i}^{z_{i+1}} f(x)dx = \sum_{i=1}^n \left(Q_i + c_m f^{(d+1)}(\eta_i) \left(\frac{z_{i+1} - z_i}{m-1} \right)^{d+2} \right)$$

By definition

$$Q_{NC(m)}^{(n)} = \sum_{i=1}^n Q_i$$

and

$$z_{i+1} - z_i = \Delta = \frac{b-a}{n}$$

Moreover, it can be shown that

$$\frac{1}{n} \sum_{i=1}^n f^{(d+1)}(\eta_i) = f^{(d+1)}(\eta)$$

for some $\eta \in [a, b]$ and so

$$\int_a^b f(x)dx = Q_{NC(m)}^{(n)} + c_m \left(\frac{b-a}{n(m-1)} \right)^{d+2} n f^{(d+1)}(\eta) \qquad (4.2.2)$$

If $|f^{(d+1)}(x)| \leq M_{d+1}$ for all $x \in [a, b]$, then

$$\left| Q_{NC(m)}^{(n)} - \int_a^b f(x)dx \right| \leq \left[|c_m| M_{d+1} \left(\frac{b-a}{m-1} \right)^{d+2} \right] \frac{1}{n^{d+1}} \qquad (4.2.3)$$

Comparing with (4.1.3), we see that the error in the composite rule is the error in the corresponding "simple" rule divided by n^{d+1}. Thus with m fixed it is possible to exercise error control by choosing n sufficiently large. For example, suppose that we want to approximate the integral with a uniformly spaced composite Simpson rule so that the error is less than a prescribed tolerance *tol*. If we know that the fourth derivative of f is bounded by M_4, then we choose n so that

$$\frac{1}{90} M_4 \left(\frac{b-a}{2} \right)^5 \frac{1}{n^4} \leq tol$$

To keep the number of function evaluations as small as possible, n should be the smallest positive integer that satisfies

$$n \geq (b-a) \sqrt[4]{\frac{M_4(b-a)}{2880 \cdot tol}}$$

The script file ShowCompQNC displays the error properties of the composite Newton-Cotes rules for three different integrands. (See Fig 4.4.)

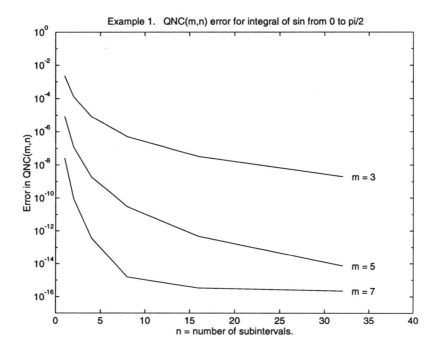

FIGURE 4.4 Error in composite Newton-Cotes rules

Problems

P4.2.1 Write a function error = errCompNC(a,b,m,DerBound,n) that returns an upper bound for the error in the uniformly spaced composite m-point Newton-Cotes quadrature rule applied to the integral of $f(x)$ from a to b. Use errNC.

P4.2.2 Complete the following function so that it performs as advertised:

```
function w = wCompNC(m,n)
%
% Pre:
%    m    integer satisfying 2<=m<=11
%    n    positive integer
% Post:
%    w    column vector of length n(m-1)+1 with the property
%         that the script
%              xvals = linspace(a,b,n*(m-1)+1)';
%              fvals = feval('f',xvals);
%              I = (b-a)*w'*fvals
%         returns  m-point, n-subinterval composite Newton-Cotes approximation
%         to the integral of f(x) from a to b.
```

You may assume the existence of the function w = wNC(m) that returns the weights of the simple m-point Newton-Cotes rule, $2 \le m \le 11$. Note that wNC(m) and wCompNC(m,1) are equivalent.

P4.2.3 Write a function: n = nBest(a,b,m,DerBound,tol) that returns an integer n such that the error bound for $Q_{NC(m)}^{(n)}$ is less than *tol*.

P4.2.4 Let $C(x)$ be the piecewise cubic Hermite interpolant of of $f(x)$ on $[a, b]$. Develop a uniformly spaced composite rule based on this quadrature rule. (See Problem 4.1.3.)

4.3   Adaptive Quadrature

Uniformly spaced composite rules that are exact for degree d polynomials are efficient if $f^{(d+1)}$ is uniformly behaved across $[a, b]$. However, if the magnitude of this derivative varies widely across the interval of integration, then the error control process discussed in §4.2 may result in an unnecessary number of function evaluations. This is because n is determined by an interval-wide derivative bound M_{d+1}. In regions where $f^{(d+1)}$ is small compared to this value, the subintervals are (possibly) much shorter than necessary. *Adaptive quadrature* methods address this problem by "discovering" where the integrand is ill behaved and shortening the subintervals accordingly.

4.3.1   An Adaptive Newton-Cotes Procedure

To obtain a good partition of $[a, b]$, we need to be able to estimate error. That way the partition can be refined if the error is not small enough. One idea is to use two different quadrature rules. The difference between the two predicted values of the integral could be taken as a measure of their inaccuracy:

```
function numI = AdaptQNC(fname,a,b,...)
    Compute the integral from a to b in two ways.  Call the values A₁ and A₂
        and assume that A₂ is better.
    Estimate the error in A₂ based on |A₁ - A₂|.
    If the error is sufficiently small, then
        numI = A₂;
    else
        mid = (a+b)/2;
        numI = AdaptQNC(fname,a,mid,...)  + AdaptQNC(fname,mid,b,...);
    end
```

This divide-and-conquer framework is similar to the one we developed for adaptive piecewise linear approximation.

The filling in of the details begins with the development of a method for estimating the error. Fix m and set $A_1 = Q_{NC(m)}^{(1)}$ and $A_2 = Q_{NC(m)}^{(2)}$. Thus A_1 is the "simple" m-point rule estimate and A_2 is the two-interval, m-point rule estimate. If these rules are exact for degree d polynomials, then it can be shown that

$$I = A_1 + \left[c_m f^{(d+1)}(\eta_1) \left(\frac{b-a}{m-1} \right)^{d+2} \right] \tag{4.3.1}$$

$$I = A_2 + \left[c_m f^{(d+1)}(\eta_2) \left(\frac{b-a}{m-1} \right)^{d+2} \right] \frac{1}{2^{d+1}} \tag{4.3.2}$$

where η_1 and η_2 are in the interval $[a, b]$. We now make the assumption $f^{(d+1)}(\eta_1) = f^{(d+1)}(\eta_2)$. This is reasonable if $f^{(d+1)}$ does not vary much on $[a, b]$. (The shorter the interval, the more likely this is to be the case.) Thus $I = A_1 + C$ and $I = A_2 + C/2^{d+1}$ where

$$C = \left[c_m f^{(d+1)}(\eta_1) \left(\frac{b-a}{m-1} \right)^{d+2} \right]$$

By subtracting these two equations for I and solving for C, we get

$$C = \frac{A_2 - A_1}{1 - \frac{1}{2^{d+1}}}$$

and so

$$|I - A_2| \approx \frac{|A_1 - A_2|}{2^{d+1}}$$

Thus the discrepancy between the two estimates divided by 2^{d+1} provides a reasonable estimate of the error in A_2. If our goal is to produce an estimate of I that has absolute error *tol* or less, then the recursion may be organized as follows:

```
function numI = AdaptQNC(fname,a,b,m,tol)
%
% Pre:
%    fname       string that names an available function of the
%                form f(x) that is defined on [a,b]. f should
%                return a column vector if x is a column vector.
%    a,b         real scalars
%       m        integer that satisfies 2 <= m <=11
%    tol         positive real
% Post:
%    numI        the composite m-point Newton-Cotes approximation of the
%                integral of f(x) from a to b, with the abscissae chosen
%                adaptively. The

% Compute the integral two ways and estimate the error.
A1 = CompQNC(fname,a,b,m,1);
A2 = CompQNC(fname,a,b,m,2);
d = 2*floor((m-1)/2)+1;
error = abs(A1-A2)/2^(d+1);
if error <= tol
   % A2 is acceptable
   numI = A2;
else
   % Sum suitably accurate left and right integrals
   mid = (a+b)/2;
   numI = AdaptQNC(fname,a,mid,m,tol/2) + AdaptQNC(fname,mid,b,m,tol/2);
end
```

If the heuristic estimate of the error is greater than *tol*, then two recursive calls are initiated to obtain estimates

$$Q_L \approx \int_a^{mid} f(x)dx = I_L$$

and

$$Q_R \approx \int_{mid}^b f(x)dx = I_R$$

that satisfy

$$|I_L - Q_L| \leq tol/2$$
$$|I_R - Q_R| \leq tol/2$$

Setting $Q = Q_L + Q_R$, we see that

$$|I - Q| = |(I_L - Q_L) + (I_R - Q_R)| \leq |I_L - Q_L| + |I_R - Q_R| \leq (tol/2) + (tol/2) = tol$$

The following script illustrates the behavior of `AdaptQNC` for various choices of tolerance and m:

```
% Script File: ShowAdapts
%
% Illustrates AdaptQNC for a range of tol values and a range of
% underlying NC rules. Uses the humps function for an integrand
% and displays information about the function evaluations.
%
  global FunEvals VecFunEvals
  close all
  x = linspace(0,1,100); y = humps(x);
  for tol = [.01   .001   .0001 .00001]
     for m=3:2:9
        figure
        plot(x,y)
        hold on
        title(sprintf('m = %2.0f   tol = %10.6f',m,tol))
        FunEvals = 0;
        VecFunEvals = 0;
        num0 = AdaptQNC('SpecHumps',0,1,m,tol);
        s = 'Scalar Evals = %3.0f   Vector Evals = %3.0f';
        xLabel(sprintf(s,FunEvals,VecFunEvals))
        hold off
        pause
     end
  end
```

The script uses MATLAB's global variable capability in order for it to report on the number of function evaluations that are required for each `AdaptQNC` call. The command

```
  global FunEvals VecFunEvals
```

	$m = 3$	$m = 5$	$m = 7$	$m = 9$
$tol = .01$	(104,26)	(98,14)	(60,6)	(26,2)
$tol = .001$	(216,54)	(154,22)	(60,6)	(26,2)
$tol = .0001$	(376,94)	(210,30)	(140,14)	(130,10)
$tol = .00001$	(696,174)	(322,46)	(260,26)	(182,14)

FIGURE 4.5 FunEvals and VecFunEvals values

designates FunEvals and VecFunEvals as global variables. They "sit" in the MATLAB workspace and are accessible by any function that also designates these two variables as global. For example,

```
    function y = SpecHumps(x)
%
% Pre:
%   x               n-vector
%   FunEvals        initialized global scalar.
%   VecFunEvals     initialized global scalar
%
% Post:
%   y     humps(x)
%
% Side Effects:
%   FunEvals is increased by length(x).
%   VecFunEvals is increased by 1.
%

    global FunEvals VecFunEvals;
    y = humps(x);
    plot(x,y,'*')
    FunEvals = FunEvals + length(x);
    VecFunEvals = VecFunEvals + 1;
```

Each call to SpecHumps updates both FunEvals and VecFunEvals. SpecHumps also plots the points at which humps is evaluated during the call. Notice that the hold toggle is on during the call so that these points are added to the current plot. The values of FunEvals and VecFunEvals computed by ShowAdapts are reported in Fig 4.5.

4.3.2   Quad and Quad8

MATLAB has two adaptive quadrature procedures, quad and quad8. The first is based on $Q_{NC(3)}$ (Simpson's rule) and the second on $Q_{NC(9)}$. We discuss quad. The calling sequence is identical for quad8. A call of the form

```
        Q = QUAD('f',a,b)
```

assigns to Q an estimation of the integral of $f(x)$ from a to b. The default relative error tolerance is 10^{-3}. Otherwise, a fourth input parameter can be used to specify the required tolerance. For example,

$$Q = QUAD('f',a,b,tol)$$

A fifth nonzero parameter can be used to produce a plot of f that reveals where it is evaluated by quad:

$$Q = QUAD('f',a,b,tol,1)$$

The number of required function evaluations can be obtained by specifying a second output parameter:

$$[Q,count] = QUAD('f',a,b,tol,1)$$

In the script file ShowQuads, these two procedures are applied to the integral of the function humps from 0 to 1 over a range of tolerances:

tol	quad	N-quad	quad8	N-quad8
0.010000	29.8576752	49	29.8582736	49
0.001000	29.8583121	109	29.8583095	65
0.000100	29.8583251	241	29.8583278	81
0.000010	29.8583254	513	29.8583254	129
0.000001	29.8583254	1081	29.8583254	161

An error message is printed if the recursion goes more than 10 deep.

Problems

P4.3.1 The one-panel midpoint rule Q_1 for the integral

$$I = \int_a^b f(x)dx$$

is defined by

$$Q_1 = (b-a)f\left(\frac{a+b}{2}\right)$$

The two-panel midpoint rule Q_2 for I is given by

$$Q_2 = \frac{b-a}{2}\left(f\left(\frac{3a+b}{4}\right) + f\left(\frac{a+3b}{4}\right)\right)$$

Using the heuristic $|I - Q_2| \leq |Q_2 - Q_1|$, write an efficient MATLAB adaptive quadrature routine of the form Adapt(a,b,fname,tol,...) that returns an estimate of I that is accurate to within the tolerance given by tol. You may extend the parameter list, and you may use nargin as required. You may ignore the possibility of infinite recursion.

P4.3.2 The MATLAB function Quad(fname,a,b,tol) returns an estimate of the integral

$$I = \int_a^b f(x)dx$$

that is usually correct to within `tol` where `fname` is a string that names the integrand function. Assume that functions `g(x)` and `h(x)` are available. Write a MATLAB script that assigns to `Idif` and `Isum` approximations to

$$I_{dif} = \int_a^b ([g(x) - h(x)]/2) dx$$

and

$$I_{sum} = \int_a^b ([g(x) + h(x)]/2) dx.$$

that are each correct to within 10^{-3}. Efficiency matters. Make effective use of `Quad`.

P4.3.3 A number of efficiency improvements can be made to `AdaptQNC`. A casual glance at `AdaptQNC` reveals two sources of redundant function evaluations: First, each function evaluation required in the assignment to `A1` is also required in the assignment to `A2`. Second, the recursive calls could (but do not) make use of previous function evaluations. In addressing these deficiencies, you are to follow these ground rules:

- A call of the form `AdaptQNC1(fname,a,b,m,tol)` must produce the same value as a call of the form `AdaptQNC(fname,a,b,m,tol)`.
- No global variables are allowed.

To "transmit" appropriate function values in the recursive calls, you will want to design `AdaptQNC1` so that it has an "optional" sixth argument `fValues`. By making this argument optional, the same five-parameter calls at the top level are permitted.

4.4 Special Topics

We briefly discuss two other approaches to the quadrature problem. Gauss quadrature is useful in certain specialized settings. When the higher derivatives of a function are smooth, fewer function evaluations are required to attain a given level of accuracy. In settings where the function evaluations are experimentally determined, spline quadrature has a certain appeal.

4.4.1 Gauss Quadrature

In the Newton-Cotes framework, the integrand is sampled at regular intervals across $[a, b]$. In the *Gauss quadrature* framework, the abscissas are positioned in such a way that the rule is correct for polynomials of maximal degree.

A simple example clarifies the main idea. Let us try to determine weights w_1 and w_2 and abscissas x_1 and x_2 so that

$$w_1 f(x_1) + w_2 f(x_2) = \int_{-1}^1 f(x) dx$$

for polynomials of degree 3 or less. This is plausible since there are four parameters to choose (w_1, w_2, x_1, x_2) and four constraints obtained by forcing the rule to be exact for the functions $1, x, x^2$, and x^3:

$$
\begin{aligned}
w_1 + w_2 &= 2 \\
w_1 x_1 + w_2 x_2 &= 0 \\
w_1 x_1^2 + w_2 x_2^2 &= 2/3 \\
w_1 x_1^3 + w_2 x_2^3 &= 0
\end{aligned}
$$

By multiplying the second equation by x_1^2 and subtracting it from the fourth equation we get $w_2 x_2 (x_1^2 - x_2^2) = 0$, and so $x_2 = -x_1$. It follows from the second equation that $w_1 = w_2$ and thus from the first equation, $w_1 = w_2 = 1$. From the third equation, $x_1^2 = 1/3$ and so $x_1 = -1/\sqrt{3}$ and $x_2 = 1/\sqrt{3}$. Thus for any $f(x)$ we have

$$\int_{-1}^{1} f(x) dx \approx f(-1/\sqrt{3}) + f(1/\sqrt{3})$$

This is the two-point *Gauss-Legendre* rule.

The m-point Gauss-Legendre rule has the form

$$Q_{\text{GL}(m)} = w_1 f(x_1) + \cdots + w_m f(x_m)$$

where the w_i and x_i are chosen to make the rule exact for polynomials of degree $2m - 1$. One way to define these $2m$ parameters is by the $2m$ *nonlinear* equations

$$w_1 x_1^k + w_2 x_2^k + \cdots + w_m x_m^k = \frac{1 - (-1)^{k+1}}{k + 1}, \qquad k = 0{:}2m - 1$$

The kth equation is the requirement that the rule

$$w_1 f(x_1) + \cdots + w_m f(x_m) = \int_{-1}^{1} f(x) dx$$

be exact for $f(x) = x^k$. It turns out that this system has a unique solution, which we encapsulate in the following function for the cases $m = 2{:}6$:

```
    function [w,x] = WGL(m);
%
% Pre:
%   m       integer that satisfies 2 <= m <= 6
%
% Post:
%   w       column m-vector consisting of the weights for the
%           m-point Gauss-Legendre rule..
%   x       column m-vector consisting of the abscissae for the
%           m-point Gauss-Legendre rule.
%
    w = ones(m,1);
    x = ones(m,1);
    if m==2
        w(1)  =   1.000000000000000; w(2)  =   w(1);
        x(1)  =  -0.577350269189626; x(2)  =  -x(1);
    elseif m==3
        w(1)  =   0.555555555555558; w(3)  =   w(1);
        w(2)  =   0.888888888888889;
        x(1)  =  -0.774596669241483; x(3)  =  -x(1);
        x(2)  =   0.000000000000000;
        :
    end
```

The Gauss-Legendre rules

$$Q_{\text{GL}(m)} = w_1 f(x_1) + \cdots + w_m f(x_m) \approx \int_{-1}^{1} f(x) dx$$

are not restrictive even though they pertain to integrals from -1 to 1. By a change of variable, we have

$$\int_a^b f(x) dx = \frac{b-a}{2} \int_{-1}^{1} g(x) dx$$

where

$$g(x) = f\left(\frac{a+b}{2} + \frac{b-a}{2} x\right)$$

and so

$$\frac{b-a}{2}\left(w_1 f\left(\frac{a+b}{2} + \frac{b-a}{2} x_1\right) + \cdots + w_m f\left(\frac{a+b}{2} + \frac{b-a}{2} x_m\right)\right) \approx \int_a^b f(x) dx$$

This gives

```
    function numI = QGL(fname,a,b,m)
%
% Pre:
%    fname      string that names an available function of the
%               form f(x) that is defined on [a,b]. f should
%               return a column vector if x is a column vector.
%       a,b     real scalars
%         m     integer that satisfies 2 <= m <=6
% Post:
%       numI    the m-point Gauss-Legendre approximation of the
%               integral of f(x) from a to b.
%
    [w,x] = WGL(m);
    fvals = feval(fname,((b-a)/2)*x + ((a+b)/2)*ones(m,1));
    numI = ((b-a)/2)*w'*fvals;
```

It can be shown that

$$\left| \int_a^b f(x) dx - Q_{\text{GL}(m)} \right| \leq \frac{(b-a)^{2m+1}(m!)^4}{(2m+1)[(2m)!]^3} M_{2m}$$

where M_{2m} is a constant that bounds $|f^{2m}(x)|$ on $[a,b]$. The script file GLvsNC compares the $Q_{\text{NC}(m)}$ and $Q_{\text{GL}(m)}$ rules when they are applied to the integral of $\sin(x)$ from 0 to $\pi/2$:

m	NC(m)	GL(m)
2	0.7853981633974483	0.9984726134041148
3	1.0022798774922104	1.0000081215555008
4	1.0010049233142790	0.9999999771971151
5	0.9999915654729927	1.0000000000395670
6	0.9999952613861668	0.9999999999999533

Notice that for this particular (easy) problem, $GL(m)$ has approximately the accuracy of $NC(2m)$.

Problems

P4.4.1 If $Q_{GL(m)}$ is the m-point Gauss-Legendre estimate for

$$I = \int_a^b f(x)dx$$

then it can be shown that

$$|I - Q_{GL(m)}| \le \frac{(b-a)^{2m+1}(m!)^4}{(2m+1)[(2m)!]^3} M_{2m} \equiv E_m$$

where the constant M_{2m} satisfies $|f^{(2m)}(x)| \le M_{2m}$ for all $x \in [a, b]$. The following questions apply to the case when $f(x) = e^{cx}$, where $c > 0$. Assume that $a < b$. **(1)** Give a good choice for M_{2m}. **(2)** Give an expression for E_{m+1}/E_m. **(3)** Write a MATLAB script that determines the smallest positive integer m so that E_m is less than tol.

P4.4.2 Write a function numI = compQGL(fname,a,b,m,n) that approximates the integral of a function from a to b by applying the m-point Gauss-Legendre rule on n equal-length subintervals of $[a, b]$.

4.4.2   Spline Quadrature

Suppose $S(x)$ is a cubic spline interpolant of (x_i, y_i), $i = 1:n$ and that we wish to compute

$$I = \int_{x_1}^{x_n} S(x)dx$$

If the ith local cubic is represented by

$$q_i(x) = \rho_{i4} + \rho_{i,3}(x - x_i) + \rho_{i,2}(x - x_i)^2 + \rho_{i,1}(x - x_i)^3$$

then

$$\int_{x_i}^{x_{i+1}} q_i(x)dx = \rho_{i,4}h_i + \frac{\rho_{i,3}}{2}h_i^2 + \frac{\rho_{i,2}}{3}h_i^3 + \frac{\rho_{i,1}}{4}h_i^4$$

where $h_i = x_{i+1} - x_i$. By summing these quantities from $i = 1:n - 1$, we obtain the sought-after spline integral:

```
    function numI = SplineQ(x,y)
%
% Pre:
%   x,y    n-vectors with x(1) < ... < x(n)
%
% Post:
%   numI    the integral from x(1) to x(n) of the not-a-knot spline
%           interpolant of (x(i),y(i)), i=1:n
%

    S = spline(x,y);
    [x,rho,L,k] = unmkpp(S);
    sum = 0;
    for i=1:L
        % Add in the integral from x(i) to x(i+1).
        h = x(i+1)-x(i);
        subI = h*(((rho(i,1)*h/4 + rho(i,2)/3)*h + rho(i,3)/2)*h + rho(i,4));
        sum = sum + subI;
    end
    numI = sum;
```

The script file `ShowSplineQ` uses this function to produce the following estimates for the integral of sine from 0 to $\pi/2$:

m	Spline Quadrature
5	1.0001345849741938
50	0.9999999990552404
500	0.9999999999998678

Here, the spline interpolates the sine function at x = `linspace(0,pi/2,m)`.

Problems

P4.4.3 Modify `SplineQ` so that a four-argument call `SplineQ(x,y,a,b)` returns the integral of the spline interpolant from a to b. Assume that $x_1 \leq a \leq b \leq x_n$.

P4.4.4 Let $a(t)$ denote the acceleration of an object at time t. If v_0 is the object's velocity at $t = 0$, then the velocity at time t is prescribed by

$$v(t) = v_0 + \int_0^t a(\tau)d\tau$$

Likewise, if x_0 is the position at $t = 0$, then the position at time t is given by

$$x(t) = x_0 + \int_0^t v(\tau)d\tau$$

Now suppose that we have snapshots $a(t_i)$ of the acceleration at times t_i, $i = 1:m$, $t_1 = 0$. Assume that we know the initial position x_0 and velocity v_0. Our goal is to estimate position from this data. Spline quadrature will be used to approximate the preceding integrals. Let $S_a(t)$ be the not-a-knot spline interpolant of the acceleration data $(t_i, a(t_i))$, $i = 1:m$, and define

$$\tilde{v}(t) = v_0 + \int_0^t S_a(\tau)d\tau$$

Let $S_v(t)$ be the not-a-knot spline interpolant of the data $(t_i, \tilde{v}(t_i))$, $i = 1:m$, and define

$$\tilde{x}(t) = x_0 + \int_0^t S_v(\tau)d\tau$$

The spline interpolant $S_x(t)$ of the data $(t_i, \tilde{x}(t_i))$ is then an approximation of the true position. Write a function

```
    function Sx = PosVel(a,t,x0,v0)}
% Pre:
%     t    an m-vector of equally spaced time values with t(1) = 0, m>=2.
%     a    m-vector of accelerations, a(i) = acceleration at time t(i).
% x0,v0    position and velocity  at t=0
%
% Post:
%     Sx the pp-representation of a spline that approximates position.
```

Try it out on the data t = linspace(0,50,500), with $a(t) = 10e^{-t/25}\sin(t)$. However, before you turn the a vector over to PosVel, contaminate it with noise: a = a + .01*randn(size(a)). Produce a plot of the exact and estimated positions across $[0,50]$ and a separate plot of $x(t) - S_x(t)$ across $[0, 50]$. Also print the value of $S_x(t)$ at $t = 50$. Repeat with $m = 50$ instead of 500. Use the MATLAB spline function.

4.5 Shared Memory Adaptive Quadrature

The quadrature problem is an excellent vehicle for introducing some of the key ideas associated with parallel computation. In this brief section we discuss the implementation of the adaptive quadrature approach in a shared memory machine.

4.5.1 The Shared Memory Paradigm

Parallel computers have more than one processor. With careful scheduling and problem subdivision, it is sometimes possible for a parallel computer to carry out a computation much more rapidly than when only a single processor is involved. An analogy to human work groups is instructive. Ten people working together can produce the average of 1000 numbers almost ten times quicker than a single person. We have to say "almost" because the very fact that a group is working together means that there will be some "waiting for others" while the 10 partial sums are collected. However, with careful coordination, the interactions between the workers can be minimized, making it possible for them to work on their part of the computation without much interference. Not all problems subdivide as nicely as the find-the-average problem, as we will see.

Parallel computers can be organized in many different ways. In this section we focus on the *shared memory* approach. In a shared memory environment, each individual processor is able to read from and write to a *shared memory*. Figure 4.6 depicts the situation. Each processor (or node) has its own *local memory* and executes its own *node program*. The act of writing a program for such a parallel computer is the act of writing a program for each of the nodes. The node programs typically reference *local variables* housed in the processor's own memory and global

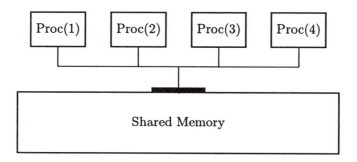

FIGURE 4.6 A shared memory multiprocessor

variables that are situated in shared memory. The idea is for all node programs to execute at the same time in a coordinated way that results in the required calculation.

To illustrate the key ideas, we outline how a four-processor shared memory parallel computer might compute an approximation Q to the integral

$$I = \int_a^b f(x)dx$$

We assume that each processor has a copy of the function AdaptQNC(fname,a,b,m,tol) developed in §4.3 and the integrand function f. Five global variables are assumed to be initialized at the outset:

$$
\begin{aligned}
\texttt{a} \ = \ & \text{the lower limit of integration (e.g., 0)} \\
\texttt{b} \ = \ & \text{the upper limit of integration, (e.g., 10)} \\
\texttt{m} \ = \ & \text{the Newton-Cotes parameter (e.g., 3)} \\
\texttt{tol} \ = \ & \text{the accuracy tolerance (e.g., .001)} \\
\texttt{p} \ = \ & \text{the number of processors in the system (e.g., 4) .}
\end{aligned}
$$

Another variable, QProc(1:4) will be used for intermediate results, and Q will be used to store the final estimate. Figure 4.7 depicts these allocations and initializations. Finally, we assume that the processors are indexed from 1 to p and that each has a local variable id.loc that houses its particular index. (Local variables will be designated with a .loc suffix.)

4.5.2 Static Scheduling

An obvious way to divide the work among the p processors is to assign the estimation of

$$I_\mu = \int_{a+(\mu-1)h}^{a+\mu h} f(x)dx \qquad h = (b-a)/p$$

to Proc(μ). Our plan is to have each processor perform this task and write the result to a location in shared memory. After each of the p processors records its estimate, the subintegrals are ready to be summed and stored in Q. If each processor executes the following node program, then these tasks will be accomplished:

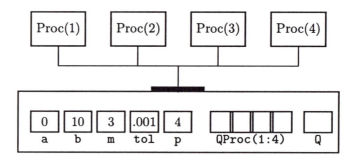

Figure 4.7 Initializations

% Node Program. The Processor's id number is stored in id.loc.

```
h.loc = (b-a)/p;
a.loc = a + (id.loc - 1)*h.loc;
b.loc = a.loc + h.loc;
Q.loc = AdaptQNC('f',a.loc,b.loc,m,tol/p);
QProc(id.loc) = Q.loc;
barrier
if id.loc == 1;
    Q = sum(QProc(1:p))
end
```

The variables id.loc, h.loc, a.loc, b.loc, and Q.loc are local while a, b, m, tol, QProc(1:p), and Q are global. Notice how shared memory variables are accessed in a manner that is no different than in uniprocessor environments. The four subintegral estimates, housed in the global array QProc(1:p), cannot be summed until they are all computed and stored by the participating processors. That is where the barrier construct comes in. It establishes a synchronization point. When a processor encounters a barrier during the execution of its node program, computation is suspended. It resumes as soon as *every* other processor reaches its barrier. In that sense, the barrier is like a stream to be traversed by p hikers. For safety, no one proceeds across the stream until *all p* hikers arrive at its edge. In our node program, the barrier ensures that the summation of the subintegral estimates does not begin until they have all been assembled in the shared memory array QProc(1:4). The order in which they are stored cannot be deduced. It depends on how long the four AdaptQNC calls take and the communication software that routes the data between these processors and the shared memory. The barrier holds up the faster processors while the slower ones catch up. Suppose the four subintegral estimates are 12, 8, 11, and 3. Eventually, QProc(1:4) is filled up. (See Fig 4.8.) At this point, each processor is "released" from the barrier and executes

```
If id.loc == 1
    Q = sum(QProc(1:p))
end
```

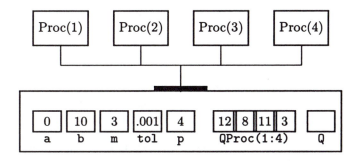

FIGURE 4.8 Subquadratures complete

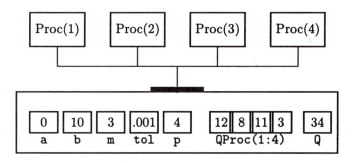

FIGURE 4.9 Overall quadrature complete

However, only Proc(1) "gets though" to carry out the summation, leaving us with the situation depicted in Fig 4.9.

Note that each node program is the same. However, the integral that is estimated by a given processor depends on its identification number.

The scheduling for the preceding parallel computation is *static*, because the "mapping" of tasks to processors is determined in advance of the computation. If the time it takes to execute a task is predictable, then static scheduling works well because the workload can be evenly distributed among the processors. Unfortunately for us, the time required to compute the ith subintegral is not generally predictable. It may be that the integrand is very rough in some small region and that one particular processor is responsible for integrating across that region. The unlucky processor will require much more time to complete its task simply because its call to AdaptQNC will require many more function evaluations.

One way to guard against this is to subdivide $[a, b]$ into rp subintervals, where r is a (possibly large) positive integer. Define as the ith task the estimation of

$$I_i = \int_{a+(i-1)h}^{a+ih} f(x)dx \qquad h = (b-a)/(rp) \qquad (4.5.1)$$

and assign tasks μ:p:rp to Proc(μ). For example, if $p = 4$ and $r = 10$, then the task assignments
are as follows:

Processor	Tasks
1	1,5,9,13,17,21,25,29,33,37
2	2,6,10,14,18,22,26,30,34,38
3	3,7,11,15,19,23,27,31,35,39
4	4,8,12,1,6,20,24,28,32,36,40

The idea behind this strategy is that with the shortened subintervals, rough integrand regions
are likely to be shared by a number of processors. Clearly, the larger the value of the parameter
r, the more likely this is to be the case. Of course, if the subintervals are too short, then an
unnecessary number of function evaluations may be required. (AdaptQNC will be "too good' for
very short intervals and will not have a chance to recur.) If a happy choice for r can be found
and placed in a global variable, then we obtain

```
% Node Program. The Processor's id number is stored in id.loc.

  h.loc = (b-a)/(r*p);
  Q.loc = 0;
  for i=id.loc:p:(r*p);
    a.loc = a + (i - 1)*h.loc;
    b.loc = a.loc + h.loc;
    Q.loc = Q.loc + AdaptQNC('f',a.loc,b.loc,m,tol/(r*p));
  end
  QProc(id.loc) = Q.loc;
  barrier
  if id.loc == 1
     Q = sum(QProc(1:p));
  end
```

Note that Q.loc is used to sum the subintegrals.

4.5.3   Dynamic Scheduling

In parallel computation we aspire to keep all the processors busy with relevant work during the
computation. When this goal is achieved, the parallel computation is said to be *load-balanced*.
An idle processor represents an squandered resource. The problem with static scheduling, as
illustrated by the examples in the previous subsection, is that the processors are locked into their
"assignments" regardless of integrand behavior. What we need is a mechanism that enables idle
processors to "pick up the slack" caused by those processors that are struggling with difficult
portions of the integral.

The *pool-of-task* approach to scheduling addresses this concern. To illustrate the main idea,
we continue to define the approximation of the integral in equation (4.5.1) as the ith task. A
special global variable NextTask is maintained in shared memory. Its value will always be the
index of the next task to be performed. At the start of execution, NextTask is assigned the value
of 1. During execution, a processor in need of something to do (1) copies the value of NextTask

into its local memory and (2) increments the value in `NextTask`. The access to `NextTask` must be carefully controlled since we do not want two different processors obtaining the same task. In particular, we must package the copy and increment operation so that only one processor at a time is dealing with `NextTask`.

To that end, we identify a *critical section* in each node program. The execution of the critical section is contingent on the value of a special kind of global variable called a `lock`. It has a 0-1 "toggle" role to play. Assume that `lock` has the value of 0 if no processor is in its critical section and the value of 1 if some processor is executing its critical section. Initially `lock` = 0.

The toggling of `lock` is under the careful control of a pair of functions `Enter` and `Leave`. These functions ensure that at most one processor is executing its critical section at any given time. The beginning of the critical section is guarded by the call `Enter(lock)`. If `lock` = 0, then it is toggled to 1 and the executing processor can enter its critical section. If `lock` = 1, then the processor will be held up (perhaps with others) because some other processor is executing its critical section. The processor executing its critical section communicates that it is done by executing a `Leave(lock)`, which restores the value of `lock` to 0. At that time, *one* of the waiting processors will seize control of its critical section. The implementation of this "semaphore" idea is at the system level and beyond the scope of our discussion. However, with these synchronization tools we are able to rewrite the node programs as follows: We suppress the details and focus on the high-level application of the idea to our problem:

```
% Node Program. The Processor's id number is stored in id.loc.

h.loc = (b-a)/(r*p);
Q.loc = 0;
i.loc = 0;
while i.loc <= r*p;
  Enter(lock)
    i.loc = NextTask;
    NextTask = NextTask +1;
  Leave(lock)
  a.loc = a + (i.loc - 1)*h.loc;
  b.loc = a.loc + h.loc;
  Q.loc = Q.loc + AdaptQNC('f',a.loc,b.loc,m,tol/(r*p));
end
QProc(id.loc) = Q.loc;
barrier
if id.loc == 1;
  Q = sum(QProc(1:p))
end
```

Recall that `NextTask` is initialized to 1. Because of the repeated incrementing of `NextTask`, `i.loc` is eventually assigned a value greater than the total number of tasks. This brings about the termination of the `while` and takes the processor to the `barrier`, where things proceed as in our earlier examples. If a processor is given an "easy" subintegral, it will return quickly and immediately request another task. Upon completion, some processors may have performed more tasks than others, but the idea is that the imbalance we saw in the statically scheduled examples is less likely.

Problems

P4.5.1 Explain why `tol` is scaled in the calls to `AdaptQNC`.

P4.5.2 Suppose function evaluations are so costly in execution time that they dominate all other costs. Suppose that $r = 10$, $p = 4$, and that tasks 2 through 40 require 100 function evaluations each. Assume that task 1 requires 1000 function evaluations. Assume that each function evaluation requires 1 minute of processor node time. Estimate how long it will take both the statically scheduled and the pool-of-task scheduled methods to execute. What if the last task required 1000 function evaluations and the first 39 just 100 evaluations?

M-Files and References

Script Files

ShowNCIdea	Displays the idea behind the Newton-Cotes rules.
ShowQNC	Computes Newton-Cotes estimates for specified integrands.
ShowQNCError	Illustrates ErrNC.
ShowCompQNC	Illustrates CompQNC on three examples.
ShowAdapts	Illustrates AdaptQNC.
ShowQuads	Illustrates Quad and Quad8.
GLvsNC	Compares Gauss-Legendre and Newton-Cotes rules.
ShowSplineQ	Illustrates SplineQ.

Function Files

WNC	Constructs the Newton-Cotes weight vector.
QNC	The simple Newton-Cotes rule.
ErrNC	Error in the simple Newton-Cotes rule.
WNCOpen	Constructs the open Newton-Cotes weight vector.
ErrNCOpen	Error in the simple open Newton-Cotes rule.
CompQNC	Equally-spaced, composite Newon-Cotes rule.
AdaptQNC	Adaptive Newton-Cotes quadrature.
SpecHumps	The humps function with function call counters.
WGL	Constructs the Gauss-Legendre weight vector.
QGL	The simple Gauss-Legendre rule.
SplineQ	Spline quadrature.

References

P. Davis and P. Rabinowitz (1984). *Methods of Numerical Integration, 2nd Ed.*, Academic Press, New York.

G.H. Golub and J.M. Ortega (1993). *Scientific Computing: An Introduction with Parallel Computing,* Academic Press, Boston.

A. Stroud (1972). *Approximate Calculation of Multiple Integrals*, Prentice Hall, Englewood Cliffs, NJ.

Chapter 5

Matrix Computations

§**5.1** Setting Up Matrix Problems

§**5.2** Matrix Operations

§**5.3** Once Again, Setting Up Matrix Problems

§**5.4** Recursive Matrix Operations

§**5.5** Distributed Memory Matrix Multiplication

The next item on our agenda is the linear equation problem $Ax = b$. However, before we get into algorithmic details, it is important to study two simpler calculations: matrix-vector multiplication and matrix-matrix multiplication. Both operations are supported in MATLAB so in that sense there is "nothing to do." However, there is much to learn by studying how these computations can be implemented. Matrix-vector multiplications arise during the course of solving $Ax = b$ problems. Moreover, it is good to uplift our ability to think at the matrix-vector level before embarking on a presentation of $Ax = b$ solvers.

The act of setting up a matrix problem also deserves attention. Often, the amount of work that is required to initialize an n-by-n matrix is as much as the work required to solve for x. We pay particular attention to the common setting when each matrix entry a_{ij} is an evaluation of a continuous function $f(x, y)$.

A theme throughout this chapter is the exploitation of structure. It is frequently the case that there is a pattern among the entries in A which can be used to reduce the amount of work. The fast Fourier transform and the fast Strassen matrix multiply algorithm are presented as examples of recursion in the matrix computations. The organization of matrix-matrix multiplication on a ring of processors is also studied and gives us a nice snapshot of what algorithm development is like in a distributed memory environment.

5.1 Setting Up Matrix Problems

Before a matrix problem can be solved, it must be set up. In many applications, the amount of work associated with the set-up phase rivals the amount of work associated with the solution phase. Therefore, it is in our interest to acquire intuition about this activity. It is also an occasion to see how many of MATLAB's vector capabilities extend to the matrix level.

5.1.1 Simple ij Recipes

If the entries in a matrix $A = (a_{ij})$ are specified by simple recipes, such as

$$a_{ij} = \frac{1}{i+j-1},$$

then a double-loop script can be used for its computation:

```
A = zeros(n,n);
for i=1:n
    for j=1:n
        A(i,j) = 1/(i+j-1);
    end
end
```

Preallocation with `zeros(n,n)` reduces memory management overhead.

Sometimes the matrix defined has patterns that can be exploited. The preceding matrix is *symmetric* since $a_{ij} = a_{ji}$ for all i and j. This means that the (i,j) recipe need only be applied half the time:

```
A = zeros(n,n);
for i=1:n
    for j=i:n
        A(i,j) = 1/1(i+j-1);
        A(j,i) = A(i,j);
    end
end
```

This particular example is a *Hilbert matrix* and it so happens that there a built-in function `A = hilbert(n)` that can be used in lieu of the preceding scripts.

The setting up of a matrix can often be made more efficient by exploiting relationships that exist between the entries. Consider the construction of the lower triangular matrix of binomial coefficients:

$$P = \begin{bmatrix} 1 & 0 & 0 & 0 & 0 \\ 1 & 1 & 0 & 0 & 0 \\ 1 & 2 & 1 & 0 & 0 \\ 1 & 3 & 3 & 1 & 0 \\ 1 & 4 & 6 & 4 & 1 \end{bmatrix}$$

The binomial coefficient "m-choose-k" is defined by

$$\begin{pmatrix} m \\ k \end{pmatrix} = \begin{cases} \dfrac{m!}{k!(m-k)!} & \text{if } 0 \le k \le m \\[2ex] 0 & \text{otherwise} \end{cases}$$

If $k \leq m$, then it specifies the number of ways that k objects can be selected from a set of m objects. The ij entry of the matrix we are setting up is defined by

$$p_{ij} = \left(\begin{array}{c} i - 1 \\ j - 1 \end{array} \right).$$

If we simply compute each entry using the factorial definition, then $O(n^3)$ flops are involved. On the other hand, from the 5-by-5 case we notice that P is lower triangular with ones on the diagonal and in the first column. An entry not in these locations is the sum of its "north" and "northwest" neighbors. That is,

$$p_{ij} = p_{i-1,j-1} + p_{i-1,j}$$

This permits the following set-up strategy:

```
P = zeros(n,n);
P(:,1) = ones(n,1);
for i=2:n
    for j=2:i
        P(i,j) = P(i-1,j-1) + P(i-1,j);
    end
end
```

This script involves $O(n^2)$ flops and is therefore an order of magnitude faster than the method that ignores the connections between the p_{ij}.

5.1.2 Matrices Defined by a Vector of Parameters

Many matrices are defined in terms of a vector of parameters. Recall the *Vandermonde* matrices from Chapter 2:

$$V = \left[\begin{array}{cccc} 1 & x_1 & x_1^2 & x_1^3 \\ 1 & x_2 & x_2^2 & x_2^3 \\ 1 & x_3 & x_3^2 & x_3^3 \\ 1 & x_4 & x_4^2 & x_4^3 \end{array} \right]$$

We developed several set-up strategies but settled on the following column-oriented technique:

```
n = length(x);
V(:,1) = ones(n,1);
for j=2:n
    % Set up column j.
    V(:,j) = x.*V(:,j-1);
end
```

The *circulant* matrices are also of this genre. They too are defined by a vector of parameters. For example

$$C = \left[\begin{array}{cccc} a_1 & a_2 & a_3 & a_4 \\ a_4 & a_1 & a_2 & a_3 \\ a_3 & a_4 & a_1 & a_2 \\ a_2 & a_3 & a_4 & a_1 \end{array} \right]$$

Each row in a circulant is a shifted version of the row above it. Here are two circulant set-up functions:

```
    function C = Circulant1(a)
%
% Pre:
%       a is a row vector
%
% Post:
%       C is a circulant matrix with C(1,:) = a
%
    n = length(a);
    C = zeros(n,n);
    for i=1:n
       for j=1:n
          C(i,j) = a( rem(n-i+j,n)+1);
       end
    end

    function C = Circulant2(a)
%
% Pre:
%       a is a row vector
%
% Post:
%       C is a circulant matrix with C(1,:) = a
%
    n = length(a);
    C = zeros(n,n);
    C(1,:) = a;
    for i=2:n
       C(i,:) = [ C(i-1,n) C(i-1,1:n-1) ];
    end
```

Circulant1 exploits the fact that $c_{ij} = a_{((n-i+j) \bmod n)+1}$ and is a scalar-level implementation. Circulant2 exploits the fact that $C(i,:)$ is a left shift of $C(i-1,:)$ and is a vector-level implementation. In Matlab the latter is much more efficient:

n	Time Circulant1 / Time Circulant2
10	4.000
20	7.583
40	13.680
80	20.862
160	30.380

See the script file CircBench.

Circulant matrices are examples of *Toeplitz* matrices. Toeplitz matrices arise in many applications and are constant along their diagonals. For example,

$$
T = \begin{bmatrix}
c_1 & r_2 & r_3 & r_4 \\
c_2 & c_1 & r_2 & r_3 \\
c_3 & c_2 & c_1 & r_2 \\
c_4 & c_3 & c_2 & c_1
\end{bmatrix}
$$

If c and r are n-vectors, then T = Toeplitz(c,r) sets up the matrix

$$
t_{ij} = \begin{cases}
c_{i-j} & i \geq j \\
\\
r_{j-i} & j > i
\end{cases}
$$

5.1.3 Band Structure

Many important classes of matrices have lots of zeros. Lower triangular matrices

$$
L = \begin{bmatrix}
\times & 0 & 0 & 0 & 0 \\
\times & \times & 0 & 0 & 0 \\
\times & \times & \times & 0 & 0 \\
\times & \times & \times & \times & 0 \\
\times & \times & \times & \times & \times
\end{bmatrix}
$$

upper triangular matrices

$$
U = \begin{bmatrix}
\times & \times & \times & \times & \times \\
0 & \times & \times & \times & \times \\
0 & 0 & \times & \times & \times \\
0 & 0 & 0 & \times & \times \\
0 & 0 & 0 & 0 & \times
\end{bmatrix}
$$

and tridiagonal matrices

$$
T = \begin{bmatrix}
\times & \times & 0 & 0 & 0 \\
\times & \times & \times & 0 & 0 \\
0 & \times & \times & \times & 0 \\
0 & 0 & \times & \times & \times \\
0 & 0 & 0 & \times & \times
\end{bmatrix}
$$

are the most important special cases. The $\times$-0 notation is a handy way to describe patterns of zeros and nonzeros in a matrix. Each "$\times$" designates a nonzero scalar.

In general, a matrix $A = (a_{ij})$ has *lower bandwidth* p if $a_{ij} = 0$ whenever $i > j + p$. Thus an upper triangular matrix has lower bandwidth 0 and a tridiagonal matrix has lower bandwidth 1. A matrix $A = (a_{ij})$ has *upper bandwidth* q if $a_{ij} = 0$ whenever $j > i + q$. Thus a lower triangular matrix has upper bandwidth 0 and a tridiagonal matrix has lower bandwidth 1. Here is a matrix with upper bandwidth 2 and lower bandwidth 3:

$$A = \begin{bmatrix} \times & \times & \times & 0 & 0 & 0 & 0 & 0 \\ \times & \times & \times & \times & 0 & 0 & 0 & 0 \\ \times & \times & \times & \times & \times & 0 & 0 & 0 \\ \times & \times & \times & \times & \times & \times & 0 & 0 \\ 0 & \times & \times & \times & \times & \times & \times & 0 \\ 0 & 0 & \times & \times & \times & \times & \times & \times \\ 0 & 0 & 0 & \times & \times & \times & \times & \times \\ 0 & 0 & 0 & 0 & \times & \times & \times & \times \end{bmatrix}$$

Diagonal matrices have upper and lower bandwidth zero and can be established using the diag function. If d = [10 20 30 40] and D = diag(d), then

$$D = \begin{bmatrix} 10 & 0 & 0 & 0 \\ 0 & 20 & 0 & 0 \\ 0 & 0 & 30 & 0 \\ 0 & 0 & 0 & 40 \end{bmatrix}$$

Two-argument calls to diag are also possible and can be used to reference to the "other" diagonals of a matrix. An entry a_{ij} is on the kth diagonal if $j - i = k$. To clarify this, here is a matrix whose entries equal the diagonal values:

$$\begin{bmatrix} 0 & 1 & 2 & 3 \\ -1 & 0 & 1 & 2 \\ -2 & -1 & 0 & 1 \\ -3 & -2 & -1 & 0 \\ -4 & -3 & -2 & -1 \end{bmatrix}$$

If v is an m-vector, then D = diag(v,k) establishes $(m + k)$-by-$(m + k)$ matrix that has kth diagonal equal to v and is zero everywhere else. Thus

$$\text{diag}([10\ 20\ 30],2) = \begin{bmatrix} 0 & 0 & 10 & 0 & 0 \\ 0 & 0 & 0 & 20 & 0 \\ 0 & 0 & 0 & 0 & 30 \\ 0 & 0 & 0 & 0 & 0 \\ 0 & 0 & 0 & 0 & 0 \end{bmatrix}$$

If A is a matrix, then v = diag(A,k) extracts the kth diagonal and assigns it (as a column vector) to v.

The functions tril and triu can be used to "punch out" a banded portion of a given matrix. If B = tril(A,k), then

$$b_{ij} = \begin{cases} a_{ij} & j \le i + k \\ 0 & j > i + k \end{cases}$$

Analogously, if B = triu(A,k), then

$$b_{ij} = \begin{cases} a_{ij} & i \le j + k \\ 0 & i > j + k \end{cases}$$

The command

```
T = -triu(tril(ones(6,6),1),-1) + 3*eye(6,6)
```

sets up the matrix

$$
T = \begin{bmatrix}
2 & -1 & 0 & 0 & 0 & 0 \\
-1 & 2 & -1 & 0 & 0 & 0 \\
0 & -1 & 2 & -1 & 0 & 0 \\
0 & 0 & -1 & 2 & -2 & 0 \\
0 & 0 & 0 & -1 & 2 & -1 \\
0 & 0 & 0 & 0 & -1 & 2
\end{bmatrix}
$$

The commands

```
T =  -diag(ones(5,1),-1) + diag(2*ones(6,1),0) - diag(ones(5,1),1)
T = Toeplitz([2; -1; zeros(4,1)],[2 ; -1; zeros(4,1)])
```

do the same thing.

5.1.4 Block Structure

The notation

$$
A = \begin{bmatrix}
a_{11} & a_{12} & a_{13} \\
a_{21} & a_{22} & a_{23}
\end{bmatrix}
$$

means that A is a 2-by-3 matrix with entries a_{ij}. The a_{ij} are understood to be scalars, and MATLAB supports the synthesis of matrices at this level (i.e., A=[a11 a12 a13; a21 a22 a23]).

The notation can be generalized to handle the case when the specified entries are *matrices* themselves. Suppose A_{11}, A_{12}, A_{13}, A_{21}, A_{22}, and A_{23} have the following shapes:

$$
A_{11} = \begin{bmatrix}
u & u & u \\
u & u & u \\
u & u & u
\end{bmatrix}
\quad
A_{12} = \begin{bmatrix}
v \\
v \\
v
\end{bmatrix}
\quad
A_{13} = \begin{bmatrix}
w & w \\
w & w \\
w & w
\end{bmatrix}
$$

$$
A_{21} = \begin{bmatrix}
x & x & x \\
x & x & x
\end{bmatrix}
\quad
A_{22} = \begin{bmatrix}
y \\
y
\end{bmatrix}
\quad
A_{23} = \begin{bmatrix}
z & z \\
z & z
\end{bmatrix}
$$

We then define the 2-by-3 *block matrix*

$$
A = \begin{bmatrix}
A_{11} & A_{12} & A_{13} \\
A_{21} & A_{22} & A_{23}
\end{bmatrix}
$$

by

$$
A = \left[
\begin{array}{ccc|c|cc}
u & u & u & v & w & w \\
u & u & u & v & w & w \\
u & u & u & v & w & w \\
\hline
x & x & x & y & z & z \\
x & x & x & y & z & z
\end{array}
\right]
$$

The lines delineate the block entries. Of course, A is also a 5-by-6 scalar matrix.

Block matrix manipulations are very important and can be effectively carried out in MATLAB . The script

```
A11 = [ 10 11 12 ; 13 14 15 ; 16 17 18 ];
A12 = [ 20 ; 21 ; 22 ];
A13 = [ 30 31 ; 32 33 ; 34 35 ];
A21 = [ 40 41 42 ; 43 44 45 ];
A22 = [ 50 ; 51 ];
A23 = [ 60 61 ; 62 63 ];
A = [ A11 A12 A13 ; A21 A22 A23 ];
```

results in the formation of

$$
A = \begin{bmatrix}
10 & 11 & 12 & 20 & 30 & 31 \\
13 & 14 & 15 & 21 & 32 & 33 \\
16 & 17 & 18 & 22 & 34 & 35 \\
40 & 41 & 42 & 50 & 60 & 61 \\
43 & 44 & 45 & 51 & 62 & 63
\end{bmatrix}
$$

The block rows of a matrix are separated by semicolons, and it is important to make sure that the dimensions are consistent. The final result must be rectangular at the scalar level. For example,

```
A = [1 zeros(1,6); ...
     zeros(2,1) [10 20;30 40] zeros(2,2); ...
     zeros(2,3) [50 60;70 80]]
```

sets up the matrix

$$
A = \begin{bmatrix}
1 & 0 & 0 & 0 & 0 \\
0 & 10 & 20 & 0 & 0 \\
0 & 30 & 40 & 0 & 0 \\
0 & 0 & 0 & 50 & 60 \\
0 & 0 & 0 & 70 & 80
\end{bmatrix}
$$

The extraction of blocks requires the colon notation. The assignment C = A(2:4,5:6) is equivalent to any of the following:

```
C = [A(2:4,5) A(2:4,6)]
C = [A(2,5:6) ; A(3,5:6) ; A(4,5:6)]
C = [A(2,5) A(2,6) ; A(3,5) A(3,6) ; A(4,5) A(4,6)]
```

Problems

P5.1.1 For n=[5 10 20 40] and for each of the two methods mentioned in §5.1.1, compute the number of flops required to set up the matrix P.

P5.1.2 Give a MATLAB one-liner using Toeplitz that sets up a circulant matrix with first row equal to a given vector a.

P5.1.3 Any Toeplitz matrix can be "embedded" in a larger circulant matrix. For example,

$$
T = \begin{bmatrix}
c & d & e \\
b & c & d \\
a & b & c
\end{bmatrix}
$$

is the leading 3-by-3 portion of

$$
C = \begin{bmatrix}
c & d & e & a & b \\
b & c & d & e & a \\
a & b & c & d & e \\
e & a & b & c & d \\
d & e & a & b & c
\end{bmatrix}
$$

Write a MATLAB function C = EmbedToep(col,row) that sets up a circulant matrix with the property that C(1:n,1:n) = Toeplitz(col,row).

P5.1.4 Write a MATLAB function A = RandBand(m,n,p,q) that returns a random m-by-n matrix that has lower bandwidth p and upper bandwidth q.

P5.1.5 Let n be a positive integer. Extrachromosomal DNA elements called *plasmids* are found in many types of bacteria. Assume that in a particular species there is a plasmid P and that exactly n copies of it appear in every cell. Sometimes the plasmid appears in two slightly different forms. These may differ at just a few points in their DNA. For example, one type might have a gene that codes for resistance of the cell to the antibiotic ampicillin, and the other could have a gene that codes for resistance to tetracycline. Let's call the two variations of the plasmid A and B.

Assume that the cells in the population reproduce in unison. Here is what happens at that time. The cell first replicates its DNA matter. Thus, if a cell has one type A plasmid and three type B plasmids, then it now has two type A plasmids and six type B plasmids. After replication, the cell divides. The two daughter cells will each receive four of the eight plasmids. There are several possibilities and to describe them we adopt a handy notation. We say that a cell is (i_A, i_B) if it has i_A type A plasmids and i_B type B plasmids. So if the parent is $(1, 3)$, then its daughter will be either $(0, 4)$, $(1, 3)$, or $(2, 2)$.

The probability that a daughter cell is (i'_A, i'_B) given its parent is (i_A, i_B) is specified by

$$
P_{i'_A, i_A} = \frac{\begin{pmatrix} 2i_A \\ i'_A \end{pmatrix} \begin{pmatrix} 2i_B \\ i'_B \end{pmatrix}}{\begin{pmatrix} 2n \\ n \end{pmatrix}}
$$

In the numerator you see the number of ways we can partition the parent's replicated DNA so that the daughter is (i'_A, i'_B). The denominator is the total number of ways we can select n plasmids from the replicated set of $2n$ plasmids.

Let $P(n)$ be the $(n+1)$-by-$(n+1)$ matrix whose (i'_A, i_A) entry is given by $P_{i'_A, i_A}$. (Note subscripting from zero.) If $n = 4$, then

$$
P = \frac{1}{\begin{pmatrix} 8 \\ 4 \end{pmatrix}}
\begin{bmatrix}
\begin{pmatrix}0\\0\end{pmatrix}\begin{pmatrix}8\\4\end{pmatrix} & \begin{pmatrix}2\\0\end{pmatrix}\begin{pmatrix}6\\4\end{pmatrix} & \begin{pmatrix}4\\0\end{pmatrix}\begin{pmatrix}4\\4\end{pmatrix} & \begin{pmatrix}6\\0\end{pmatrix}\begin{pmatrix}2\\4\end{pmatrix} & \begin{pmatrix}8\\0\end{pmatrix}\begin{pmatrix}0\\4\end{pmatrix} \\[2ex]
\begin{pmatrix}0\\1\end{pmatrix}\begin{pmatrix}8\\3\end{pmatrix} & \begin{pmatrix}2\\1\end{pmatrix}\begin{pmatrix}6\\3\end{pmatrix} & \begin{pmatrix}4\\1\end{pmatrix}\begin{pmatrix}4\\3\end{pmatrix} & \begin{pmatrix}6\\1\end{pmatrix}\begin{pmatrix}2\\3\end{pmatrix} & \begin{pmatrix}8\\1\end{pmatrix}\begin{pmatrix}0\\3\end{pmatrix} \\[2ex]
\begin{pmatrix}0\\2\end{pmatrix}\begin{pmatrix}8\\2\end{pmatrix} & \begin{pmatrix}2\\2\end{pmatrix}\begin{pmatrix}6\\2\end{pmatrix} & \begin{pmatrix}4\\2\end{pmatrix}\begin{pmatrix}4\\2\end{pmatrix} & \begin{pmatrix}6\\2\end{pmatrix}\begin{pmatrix}2\\2\end{pmatrix} & \begin{pmatrix}8\\2\end{pmatrix}\begin{pmatrix}0\\2\end{pmatrix} \\[2ex]
\begin{pmatrix}0\\3\end{pmatrix}\begin{pmatrix}8\\1\end{pmatrix} & \begin{pmatrix}2\\3\end{pmatrix}\begin{pmatrix}6\\1\end{pmatrix} & \begin{pmatrix}4\\3\end{pmatrix}\begin{pmatrix}4\\1\end{pmatrix} & \begin{pmatrix}6\\3\end{pmatrix}\begin{pmatrix}2\\1\end{pmatrix} & \begin{pmatrix}8\\3\end{pmatrix}\begin{pmatrix}0\\1\end{pmatrix} \\[2ex]
\begin{pmatrix}0\\4\end{pmatrix}\begin{pmatrix}8\\0\end{pmatrix} & \begin{pmatrix}2\\4\end{pmatrix}\begin{pmatrix}6\\0\end{pmatrix} & \begin{pmatrix}4\\4\end{pmatrix}\begin{pmatrix}4\\0\end{pmatrix} & \begin{pmatrix}6\\4\end{pmatrix}\begin{pmatrix}2\\0\end{pmatrix} & \begin{pmatrix}8\\4\end{pmatrix}\begin{pmatrix}0\\0\end{pmatrix}
\end{bmatrix}
$$

Call this matrix the plasmid transition matrix. Write a MATLAB function SetUp(n) that computes the $(n+1)$-by-$(n+1)$ plasmid transition matrix $P(n)$. (You'll have to adapt the preceding discussion to conform to MATLAB's subscripting from one requirement.) Exploit structure. If successful, you should find that the the number of flops required is quadratic in n. Print a table that indicates the number of flops required to construct $P(n)$ for $n = 5,6,10,11,20,21,40,41$. (See F.C. Hoppenstadt and C.Peskin (1992), *Mathematics in Medicine and the Life Sciences*, Springer-Verlag, New York, p.50.)

5.2 Matrix Operations

Once a matrix it is set up, it can participate in matrix-vector and matrix-matrix products. Although these operations are MATLAB one-liners, it is instructive to examine the different ways that they can be implemented.

5.2.1 Matrix-Vector Multiplication

Suppose $A \in \mathbb{R}^{m \times n}$, and we wish to compute the matrix-vector product $y = Ax$, where $x \in \mathbb{R}^n$. The usual way this computation proceeds is to compute the dot products

$$y_i = \sum_{j=1}^{n} a_{ij} x_j$$

one at a time for $i = 1{:}m$. This leads to the following algorithm:

```
[m,n] = size(A);
y = zeros(m,1);
for i = 1:m
    for j = 1:n
        y(i) = y(i) + A(i,j)*x(j);
    end
end
```

The one-line assignment y = A*x is equivalent and requires $2mn$ flops.

Even though it is not necessary to hand-code matrix-vector multiplication in MATLAB , it is instructive to reconsider the preceding double loop. In particular, recognizing that the j-loop oversees an inner product of the ith row of A and the x vector, we have

```
function y = MatVecRO(A,x)
%
% Pre:
%    A       m-by-n matrix.
%    x       n-by-1
% Post:
%    y       A*x (row-oriented method)
%
   [m,n] = size(A);
   y = zeros(m,1);
   for i=1:m
       y(i) = A(i,:)*x;
   end
```

The colon notation has the effect of highlighting the dot products that make up Az. The procedure is *row oriented* because A is accessed by row.

A *column-oriented* algorithm for matrix-vector products can also be developed. We start with a 3-by-2 observation:

$$y = Ax = \begin{bmatrix} 1 & 2 \\ 3 & 4 \\ 5 & 6 \end{bmatrix} \begin{bmatrix} 7 \\ 8 \end{bmatrix} = \begin{bmatrix} 1 \cdot 7 + 2 \cdot 8 \\ 3 \cdot 7 + 4 \cdot 8 \\ 5 \cdot 7 + 6 \cdot 8 \end{bmatrix} = 7 \begin{bmatrix} 1 \\ 3 \\ 5 \end{bmatrix} + 8 \begin{bmatrix} 2 \\ 4 \\ 6 \end{bmatrix} = \begin{bmatrix} 23 \\ 53 \\ 83 \end{bmatrix}$$

In other words, y is a linear combination of A's columns with the x_j being the coefficients. This leads us to the following reorganization of `MatVecRO`:

```
function y = MatVecCO(A,x)
%
% Pre:
%    A        m-by-n matrix.
%    x        n-by-1
%
% Post:
%    y        A*x (column-oriented method)
%
   [m,n] = size(A);
   y = zeros(m,1);
   for j=1:n
      y = y + A(:,j)*x(j);
   end
```

In terms of program transformation, this function is just `MatVecRO` with the i and j loops swapped. The inner loop now oversees an operation of the form

$$\text{vector} \leftarrow \text{vector} + \text{vector} \cdot \text{scalar}$$

This is known as the *saxpy* operation. Along with the dot product, it is a key player in matrix computations. Here is an expanded view of the saxpy operation in `MatVecCO`:

$$\begin{bmatrix} \mathtt{y(1)} \\ \mathtt{y(2)} \\ \vdots \\ \mathtt{y(m)} \end{bmatrix} = \begin{bmatrix} \mathtt{A(1,j)} \\ \mathtt{A(2,j)} \\ \vdots \\ \mathtt{A(m,j)} \end{bmatrix} \mathtt{x(j)} + \begin{bmatrix} \mathtt{y(1)} \\ \mathtt{y(2)} \\ \vdots \\ \mathtt{y(m)} \end{bmatrix}$$

`MatVecCO` requires $2mn$ flops just like `MatVecRO`. However, to stress once again the limitations of flop counting, we point out that in certain powerful computing environments our two matrix-vector product algorithms may execute at radically different rates. For example, if the matrix entries entries a_{ij} are stored column by column in memory, then the saxpy version accesses A-entries that are contiguous in memory. In contrast, the row-oriented algorithm accesses non-contiguous a_{ij}. As a result of that inconvenience, it may require much more time to execute.

5.2.2   Exploiting Structure

In many matrix computations the matrices are structured with lots of zeros. In such a context it may be possible to streamline the computations. As a first example of this, we examine the matrix-vector product problem $y = Az$, where $A \in \mathbb{R}^{n \times n}$ is upper triangular. The product looks like this in the $n = 4$ case:

$$
\begin{bmatrix} \times \\ \times \\ \times \\ \times \end{bmatrix} = \begin{bmatrix} \times & \times & \times & \times \\ 0 & \times & \times & \times \\ 0 & 0 & \times & \times \\ 0 & 0 & 0 & \times \end{bmatrix} \begin{bmatrix} \times \\ \times \\ \times \\ \times \end{bmatrix}
$$

The derivation starts with an examination of `MatVecRO`. Observe that the inner products in the loop

```
for i = 1:n
    y(i) = A(i,:)*x
end
```

involve long runs of zeros when `A` is upper triangular. For example, if $n = 8$, then the inner product `A(5,:)*x` looks like

$$
\begin{bmatrix} 0 & 0 & 0 & 0 & 0 & \times & \times & \times \end{bmatrix} \begin{bmatrix} \times \\ \times \\ \times \\ \times \\ \times \\ \times \\ \times \\ \times \end{bmatrix}
$$

and requires a reduced number of flops because of all the zeros. Thus we must "shorten" the inner products so that they only include the nonzero portion of the row.

From the observation that the first i entries in `A(i,:)` are zero, we see that `A(i,i:n)*x(i:n)` is the nonzero portion of the full inner product `A(i,:)*x` that we need. It follows that

```
[n,n] = size(A);
y = zeros(n,1);
for i = 1:n
    y(i) = A(i,1:i)*z(1:i)
end
```

is a structure-exploiting upper triangular version of `MatVecRO`. The assignment to `y(i)` requires $2i$ flops, and so overall

$$
\sum_{i=1}^{n}(2i) = 2(1 + 2 + \cdots + n) = n(n+1)
$$

flops are required. However, in keeping with the philosophy of flop counting, we do not care about the $O(n)$ term and so we merely state that the algorithm requires n^2 flops. Our streamlining halved the number of floating point operations.

MatVecCO can also be abbreviated. Note that A(:,j) is zero in components $j+1$ through n, and so the "essential" saxpy to perform in the jth step is

$$
\begin{bmatrix} y(1) \\ y(2) \\ \vdots \\ y(j) \end{bmatrix} = \begin{bmatrix} A(1,j) \\ A(2,j) \\ \vdots \\ A(j,j) \end{bmatrix} x(j) + \begin{bmatrix} y(1) \\ y(2) \\ \vdots \\ y(j) \end{bmatrix}
$$

rendering

```
[n,n] = size(A);
y = zeros(n,1);
for j = 1:n
    y(j:n) = A(j:n,j)*x(j) + y(j:n);
end
```

Again, the number of required flops is halved.

5.2.3 Matrix-Matrix Multiplication

If $A \in \mathbb{R}^{m \times r}$ and $B \in \mathbb{R}^{r \times n}$, then the product $C = AB$ is defined by

$$
c_{ij} = \sum_{k=1}^{r} a_{ik} b_{kj}
$$

for all i and j that satisfy $1 \le i \le m$ and $1 \le j \le n$. In other words, each entry in C is the inner product of a row in A and a column in B. Thus the fragment

```
C = zeros(m,n);
for j=1:n
    for i=1:m
        for k=1:r
            C(i,j) = C(i,j) + A(i,k)*B(k,j);
        end
    end
end
```

computes the product AB and assigns the result to C. MATLAB supports matrix-matrix multiplication, and so this can be implemented with the one-liner

```
C = A*B
```

However, there are a number of different ways to look at matrix multiplication, and we shall present four distinct versions.

We start with the recognition that the innermost loop in the preceding script oversees a the dot product between row i of A and column j of B:

```
    function C = MatMatDot(A,B)
%
% Pre:
%    A        m-by-p matrix.
%    B        p-by-n matrix
%
% Post:
%    C        A*B (Dot Product Version)
%

   [m,p] = size(A);
   [p,n] = size(B);
   C = zeros(m,n);
   for j=1:n
      % Compute j-th column of C.
      for i=1:m
         C(i,j) = A(i,:)*B(:,j);
      end
   end
```

On other hand, we know that the jth column of C equals A times the jth column of B. If we apply MatVecCO to each of these matrix vector products, we obtain

```
    function C = MatMatSax(A,B)
%
% Pre:
%    A        m-by-p matrix.
%    B        p-by-n matrix
%
% Post:
%    C        A*B (Saxpy Version)
%

   [m,p] = size(A);
   [p,n] = size(B);
   C = zeros(m,n);
   for j=1:n
      % Compute j-th column of C.
      for k=1:m
         C(:,j) = C(:,j) + A(:,k)*B(k,j);
      end
   end
```

This version of matrix multiplication highlights the saxpy operation. By replacing the inner loop in this with a single matrix-vector product we obtain

```
   function C = MatMatVec(A,B)
%
% Pre:
%    A        m-by-p matrix.
%    B        p-by-n matrix
%
% Post:
%    C        A*B (Matrix-Vector Version)
%
   [m,p] = size(A);
   [p,n] = size(B);
   C = zeros(m,n);
   for j=1:n
      % Compute j-th column of C.
      C(:,j) = A*B(:,j);
   end
```

Finally, we observe that a matrix multiplication is a sum of outer products. The *outer product* between a column m-vector u and a row n-vector v is given by

$$uv^T = \begin{bmatrix} u_1 \\ u_2 \\ \vdots \\ u_m \end{bmatrix} \begin{bmatrix} v_1 & v_2 & \cdots & v_n \end{bmatrix} = \begin{bmatrix} u_1v_1 & u_1v_2 & \cdots & u_1v_n \\ u_2v_1 & u_2v_2 & \cdots & u_2v_n \\ \vdots & \vdots & \ddots & \vdots \\ u_mv_1 & u_mv_2 & \cdots & u_mv_n \end{bmatrix}$$

Appreciate this as just the ordinary matrix multiplication of an m-by-1 matrix and a 1-by-n matrix:

$$\begin{bmatrix} 10 \\ 15 \\ 20 \end{bmatrix} \begin{bmatrix} 1 & 2 & 3 & 4 \end{bmatrix} = \begin{bmatrix} 10 & 20 & 30 & 40 \\ 15 & 30 & 45 & 60 \\ 20 & 40 & 60 & 80 \end{bmatrix}$$

Returning to the matrix multiplication problem,

$$C = AB = \begin{bmatrix} A(:,1) & | & A(:,2) & | & \cdots & | & A:,p) \end{bmatrix} \begin{bmatrix} B(1,:) \\ B(2,:) \\ \vdots \\ B(p,:) \end{bmatrix} = \sum_{k=1}^{p} A(:,k)B(k,:)$$

Thus

$$\begin{bmatrix} 1 & 2 \\ 3 & 4 \\ 5 & 6 \end{bmatrix} \begin{bmatrix} 10 & 20 \\ 30 & 40 \end{bmatrix} = \begin{bmatrix} 1 \\ 3 \\ 5 \end{bmatrix} \begin{bmatrix} 10 & 20 \end{bmatrix} + \begin{bmatrix} 2 \\ 4 \\ 6 \end{bmatrix} \begin{bmatrix} 30 & 40 \end{bmatrix}$$

$$= \begin{bmatrix} 10 & 20 \\ 30 & 60 \\ 50 & 100 \end{bmatrix} + \begin{bmatrix} 60 & 80 \\ 120 & 160 \\ 180 & 240 \end{bmatrix}$$

$$= \begin{bmatrix} 70 & 100 \\ 150 & 220 \\ 230 & 340 \end{bmatrix}$$

This leads to the outer product version of matrix multiplication:

```
function C = MatMatOuter(A,B)
%
% Pre:
%    A       m-by-p matrix.
%    B       p-by-n matrix
%
% Post:
%    C       A*B (Outer Product Version)
%
   [m,p] = size(A);
   [p,n] = size(B);
   C = zeros(m,n);
   for k=1:p
      % Add in k-th outer product
      C = C + A(:,k)*B(k,:);
   end;
```

The script file `MatBench` benchmarks the four various matrix-multiply functions that we have developed along with the direct, one-liner `C = A*B`.

n	Dot	Saxpy	MatVec	Outer	Direct
10	0.400	0.617	0.067	0.100	0.017
20	1.583	2.333	0.150	0.500	0.050
40	6.567	10.067	0.550	3.333	0.367
80	29.083	49.733	3.033	25.517	2.617

The most important thing about the preceding table is not the actual values reported but that it shows the weakness of flop counting. Methods for the same problem that involve the same number of flops can perform very differently. The nature of the kernel operation (saxpy, dot product, matrix-vector product, outer product, etc.) is more important than the amount of arithmetic involved.

5.2.4   Error and Norms

We conclude this section with a brief look at how errors are quantified in the matrix computation area. The concept of a *norm* is required. Norms are a vehicle for measuring distance in a vector space. For vectors $x \in \mathbb{R}^n$, the 1, 2, and infinity norms are of particular importance:

$$\| x \|_1 = |x_1| + \cdots + |x_n|$$
$$\| x \|_2 = \sqrt{x_1^2 + \cdots + x_n^2}$$
$$\| x \|_\infty = \max\{|x_1|, \ldots, |x_n|\}$$

A norm is just a generalization of absolute value. Whenever we think about vectors of errors in an order-of-magnitude sense, then the choice of norm is generally not important. It is possible to show that

$$\| x \|_\infty \;\leq\; \| x \|_1 \;\leq\; n \, \| x \|_\infty$$

$$\| x \|_\infty \;\leq\; \| x \|_2 \;\leq\; \sqrt{n} \, \| x \|_\infty$$

Thus the 1-norm cannot be particularly small without the others following suit.

In MATLAB, if x is a vector, norm(x,1), norm(x,2), and norm(x,inf) can be used to ascertain these quantities. A single-argument call to norm returns the 2-norm (e.g., norm(x)). The script AveNorms explores the ratios $\| x \|_1 / \| x \|_\infty$ and $\| x \|_2 / \| x \|_\infty$ for large collections of random n-vectors.

The idea of a norm extends to matrices and, as in the vector case, there are number of important special cases. If $A \in \mathbb{R}^{m \times n}$, then

$$\| A \|_1 \;=\; \max_{1 \leq j \leq n} \sum_{i=1}^{m} |a_{ij}| \qquad\qquad \| A \|_2 = \max_{\| x \|_2 = 1} \| Ax \|_2$$

$$\| A \|_\infty \;=\; \max_{1 \leq i \leq m} \sum_{j=1}^{n} |a_{ij}| \qquad\qquad \| A \|_F = \sqrt{\sum_{i=1}^{m} \sum_{j=1}^{n} |a_{ij}|^2}$$

In MATLAB if A is a matrix, then norm(A,1), norm(A,2), norm(A,inf), and norm(A,'fro') can be used to compute these values. As a simple illustration of how matrix norms can be used to quantify error at the matrix level, we prove a result about the roundoff errors that arise when an m-by-n matrix is stored.

Theorem 5 *If $\hat{A}$ is the stored version of $A \in \mathbb{R}^{m \times n}$, then $\hat{A} = A + E$ where $E \in \mathbb{R}^{n \times n}$ and*

$$\| E \|_1 \leq \texttt{eps} \; \| A \|_1$$

Proof. From Theorem 1, if $\hat{A} = (\hat{a}_{ij})$, then

$$\hat{a}_{ij} = fl(a_{ij}) = a_{ij}(1 + \epsilon_{ij})$$

where $|\epsilon_{ij}| \leq$ eps . Thus

$$\| E \|_1 \;=\; \| \hat{A} - A \|_1 \;=\; \max_{1 \leq j \leq n} \sum_{i=1}^{m} |\hat{a}_{ij} - a_{ij}|$$

$$\leq \; \max_{1 \leq j \leq n} \sum_{i=1}^{m} |a_{ij}\epsilon_{ij}| \;\leq\; \texttt{eps} \max_{1 \leq j \leq n} \sum_{i=1}^{m} |a_{ij}| \;=\; \texttt{eps} \, \| A \|_1 \qquad \square$$

This says that errors of order $\texttt{eps}\| A \|_1$ arise when a real matrix A is stored in floating point. There is nothing special about our choice of the 1-norm. Similar results apply for the other norms defined earlier.

When the effect of roundoff error is the issue, we will be content with order-of-magnitude approximation. For example, it can be shown that if A and B are matrices of floating point numbers, then

$$\| \, \mathrm{fl}(AB) - AB \, \| \approx \mathbf{eps} \| \, A \, \| \| \, B \, \|$$

By $\mathrm{fl}(AB)$ we mean the computed, floating point product of A and B. The result says that the errors in the computed product are roughly the product of the unit roundoff **eps**, the size of the numbers in A, and the size of the numbers in B. The running of the following script affirms this result:

```
% Script File: ProdBound
%
% Examines the error in 3-digit matrix multiplication.
%

    eps3 = .005;        % 3-digit machine precision
    disp('    n    1-norm factor ')
    disp('-----------------------')
    for n = 2:10
        s = 0;
        for r=1:10
            A = randn(n,n);
            B = randn(n,n);
            C = Prod3Digit(A,B);
            E = C - A*B;
            s = s+ norm(E,1)/(eps3*norm(A,1)*norm(B,1));
        end
        disp(sprintf('%4.0f      %8.3f   ',n,s/10))
    end
```

The function `Prod3Digit(A,B)` returns the product AB computed using simulated three-digit floating point arithmetic developed in §1.5.3. The result is compared to the "exact" product obtained by using the prevailing, full machine precision. Sample results are given in Fig 5.1.

Problems

P5.2.1 Suppose $A \in \mathbf{R}^{n \times n}$ has the property that a_{ij} is zero whenever $i > j + 1$. Write an efficient, row-oriented dot product algorithm that computes $y = Az$.

P5.2.2 Suppose $A \in \mathbf{R}^{m \times n}$ is upper triangular and that $x \in \mathbf{R}^n$. Write a MATLAB fragment for computing the product $y = Ax$. (Do not make assumptions like $m \geq n$ or $m \leq n$.)

P5.2.3 Modify `MatMatSax` so that it efficiently handles the case when B is upper triangular.

P5.2.4 Modify `MatMatSax` so that it efficiently handles the case when both A and B are upper triangular and n-by-n.

P5.2.5 Modify `MatMatDot` so that it efficiently handles the case when A is lower triangular and B are upper triangular and both are n-by-n.

P5.2.6 Modify `MatMatSax` so that it efficiently handles the case when B has upper and lower bandwidth p. Assume that both A and B are n-by-n.

n	$\dfrac{\parallel \mathrm{fl}(AB) - AB \parallel_1}{\mathbf{eps}\parallel A \parallel_1 \parallel B \parallel_1}$
2	0.417
3	0.364
4	0.268
5	0.259
6	0.246
7	0.230
8	0.225
9	0.241
10	0.221

FIGURE 5.1 Sample roundoff error in matrix-matrix multiplication

P5.2.7 Write a function `B = MatPower(A,k)` so that computes $B = A^k$, where A is a square matrix and k is a nonnegative integer. Hint: First consider the case when k is a power of 2. Then consider binary expansions (e.g., $A^{29} = A^{16}A^8 A^4 A$).

P5.2.8 Develop a nested multiplication for the product $y = (c_1 I + c_2 A + \cdots + c_k A^{k-1})v$, where the c_i are scalars, A is a square matrix, and v is a vector.

P5.2.9 Write a MATLAB function that returns the matrix

$$C = \left[\ B\ |\ AB\ |\ A^2 B\ |\ \cdots\ |\ A^{p-1}B\ \right]$$

where A is n-by-n, B is n-by-t, and p is a positive integer.

P5.2.10 Write a MATLAB function `ScaleRows(A,d)` that assumes A is m-by-n and d is m-by-1 and multiplies the ith row of A by $d(i)$.

P5.2.11 Let $P(n)$ be the matrix of P5.1.2. If v is a plasmid state vector, then after one reproductive cycle, $P(n)v$ is the state vector for the population of daughters. A vector $v(0:n)$ is *symmetric* if $v(n:-1:0) = v$. Thus [2;5;6;5;2] and [3;1;1;3] are symmetric.) Write a MATLAB function `V = Forward(P,v0,k)` that sets up a matrix V with k rows. The kth row of V should be the transpose of the vector $P^k v_0$. Assume v_0 is symmetric.

5.3 Once Again, Setting Up Matrix Problems

On numerous occasions we have been required to evaluate a continuous function $f(x)$ on a vector of values (e.g., `sqrt(linspace(0,9))`). The analog of this in two dimensions is the evaluation of a function $f(x,y)$ on a pair of vectors x and y.

5.3.1 Two-Dimensional Tables of Function Values

Suppose $f(x,y) = \exp^{-(x^2+3y^2)}$ and that we want to set up an n-by-n matrix F with the property that

$$f_{ij} = e^{-(x_i^2 + 3y_j^2)}$$

where $x_i = (i-1)/(n-1)$ and $y_j = (j-1)/(n-1)$. We can proceed at the scalar, vector, or matrix level. At the scalar level we evaluate **exp** at each entry:

```
v = linspace(0,1,n);
F = zeros(n,n);
for i=1:n
   for j=1:n
      F(i,j) = exp(-(v(i)^2 + 3*v(j)^2));
   end
end
```

At the vector level can set F up by column:

```
v = linspace(0,1,n)';
F = zeros(n,n);
for j=1:n
   F(:,j) = exp(-( v.^2 + 3*v(j)^2));
end
```

Finally, we can even evaluate **exp** on the matrix of arguments:

```
v = linspace(0,1,n);
A = zeros(n,n);
for i=1:n
   for j=1:n
      A(i,j) = -(v(i)^2 + 3*v(j)^2);
   end
end
F = exp(A);
```

Many of MATLAB's built-in functions, like **exp**, accept matrix arguments. The assignment **F = exp(A)** sets F to be a matrix that is the same size as A with $f_{ij} = e^{a_{ij}}$ for all i and j.

In general, the most efficient approach depends on the structure of the matrix of arguments, the nature of the underlying function $f(x,y)$, and what is already available through M-files. Regardless of these details, it is best to be consistent with MATLAB's vectorizing philosophy designing all functions so that they can accept vector arguments. For example,

```
   function F = SampleF(x,y)
%
% Pre:
%   x      column n-vector
%   y      column m-vector
% Post:
%   F      m-by-n matrix with F(i,j) = exp(-(x(j)^2 + 2y(i)^2)/4)
%
   n = length(x); m = length(y);
   A = -((2*y.^2)*ones(1,n) + ones(m,1)*(x.^2)')/4;
   F = exp(A);
```

Notice that the matrix A is the sum of two outer products and that $a_{ij} = -(x_j^2 + 2y_i^2)/4$. The setting up of this grid of points allows for a single (matrix-valued) call to **exp**.

5.3.2   Contour Plots

While the discussion of tables is still fresh, we introduce the MATLAB's contour plotting capability. If $f(x, y)$ is a function of two real variables, then a curve in the xy-plane of the form $f(x, y) = c$ is a *contour*. The function contour can be used to display such curves. Here is a script that displays various contour plots of the function SampleF:

```
% Script File: ShowContour
%
% Illustrates various contour plots.

  close all
% Set up array of function values.
  x = linspace(-2,2,50)';
  y = linspace(-1.5,1.5,50)';
  F = SampleF(x,y);

% Number of contours set to default value:
  figure
  Contour(x,y,F)
  axis('equal')

% Five contours:
  figure
  contour(x,y,F,5);
  axis('equal')

% Five contours with specified values:
  figure
  contour(x,y,F,[1 .8  .6  .4  .2])
  axis('equal')

% Four contours with manual labeling:
  figure
  c = contour(x,y,F,4);
  clabel(c,'manual');
  axis('equal')
```

Contour(x,y,F) assumes that F(i,j) is the value of the underlying function f at (x_j, y_i). Clearly, the length of x and the length of y must equal the column and row dimension of F. The argument after the array is used to supply information about the number of contours and the associated "elevations." Contour(x,y,F,N) specifies N contours. Contour(x,y,F,v), where v is a vector, specifies elevations v(i), where i=1:length(v). The contour elevations can be labeled using the mouse by the command sequence of the form

```
  c = contour(x,y,F,...);
  clabel(c,'manual');
```

Type `help clabel` for more details. A sample labeled contour plot of the function `SampleF` is shown in Fig 5.2. See the script `ShowContour`.

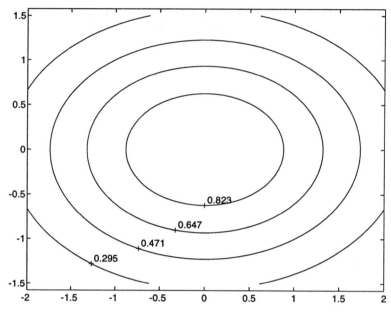

FIGURE 5.2 A contour plot

5.3.3 Spotting Matrix-Vector Products

Let us consider the problem of approximating the double integral

$$I = \int_a^b \int_c^d f(x,y)dy \, dx$$

using a quadrature rule of the form

$$\int_a^b g(x)dx \approx (b-a) \sum_{i=1}^{N_x} \omega_i g(x_i) \; \equiv \; Q_x$$

in the x-direction and a quadrature rule of the form

$$\int_c^d g(y)dy \approx (d-c) \sum_{j=1}^{N_y} \mu_j g(y_j) \; \equiv \; Q_y$$

in the y-direction. Doing this, we obtain

$$I \; = \; \int_a^b \left(\int_c^d f(x,y)dy \right) dx \; \approx \; (b-a) \sum_{i=1}^{N_x} \omega_i \left(\int_c^d f(x_i,y)dy \right)$$

$$\approx (b-a)\sum_{i=1}^{N_x}\omega_i\left((d-c)\sum_{j=1}^{N_y}\mu_j f(x_i,y_j)\right)$$

$$= (b-a)(d-c)\sum_{i=1}^{N_x}\omega_i\left(\sum_{j=1}^{N_y}\mu_j f(x_i,y_j)\right) \equiv Q$$

Observe that the quantity in parentheses is the ith component of the vector $F\mu$ where

$$F=\begin{bmatrix} f(x_1,y_1) & \cdots & f(x_1,y_{N_y}) \\ \vdots & \ddots & \vdots \\ f(x_{N_x},y_1) & \cdots & f(x_{N_x},y_{N_y)} \end{bmatrix} \quad\text{and}\quad \mu=\begin{bmatrix}\mu_1 \\ \vdots \\ \mu_{N_y}\end{bmatrix}$$

It follows that

$$Q=(b-a)(d-c)\omega^T(F\mu)$$

where

$$\omega^T=\begin{bmatrix}\omega_1 \\ \vdots \\ \omega_{N_x}\end{bmatrix}$$

If Q_x and Q_y are taken to be composite Newton-Cotes rules, then we obtain the following implementation:

```
function numI2D = CompQNC2D(fname,a,b,c,d,mx,nx,my,ny)
%
%  Pre:
%     fname    string that names a function of the form f(x,y).
%              If x and y are vectors, then it should return a
%              matrix F with the property that F(i,j) =
%              f(x(j),y(i)), i=1:length(y), j=1:length(x).
%
%  a,b,c,d    real scalars
%     mx,my    integers that satisfy 2<=mx<=11, 2<=my<=11.
%     nx,ny    positive integers
%
%  Post:
%     numI2D   approximation to the integral of f(x,y) over the
%              rectangle [a,b]x[c,d]. The compQNC(mx,nx) rule is
%              used in the x-direction and the compQNC(my,ny)
%              rule is used in the y-direction.

   [omega,x] = wCompNC(a,b,mx,nx);
   [mu,y]    = wCompNC(c,d,my,ny);
   F = feval(fname,x,y);
   numI2D = (b-a)*(d-c)*(mu'*F*omega);
```

The function wCompNC uses wNC from Chapter 4 and sets up the vector of weights and the vector of abscissas for the composite m-point Newton-Cotes rule. The script Show2DQuad benchmarks this function when it is applied to

$$I = \int_0^2 \int_0^2 \left(\frac{1}{((x-.3)^2+.1)((y-.4)^2+.1)} + \frac{1}{((x-.7)^2+.1)((y-.3)^2+.1)} \right) dy\, dx$$

The integrand function is implemented in SampleF2.

```
% Script File: Show2DQuad
%
% Integral of SampleF2 over [0,2]x[0,2] for various 2D composite QNC(7) rules.

    clc; home; m=7;
    disp(' Subintervals     Integral          Relative Time')
    disp('------------------------------------------------------')
    for n = [1 2 4 8 16 32 64]
       t1 = clock;
       numI2D = CompQNC2D('SampleF2',0,2,0,2,m,n,m,n);
       t2 = etime(clock,t1);
       if n==1;
          base = t2;
          time = 1;
       else
          time = t2/base;
       end
       disp(sprintf(' %7.0f  %17.15f  %11.1f',n,numI2D,time))
    end;
```

Here are the results:

Subintervals	Integral	Relative Time
1	51.916084880415639	1.0
2	45.724380776263146	1.1
4	46.219888856629716	1.9
8	46.220375502796415	3.7
16	46.220349664520825	9.2
32	46.220349653054726	24.5
64	46.220349653055095	86.0

For larger values of n we would find that the amount of computation increases by a factor of 4 with a doubling of n reflecting the $O(n^2)$ nature of the calculation.

Problems

P5.3.1 Modify CompQNC2D so that it computes and uses F row at a time. The implementation should not require a two-dimensional array.

P5.3.2 Suppose we are given an interval $[a, b]$, a *Kernel function* $K(x, y)$ defined on $[a, b] \times [a, b]$, and another function $g(x)$ defined on $[a, b]$. Our goal is to find a function $f(x)$ with the property that

$$\int_a^b K(x, y)f(y)dy = g(x) \qquad a \leq x \leq b$$

Suppose Q is a quadrature rule of the form

$$Q = (b - a) \sum_{j=1}^N \omega_j s(x_j)$$

that approximates integrals of the form

$$I = \int_a^b s(x)dx$$

(The ω_j and x_j are the weights and abscissas of the rule.) We can then replace the integral in our problem with

$$(b - a) \sum_j^N \omega_j K(x, x_j)f(x_j) = g(x).$$

If we force this equation to hold at $x = x_1, \ldots, x_N$, then we obtain an N-by-N linear system:

$$(b - a) \sum_{j=1}^N \omega_j K(x_i, x_j)f(x_j) = g(x_i) \qquad i = 1{:}N$$

in the N unknowns $f(x_j)$, $j = 1{:}N$.

Write a MATLAB function Kmat = Kernel(a,b,m,n,sigma) that returns the matrix of coefficients defined by this method, where Q is the composite m-point Newton-Cotes rule with n equal subintervals across $[a, b]$ and $K(x, y) = e^{-(x-y)^2/\sigma}$. Be as efficient as possible, avoiding redundant exp evaluations.

Test this solution method with $a = 0$, $b = 5$, and

$$g(x) = \frac{1}{(x - 2)^2 + .1} + \frac{1}{(x - 4)^2 + .2}$$

For $\sigma = .01$, plot the not-a-knot spline interpolant of the computed solution (i.e., the $(x_i, f(x_i))$. Do this for the four cases $(m, n) = (3, 5), (3, 10), (5, 5), (5, 10)$. Use subplot(2,2,*). For each subplot, print the time required by your computer to execute Kernel. Repeat with $\sigma = .1$.

5.4 Recursive Matrix Operations

Some of the most interesting algorithmic developments in matrix computations are recursive. Two examples are given in this section. The first is the fast Fourier transform, a super-quick way of computing a special, very important matrix-vector product. The second is a recursive matrix multiplication algorithm that involves markedly fewer flops than the conventional algorithm.

5.4.1 The Fast Fourier Transform

The discrete Fourier transform (DFT) matrix is a complex Vandermonde matrix. Complex numbers have the form $a + i \cdot b$, where $i = \sqrt{-1}$. If we define

$$\omega_4 = \exp(-2\pi i/4) = \cos(2\pi/4) - i \cdot \sin(2\pi/4) = -i$$

then the 4-by-4 DFT matrix is given by

$$F_4 = \begin{bmatrix} 1 & 1 & 1 & 1 \\ 1 & \omega_4 & \omega_4^2 & \omega_4^3 \\ 1 & \omega_4^2 & \omega_4^4 & \omega_4^6 \\ 1 & \omega_4^3 & \omega_4^6 & \omega_4^9 \end{bmatrix}$$

The parameter ω_4 is a fourth root of unit, meaning that $\omega_4 = 1$. It follows that

$$F_4 = \begin{bmatrix} 1 & 1 & 1 & 1 \\ 1 & -i & -1 & i \\ 1 & -1 & 1 & -1 \\ 1 & i & -1 & -i \end{bmatrix}$$

MATLAB supports complex matrix manipulation. The commands

```
i = sqrt(-1);
F = [1 1 1 1; 1 -i -1 i; 1 -1 1 -1; 1 i -1 -i]
```

assign the 4-by-4 DFT matrix to F.

For general n, the DFT matrix is defined in terms of

$$\omega_n = \exp(-2\pi i/n) = \cos(2\pi/n) - i \cdot \sin(2\pi/n)$$

In particular, the n-by-n DFT matrix is defined by

$$F_n = (f_{pq}) \qquad f_{pq} = \omega_n^{(p-1)(q-1)}$$

Setting up the DFT matrix gives us an opportunity to sample MATLAB's complex arithmetic capabilities:

```
F = ones(n,n);
F(:,2) = exp(-2*pi*sqrt(-1)/n).^(0:n-1)';
for k=3:n
    F(:,k) = F(:,2).*F(:,k-1);
end
```

There is really nothing new here except that the generated matrix entries are complex with the involvement of $\sqrt{-1}$. The real and imaginary parts of a matrix can be extracted using the functions real and imag. Thus, if $F = F_R + i \cdot F_I$ and $x = x_R + i \cdot x_I$, where F_R, F_I, x_R, and x_I are real, then

```
y = F*x
```

is equivalent to

```
FR = real(F); FI = imag(F);
xR = real(x); xI = imag(x);
y = (FR*xR - FI*xI) + sqrt(-1)*(FR*xI + FI*xR);
```

because

$$y = (F_R x_R - F_I x_I) + i \cdot (F_R x_I + F_I * x_R)$$

Many of MATLAB's built-in functions like **exp**, accept complex matrix arguments. When complex computations are involved, **flops** counts *real* flops. Note that complex addition requires two real flops and that complex multiplication requires six real flops.

Returning to the DFT, it is possible to compute $y = F_n x$ without explicitly forming the DFT matrix F_n:

```
n = length(x);
y = x(1)*ones(n,1);
for k=2:n
    y = y + exp(-2*pi*sqrt(-1)*(k-1)*(0:n-1)') *x(k);
end
```

The update carries out the saxpy computation

$$y = \begin{bmatrix} 1 \\ \omega_n^{k-1} \\ \omega_n^{2(k-1)} \\ \vdots \\ \omega_n^{(n-1)(k-1)} \end{bmatrix} x_k$$

Notice that since $\omega_n^n = 1$, all powers of ω_n are in the set $\{1, \omega_n, \omega_n^2, \ldots, \omega_n^{n-1}\}$. In particular, $\omega_n^m = \omega_n^{m \bmod n}$. Thus, if

```
v = exp(-2*pi*sqrt(-1)/n)^(0:n-1)';
z = rem((k-1)*(0:n-1)',n ) +1;
```

then v(z) equals the kth column of F_n and we obtain

```
    function y = DFT(x)
%
% Pre:
%     x a column vector with power-of-two length.
% Post
%     y     the DFT of x.
%
    n = length(x);
    y = x(1)*ones(n,1);
    if n > 1
        v = exp(-2*pi*sqrt(-1)/n)^(0:n-1)';
        for k=2:n
            z = rem((k-1)*(0:n-1)',n ) +1;
            y = y + v(z)*x(k);
        end
    end
```

This is an $O(n^2)$ algorithm. We now show how to obtain an $O(n\log_2 n)$ implementation by exploiting the structure of F_n.

The starting point is to look at an even order DFT matrix when we permute its columns so that the odd-indexed columns come first. Consider the case $n = 8$. Noting that $\omega_8^m = \omega_n^{m\bmod 8}$, we have

$$F_8 = \begin{bmatrix} 1 & 1 & 1 & 1 & 1 & 1 & 1 & 1 \\ 1 & \omega & \omega^2 & \omega^3 & \omega^4 & \omega^5 & \omega^6 & \omega^7 \\ 1 & \omega^2 & \omega^4 & \omega^6 & 1 & \omega^2 & \omega^4 & \omega^6 \\ 1 & \omega^3 & \omega^6 & \omega & \omega^4 & \omega^7 & \omega^2 & \omega^5 \\ 1 & \omega^4 & 1 & \omega^4 & 1 & \omega^4 & 1 & \omega^4 \\ 1 & \omega^5 & \omega^2 & \omega^7 & \omega^4 & \omega & \omega^6 & \omega^3 \\ 1 & \omega^6 & \omega^4 & \omega^2 & 1 & \omega^6 & \omega^4 & \omega^2 \\ 1 & \omega^7 & \omega^6 & \omega^5 & \omega^4 & \omega^3 & \omega^2 & \omega \end{bmatrix}$$

where $\omega = \omega_8$. If $cols = [1\,3\,5\,7\,2\,4\,6\,8]$, then

$$F_8(:,cols) = \left[\begin{array}{cccc|cccc} 1 & 1 & 1 & 1 & 1 & 1 & 1 & 1 \\ 1 & \omega^2 & \omega^4 & \omega^6 & \omega & \omega^3 & \omega^5 & \omega^7 \\ 1 & \omega^4 & 1 & \omega^4 & \omega^2 & \omega^6 & \omega^2 & \omega^6 \\ 1 & \omega^6 & \omega^4 & \omega^2 & \omega^3 & \omega & \omega^7 & \omega^5 \\ \hline 1 & 1 & 1 & 1 & -1 & -1 & -1 & -1 \\ 1 & \omega^2 & \omega^4 & \omega^6 & -\omega & -\omega^3 & -\omega^5 & -\omega^7 \\ 1 & \omega^4 & 1 & \omega^4 & -\omega^2 & -\omega^6 & -\omega^2 & -\omega^6 \\ 1 & \omega^6 & \omega^4 & \omega^2 & -\omega^3 & -\omega & -\omega^7 & -\omega^5 \end{array}\right]$$

The lines through the matrix help us think of the matrix as a 2-by-2 matrix with 4-by-4 "blocks." Noting that $\omega^2 = \omega_8^2 = \omega_4$ we see that

$$F_8(:,cols) = \left[\begin{array}{c|c} F_4 & DF_4 \\ \hline F_4 & -DF_4 \end{array}\right]$$

where

$$D = \begin{bmatrix} 1 & 0 & 0 & 0 \\ 0 & \omega & 0 & 0 \\ 0 & 0 & \omega^2 & 0 \\ 0 & 0 & 0 & \omega^3 \end{bmatrix}$$

It follows that if x is an 8-vector, then

$$F_8 x = F(:,cols)x(cols) = \left[\begin{array}{c|c} F_4 & DF_4 \\ \hline F_4 & -DF_4 \end{array}\right]\left[\begin{array}{c} x(1{:}2{:}8) \\ x(2{:}2{:}8) \end{array}\right] = \left[\begin{array}{c|c} I & D \\ \hline I & -D \end{array}\right]\left[\begin{array}{c} F_4 x(1{:}2{:}8) \\ F_4 x(2{:}2{:}8) \end{array}\right].$$

Thus, by simple scalings we can obtain the eight-point DFT $y = F_8 x$ from the four-point DFTs $y_T = F_4 x(1{:}2{:}8)$ and $y_B = F_4 x(2{:}2{:}8)$:

$$\begin{aligned} y(1{:}4) &= y_T + d\,.{*}\,y_B \\ y(5{:}8) &= y_T - d\,.{*}\,y_B \end{aligned}$$

where d is the following "vector of weights":

$$d = \begin{bmatrix} 1 \\ \omega \\ \omega^2 \\ \omega^3 \end{bmatrix}$$

In general, if $n = 2m$, then $y = F_n x$ is given by

$$\begin{aligned} y(1:m) &= y_T + d.* y_B \\ y(m+1:n) &= y_B - d.* y_B \end{aligned}$$

where

$$\begin{aligned} y_T &= F_m x(1:2:n) \\ y_B &= F_m x(2:2:n) \end{aligned}$$

and

$$d = \begin{bmatrix} 1 \\ \omega \\ \vdots \\ \omega_n^{m-1} \end{bmatrix}$$

For $n = 2^t$ we can recur on this process until $n = 1$. (The one-point DFT of a one-vector is itself). This gives

```
function y = FFTRecur(x)
%
% Pre:
%    x    a column vector with power-of-two length.
% Post:
%    y    the DFT of x.
%
   n = length(x);
   if n ==1
      y = x;
   else
      m = n/2;
      yT = FFTRecur(x(1:2:n));
      yB = FFTRecur(x(2:2:n));
      d = exp(-2*pi*sqrt(-1)/n).^(0:m-1)';
      z = d.*yB;
      y = [ yT+z ; yT-z ];
   end
```

This is a member of the *fast Fourier transform* (FFT) family of algorithms. They involve $O(n \log_2 n)$ flops. We have illustrated a radix-2 FFT. It requires n to be a power of 2. Other radices are possible. MATLAB includes a radix-2 fast Fourier transform fft. The script fftwork tabulates the number of flops required by DFT, FFTRecur, and FFT:

n	DFT Flops	FFTRecur Flops	FFT Flops
2	81	40	29
4	233	148	84
8	777	420	201
16	2825	1076	462
32	10761	2612	1051
64	41993	6132	2376
128	165897	14068	5333
256	659465	31732	11874

The reason that `FFTRecur` involves more flops than `FFT` concerns the computation of the "weight vector" `d`. As it stands, there is a considerable amount of redundant computation with respect to the exponential values that are required. This can be avoided by precomputing the weights, storing them in a vector, and then merely "looking up" the values as they are needed during the recursion. With care, the amount of work required by a radix-2 FFT is $5n \log_2 n$ flops.

5.4.2   Strassen Multiplication

Ordinarily, 2-by-2 matrix multiplication $C = AB$ requires eight multiplications and four additions:

$$\left[\begin{array}{cc} C_{11} & C_{12} \\ C_{21} & C_{22} \end{array} \right] = \left[\begin{array}{cc} C_{11} & A_{12} \\ A_{21} & A_{22} \end{array} \right] \left[\begin{array}{cc} B_{11} & B_{12} \\ B_{21} & B_{22} \end{array} \right] = \left[\begin{array}{cc} A_{11}B_{11} + A_{12}B_{21} & A_{11}B_{12} + A_{12}B_{22} \\ A_{21}b_{11} + A_{22}B_{21} & A_{21}b_{12} + A_{22}B_{22} \end{array} \right]$$

In the *Strassen* multiplication scheme, the computations are rearranged so that they involve seven multiplications and 18 additions:

$$
\begin{aligned}
P_1 &= (A_{11} + A_{22})(B_{11} + B_{22}) \\
P_2 &= (A_{21} + A_{22})B_{11} \\
P_3 &= A_{11}(B_{12} - B_{22}) \\
P_4 &= A_{22}(B_{21} - B_{11}) \\
P_5 &= (A_{11} + A_{12})B_{22} \\
P_6 &= (A_{21} - A_{11})(B_{11} + B_{12}) \\
P_7 &= (A_{12} - A_{22})(B_{21} + B_{22}) \\
C_{11} &= P_1 + P_4 - P_5 + P_7 \\
C_{12} &= P_3 + P_5 \\
C_{21} &= P_2 + P_4 \\
C_{22} &= P_1 + P_3 - P_2 + P_6
\end{aligned}
$$

It is easy to verify that these recipes correctly define the product AB. However, why go through these convoluted formulas when ordinary 2-by-2 multiplication involves just eight multiplies and four additions? To answer this question, we first observe that the Strassen specification holds when the A_{ij} and B_{ij} are square matrices themselves. In this case, it amounts to a special method for computing block 2-by-2 matrix products. The seven multiplications are now m-by-m matrix multiplications and require $2(7m^3)$ flops. The 18 additions are matrix additions and they involve $18m^2$ flops. Thus, for this block size the Strassen multiplication requires

$$2(7m^3) + 18m^2 = \frac{7}{8}(2n^3) + \frac{9}{2}n^2$$

flops while the corresponding figure for the conventional algorithm is given by $2n^3 + n^2$. We see that for large enough n, the Strassen approach involves less arithmetic.

The idea can obviously be applied recursively. In particular, we can apply the Strassen algorithm to each of the half-sized block multiplications associated with the P_i. Thus, if the original A and B are n-by-n and $n = 2^q$, then we can recursively apply the Strassen multiplication algorithm all the way to the 1-by-1 level. However, for small n the Strassen approach involves more flops than the ordinary matrix multiplication algorithm, Therefore, for some $n_{min} \geq 1$ it makes sense to "switch over" to the standard algorithm. In the following implementation, $n_{min} = 16$:

```
    function C = strass(A,B)
%
% Pre:
%    A,B  n-by-n matrices, n a power of 2.
%    nmin  positive integer
% Post:
%       C  n-by-n matrix equal to A*B.
%
    nmin = 16;
    [n,n] = size(A);
    if n <= nmin
       C = A*B;
    else
       m = n/2; u = 1:m;  v = m+1:n;
       P1 = strass(A(u,u)+A(v,v),B(u,u)+B(v,v));
       P2 = strass(A(v,u)+A(v,v),B(u,u));
       P3 = strass(A(u,u),B(u,v)-B(v,v));
       P4 = strass(A(v,v),B(v,u)-B(u,u));
       P5 = strass(A(u,u)+A(u,v),B(v,v));
       P6 = strass(A(v,u)-A(u,u),B(u,u) + B(u,v));
       P7 = strass(A(u,v)-A(v,v),B(v,u)+B(v,v));
       C = [ P1+P4-P5+P7    P3+P5; P2+P4 P1+P3-P2+P6];
    end
```

The script StrassFlops tabulates the number of flops required by this function for various values of n:

n	Strass Flops	Ordinary Flops
2	16	16
4	128	128
8	1024	1024
16	8192	8192
32	61954	65536
64	452112	524288
128	3238514	4194304

Problems

P5.4.1 Modify `FFTRecur` so that it does not involve redundant weight computation.

P5.4.2 Modify `strass` so that it can handle general n. Hint: You will have to figure out how to partition the matrix multiplication if n is odd.

P5.4.3 Let $f(q, r)$ be the number of flops required by `strass` if $n = 2^q$ and $n_{min} = 2^r$. Assume that $q \geq r$. Develop an analytical expression for $f(q, r)$. For $q = 1:20$, compare $\min_{r \leq s} f(q, r)$ and $2 \cdot (2^q)^3$.

5.5 Distributed Memory Matrix Multiplication

In a shared memory machine, individual processors are able to read and write data to a (typically large) global memory. A snapshot of what it is like to compute in a shared memory environment is given at the end of Chapter 4, where we used the quadrature problem as an example.

In contrast to the shared memory paradigm is the *distributed memory* paradigm. In a distributed memory computer, there is an *interconnection network* that links the processors and is the basis for communication. In this section we discuss the parallel implementation of matrix multiplication, a computation that is highly parallelizable and provides a nice opportunity to contrast the shared and distributed memory approaches.[1]

5.5.1 Networks and Communication

We start by discussing the general set-up in a distributed memory environment. Popular interconnection networks include the ring and the mesh. See Figs. 5.3 and 5.4.

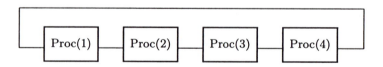

FIGURE 5.3 A ring multiprocessor

The individual processors come equipped with their own processing units, local memory, and input/output ports. The act of designing a parallel algorithm for a distributed memory system is the act of designing a *node program* for each of the participating processors. As we shall see, these programs look like ordinary programs with occasional `send` and `recv` commands that are used for the sending and receiving of messages. For us, a message is a matrix. Since vectors and scalars are special matrices, a message may consist of a vector or a scalar.

In the following we suppress very important details such as (1) how data and programs are downloaded into the nodes, (2) the subscript computations associated with local array access,

---

[1]The distinction between shared memory multiprocessors and distributed memory multiprocessors is fuzzy. A shared memory can be physically distributed. In such a case, the programmer has all the convenience of being able to write node programs that read and write directly into the shared memory. However, the physical distribution of the memory means that either the programmer or the compiler must strive to access "nearby" memory as much as possible.

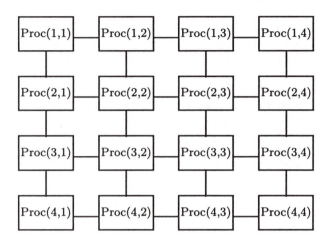

FIGURE 5.4 A mesh multiprocessor

and (3) the formatting of messages. We designate the μth processor by $\text{Proc}(\mu)$. The μth node program is usually a function of μ.

The distributed memory paradigm is quite general. At the one extreme we can have networks of workstations. On the other hand, the processors may be housed on small boards that are stacked in a single cabinet that sits on a desk.

The kinds of algorithms that can be efficiently solved on a distributed memory multiprocessor are defined by a few key properties:

- *Local memory size.* An interesting aspect of distributed memory computing is that the amount of data for a problem may be so large that it cannot fit inside a single processor. For example, a 10,000-by-10,000 matrix of floating point numbers requires almost a thousand megabytes of storage. The storage of such an array may require the local memories from hundreds of processors. Local memory often has a hierarchy, as does the memory of conventional computers (e.g., disks $\rightarrow$ slow random access memory $\rightarrow$ fast random access memory, etc.)

- *Processor speed.* The speed at which the individual processing units execute is of obvious importance. In sophisticated systems, the individual nodes may have vector capabilities, meaning that the extraction of peak performance from the system requires algorithms that vectorize at the node level. Another complicating factor may be that system is made up of different kinds of processors. For example, maybe every fourth node has a vector processing accelerator.

- *Message passing overhead.* The time it takes for one processor to send another processor a message determines how often a node program will want to break away from productive calculation to receive and send data. It is typical to model the time it takes to send or receive an n-byte message by

$$T(n) = \alpha + \beta n \qquad (5.5.1)$$

Here, α is the *latency* and β is the *bandwidth*. The former measures how long it takes to "get ready" for a send/receive while the latter is a reflection of the "size" of the wires that connect the nodes. This model of communication provides some insight into performance, but it is seriously flawed in at least two regards. First, the proximity of the receiver to the sender is usually an issue. Clearly, if the sender and receiver are neighbors in the network, then the system software that routes the message will not have so much to do. Second, the message passing software or hardware may require the breaking up of large messages into smaller packets. This takes time and magnifies the effect of latency.

The examples that follow will clarify some of these issues.

5.5.2   A Two-processor Matrix Multiplication

Consider the matrix-matrix multiplication problem

$$C \leftarrow C + AB$$

where the three matrices $A, B, C \in \mathbb{R}^{n \times n}$ are distributed in a two-processor distributed memory network. To be specific, assume that $n = 2m$ and that Proc(1) houses the "left halves"

$$A_L = A(:, 1{:}m) \quad B_L = B(:, 1{:}m) \quad C_L = C(:, 1{:}m)$$

and that Proc(2) houses the "right halves"

$$A_R = A(:, m+1{:}n) \quad B_R = B(:, m+1{:}n) \quad C_R = C(:, m+1{:}n)$$

Pictorially we have

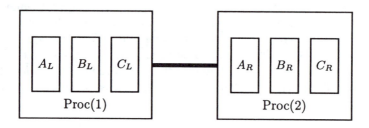

We assign to Proc(1) and Proc(2) the computation of the new C_L and C_R, respectively. Let's see where the data for these calculations come from. We start with a complete specification of the overall computation:

```
for j=1:n
   C(:,j) = C(:,j) + A*B(:,j);
end
```

Note that the jth column of the updated C is the jth column of the original C plus A times the jth column of B. The matrix-vector product $A*B(:,j)$ can be expressed as a linear combination of A-columns:

$$A * B(:,j) = A(:,1) * B(1,j) + A(:,2) * B(2,j)+ \cdots +A(:,n) * B(n,j).$$

Thus the preceding fragment expands to

```
for j=1:n
   for k=1:n
      C(:,j) = C(:,j) + A(:,k)*B(k,j);
   end
end
```

Note that Proc(1) is in charge of

```
for j=1:m
   for k=1:n
      C(:,j) = C(:,j) + A(:,k)*B(k,j);
   end
end
```

while Proc(2) must carry out

```
for j=m+1:n
   for k=1:n
      C(:,j) = C(:,j) + A(:,k)*B(k,j);
   end
end
```

From the standpoint of communication, there is both good news and bad news. The good news is that the B-data and C-data required are local. Proc(1) requires (and has) the left portions of C and B. Likewise, Proc(2) requires (and has) the right portions of these same matrices. The bad news concerns A. Both processors must "see" all of A during the course of execution. Thus, for each processor exactly one half of the A-columns are nonlocal, and these columns must somehow be acquired during the calculation. To highlight the local and nonlocal A-data, we pair off the left and the right A-columns:

```
Proc(1) does this:                      Proc(2) does this:

for j=1:m                               for j=m+1:n
   for k=1:m                               for k=1:m
      C(:,j) = C(:,j) + A(:,k)*B(k,j);        C(:,j) = C(:,j) + A(:,k+m)*B(k+m,j);
      C(:,j) = C(:,j) + A(:,k+m)*B(k+m,j);    C(:,j) = C(:,j) + A(:,k)*B(k,j);
   end                                     end
end                                     end
```

In each case, the second update of $C(:,j)$ requires a nonlocal A-column. Somehow, Proc(1) has to get hold of $A(:,k+m)$ from Proc(2). Likewise, Proc(2) must get hold of $A(:,k)$ from Proc(1).

The primitives **send** and **recv** are to be used for this purpose. They have the following syntax:

$$\textbf{send}(\{matrix\} , \{destination\ node\}) \qquad \textbf{recv}(\{matrix\} , \{source\ node\})$$

If a processor invokes a **send**, then we assume that execution resumes immediately after the message is sent. If a **recv** is encountered, then we assume that execution of the node program is suspended until the requested message arrives. We also assume that messages arrive in the same order that they are sent.[2]

With **send** and **recv**, we can solve the nonlocal data problem in our two-processor matrix multiply:

```
Proc(1) does this:                      Proc(2) does this:

for j=1:m                               for j=m+1:n
   for k=1:m                               for k=1:m
      send(A(:,k),2);                          send(A(:,k+m),1);
      C(:,j) = C(:,j) + A(:,k)*B(k,j);         C(:,j) = C(:,j) + A(:,k+m)*B(k+m,j);
      recv(v,2);                               recv(v,1);
      C(:,j) = C(:,j) + v*B(k+m,j);            C(:,j) = C(:,j) + v*B(k,j);
   end                                      end
end                                     end
```

Each processor has a local work vector v that is used to hold an A-column solicited from its neighbor. The correctness of the overall process follows from the assumption that the messages arrive in the order that they are sent. This ensures that each processor "knows" the index of the incoming columns. Of course, this is crucial because incoming A-columns have to scaled by the appropriate B entries.

Although the algorithm is correct and involves the expected amount of floating point arithmetic, it is inefficient from the standpoint of communication. Each processor sends a given local A-column to its neighbor m times. A better plan is to use each incoming A-column in all the $C(:,j)$ updates "at once." To do this, we merely reorganize the order in which things are done in the node programs:

```
Proc(1) does this:                      Proc(2) does this:

for k=1:m                               for k=1:m
   send(A(:,k),2);                          send(A(:,k+m),1);
   for j=1:m                                for j=m+1:n
      C(:,j) = C(:,j) + A(:,k)*B(k,j);         C(:,j) = C(:,j) + A(:,k+m)*B(k+m,j);
   end;                                    end
   recv(v,2);                               recv(v,1);
   for j=1:m                                for j=m+1:n
      C(:,j) = C(:,j) + v*B(k+m,j);            C(:,j) = C(:,j) + v*B(k,j);
   end                                    end
end                                     end
```

Although there is "perfect symmetry" between the two node programs, we cannot assume that they proceed in lock-step fashion. However, the program is *load balanced* because each processor has roughly the same amount of arithmetic and communication.

---

[2]This need not be true in practice. However, a system of tagging messages can be incorporated that can be used to remove ambiguities.

In the preceding, the matrices A, B, and C are accessed as if they were situated in a single memory. Of course, this will not be the case in practice. For example, Proc(2) will have a local n-by-m array to house its portion of C. Let's call this array C.loc and assume that C.loc(i,j) houses matrix element $C(i, j + m)$. Similarly, there will be local array A.loc and B.loc that house their share of A and B, respectively. If we rewrite Proc(2)'s node program in the true "language" of its local arrays, then it becomes

```
for k=1:m
   send(A.loc(:,k),1);
   for j=1:m
      C.loc(:,j) = C.loc(:,j) + A.loc(:,k)*B.loc(k,j);
   end
   recv(v,1);
   for j=1:m
      C.loc(:,j) = C(:,j) + v*B.loc(k+m,j);
   end
end
```

We merely mention this to stress that what may be called "subscript reasoning" undergoes a change when working in distributed memory environments.

5.5.3 Performance Analysis

Let's try to anticipate the time required to execute the communication-efficient version of the two-processor matrix multiply. There are two aspects to consider: computation and communication. With respect to computation, we note first that overall calculation involves $2n^3$ flops. This is because there are n^2 entries in C to update and each update requires the execution of a length n inner product (e.g., $C(i, j) = C(i, j) + A(i, :) * B(:, j)$). Since an inner product of that length involves n adds and n multiplies, $n^2(2n)$ flops are required in total. These flops are distributed equally between the two processors. If computation proceeds at a uniform rate of R flops per second, then

$$T_{comp} = n^3/R$$

seconds are required to take care of the arithmetic.[3]

With respect to communication costs, we use the model (5.5.1) and conclude that each processor spends

$$T_{comm} = n(\alpha + 8\beta n)$$

seconds sending and receiving messages. (We assume that each floating point number has an 8-byte representation.) Note that this is not the whole communication overhead story because we are not taking into account the idle wait times associated with the receives. (The required vector may not have arrived at time of the recv.) Another factor that complicates performance evaluation is that each node may have a special input/output processor that more or less handles communication making it possible to overlap computation and communication.

---

[3]Recall the earlier observation that with many advanced architectures, the execution rate varies with the operation performed and whether or not it is vectorized.

Ignoring these possibly significant details leads us to predict an overall execution time of $T = T_{comp} + T_{comm}$ seconds. It is instructive to compare this time with what would be required by a single-processor program. Look at the ratio

$$S = \frac{2n^3/R}{(n^3/R) + n(\alpha + 8\beta n)} = \frac{2}{1 + R\left(\frac{\alpha}{n^2} + \frac{8\beta}{n}\right)}$$

S represents the *speed-up* of the parallel program. We make two observations: (1) Communication overheads are suppressed as n increases, and (2) if α and β are fixed, then speed-up decreases as the rate of computation R improves.

In general, the speed-up of a parallel program executing on p processors is a ratio:

$$\text{Speed-up} = \frac{\text{Time required by the best single-processor program}}{\text{Time required for the } p\text{-processor implementation}}$$

In this definition we do not just set the numerator to be the $p = 1$ version of the parallel code because the best uniprocessor algorithm may not parallelize. Ideally, one would like the speed-up for an algorithm to equal p.

Problems

P5.5.1 Assume that $n = 3m$ and that the n-by-n matrices A, B, C are distributed around a three-processor ring. In particular, assume that processors 1, 2, and 3 house the left third, middle third, and right third of these matrices. For example, $B(:, m + 1:2m)$ would be housed in Proc(2). Write a parallel program for the computation $C \leftarrow C + AB$.

P5.5.2 Suppose we have a two-processor distributed memory system in which floating point arithmetic proceeds at R flops per second. Assume that when one processor sends or receives a message of k floating point numbers, then $\alpha + \beta k$ seconds are required. Proc(1) houses an n-by-n matrix A, and each processor houses a copy of an n-vector x. The goal is to store the vector $y = Ax$ in Proc(1)'s local memory. You may assume that n is even. **(1)** How long would this take if Proc(1) handles the entire computation itself? **(2)** Describe how the two processors can share the computation. Indicate the data that must flow between the two processors and what they must each calculate. You do not have to write formal node programs. Clear concise English will do. **(3)** Does it follow that if n is large enough, then it is more efficient to distribute the computation? Justify your answer.

M-Files and References

Script Files

CircBench	Benchmarks Circulant1 and Circulant2.
MatBench	Benchmarks various matrix-multiply methods.
AveNorms	Compares various norms on random vectors.
ProdBound	Examines the error in three-digit matrix multiplication.
ShowContour	Displays various contour plots of SampleF.
Show2dQuad	Illustrates CompQNC2D.
FFTflops	Compares FFT and DFT flops.
StrassFlops	Examines Strassen multiply.

Function Files

`Circulant1`	Scalar-level circulant matrix set-up.
`Circulant2`	Vector-level circulant matrix set-up.
`MatVecRO`	Row-oriented matrix-vector product.
`MatVecCO`	Column-Oriented matrix-vector product.
`MatMatDot`	Dot-product matrix-matrix product.
`MatMatSax`	Saxpy matrix-matrix product.
`MatMatVec`	Matrix-vector matrix-matrix product.
`MatMatOuter`	Outer product matrix-matrix product.
`Prod3Digit`	Three-digit matrix-matrix product.
`SampleF`	A Gaussian type function of two variables.
`CompQNC2D`	Two-dimensional Newton-Cotes rules.
`wCompNC`	Weight vector for composite Newton-Cotes rules.
`SampleF2`	A function of two variables with strong local maxima.
`DFT`	Discrete Fourier transform.
`FFTRecur`	A recursive radix-2 FFT.
`Strass`	Recursive Strassen matrix multiply.

References

T.F. Coleman and C.F. Van Loan (1988). *Handbook for Matrix Computations*, SIAM Publications, Philadelphia, PA.

G.H. Golub and C.F. Van Loan (1996). *Matrix Computations, Third Edition*, Johns Hopkins University Press, Baltimore, MD.

C.F. Van Loan (1992). *Computational Frameworks for the Fast Fourier Transform*, SIAM Publications, Philadelphia, PA.

Chapter 6

Linear Systems

§**6.1** Triangular Problems

§**6.2** Banded Problems

§**6.3** Full Problems

§**6.4** Analysis

The linear equation problem involves finding a vector $x \in \mathbb{R}^n$ so that $Ax = b$, where $b \in \mathbb{R}^n$ and $A \in \mathbb{R}^{n \times n}$ is nonsingular. This problem is at the heart of many problems in scientific computation. We have already seen that the vector of coefficients that define a polynomial interpolant is a solution to a linear equation problem. In Chapter 3 the problem of spline interpolation led to a tridiagonal system. Most numerical techniques in optimization and differential equations involve repeated linear equation solving. Hence it is extremely important that we know how to solve this problem efficiently and that we fully understand what can be expected in terms of precision.

In this chapter the well-known process of Gaussian elimination is related to the factorization $A = LU$, where L is lower triangular and U is upper triangular. We arrive at the general algorithm in stages, discussing triangular, tridiagonal, and Hessenberg systems first. The need for pivoting is established, and this prompts a discussion of permutation matrices and how they can be manipulated in MATLAB. Finally, we explore the issue of linear system sensitivity and identify the important role that the condition number plays.

6.1 Triangular Problems

At one level, the goal of Gaussian elimination is to convert a given linear system into an equivalent, easy-to-solve triangular system. Triangular system solving is easy because the unknowns can be resolved without any further manipulation of the matrix of coefficients. Consider the following 3-by-3 lower triangular case:

$$\begin{bmatrix} \ell_{11} & 0 & 0 \\ \ell_{21} & \ell_{22} & 0 \\ \ell_{31} & \ell_{32} & \ell_{33} \end{bmatrix} \begin{bmatrix} x_1 \\ x_2 \\ x_3 \end{bmatrix} = \begin{bmatrix} b_1 \\ b_2 \\ b_3 \end{bmatrix}$$

The unknowns can be determined as follows:

$$
\begin{aligned}
x_1 &= b_1/\ell_{11} \\
x_2 &= (b_2 - \ell_{21}x_1)/\ell_{22} \\
x_3 &= (b_3 - \ell_{31}x_1 - \ell_{32}x_2)/\ell_{33}
\end{aligned}
$$

This is the 3-by-3 version of an algorithm known as *forward substitution*. Notice that the process requires $\det(A) = \ell_{11}\ell_{22}\ell_{33}$ to be nonzero.

6.1.1 Forward Substitution

Let's look at the general $Lx = b$ problem when L is lower triangular. To derive a specification for x_i, we merely rearrange the ith equation

$$
\ell_{i1}x_1 + \cdots + \ell_{ii}x_i = b_i
$$

to obtain

$$
x_i = \left(b_i - \sum_{j=1}^{i-1} \ell_{ij}x_j \right) \Big/ \ell_{ii}
$$

If this is evaluated for $i = 1{:}n$, then a complete specification for x is obtained:

```
for i = 1:n
    x(i) = b(i);
    for j=1:i-1
        x(i) = x(i) - L(i,j)*x(j);
    end
    x(i) = x(i)/L(i,i);
end
```

Note that the j-loop effectively subtracts the inner product

$$
\sum_{j=1}^{i-1} \ell_{ij}x_j = L(i, 1{:}i-1) * x(1{:}i-1)
$$

from b_i, so we can vectorize as follows:

```
x(1) = b(1)/L(1,1);
for i = 2:n
    x(i) = ( b(i) - L(i,1:i-1)*x(1:i-1) ) /L(i,i);
end
```

Since the computation of x_i involves about $2i$ flops, the entire process requires about

$$
2\,(1 + 2 + \cdots + n) \approx n^2
$$

flops. The forward substitution algorithm that we have derived is row oriented. At each stage an inner product must be computed that involves part of a row of L and the previously computed portion of x.

A column-oriented version that features the saxpy operation can also be obtained. Consider the $n = 3$ case once again. Once x_1 is resolved, it can be removed from equations 2 and 3, leaving us with a reduced, 2-by-2 lower triangular system. For example, to solve

$$
\begin{bmatrix} 2 & 0 & 0 \\ 1 & 5 & 0 \\ 7 & 9 & 8 \end{bmatrix} \begin{bmatrix} x_1 \\ x_2 \\ x_3 \end{bmatrix} = \begin{bmatrix} 6 \\ 2 \\ 5 \end{bmatrix}
$$

we find that $x_1 = 3$ and then deal with the 2-by-2 system

$$
\begin{bmatrix} 5 & 0 \\ 9 & 8 \end{bmatrix} \begin{bmatrix} x_2 \\ x_3 \end{bmatrix} = \begin{bmatrix} 2 \\ 5 \end{bmatrix} - 3 \begin{bmatrix} 1 \\ 7 \end{bmatrix} = \begin{bmatrix} -1 \\ -16 \end{bmatrix}
$$

This implies that $x_2 = -1/5$. The system is then reduced to

$$
8x_3 = -16 - 9(-1/5)
$$

from which we conclude that $x_3 = -71/40$. In general, at the jth step we solve for x_j and then remove it from equations $j + 1$ through n. At the start, $x_1 = b_1/\ell_{11}$ and equations 2 through n transform to

$$
\begin{bmatrix} \ell_{22} & 0 & \cdots & 0 \\ \ell_{32} & \ell_{33} & \cdots & 0 \\ \vdots & \vdots & \ddots & \vdots \\ \ell_{n2} & \ell_{n3} & \cdots & \ell_{nn} \end{bmatrix} \begin{bmatrix} x_2 \\ x_3 \\ \vdots \\ x_n \end{bmatrix} = \begin{bmatrix} b_2 - x_1\ell_{21} \\ b_3 - x_1\ell_{31} \\ \vdots \\ b_n - x_1\ell_{n1} \end{bmatrix} = b(2{:}n) - x_1 L(2{:}n, 1)
$$

In general the j-th step computes $x_j = b_j/\ell_{jj}$ and then performs the saxpy update

$$
b(j + 1{:}n) \leftarrow b(j + 1{:}n) - x_j L(j + 1{:}n, j)
$$

Putting it all together, we obtain

```
function x = LTriSol(L,b)
%
% Pre:
%   L    n-by-n nonsingular lower triangular matrix
%   b    n-by-1
%
% Post:
%   x    Lx = b

n = length(b);
x = zeros(n,1);
for j=1:n-1
    x(j) = b(j)/L(j,j);
    b(j+1:n) = b(j+1:n) - x(j)*L(j+1:n,j);
end
x(n) = b(n)/L(n,n);
```

This version involves n^2 flops, just like the row-oriented, dot product version developed earlier.

6.1.2   Backward Substitution

The upper triangular case is analogous. The only difference is that the unknowns are resolved in reverse order. Thus to solve

$$
\begin{bmatrix}
u_{11} & u_{12} & u_{13} \\
0 & u_{22} & u_{23} \\
0 & 0 & u_{33}
\end{bmatrix}
\begin{bmatrix}
x_1 \\ x_2 \\ x_3
\end{bmatrix}
=
\begin{bmatrix}
b_1 \\ b_2 \\ b_3
\end{bmatrix}
$$

we work from the bottom to the top:

$$
\begin{aligned}
x_3 &= b_3/u_{33} \\
x_2 &= (b_2 - u_{23}x_3)/u_{22} \\
x_1 &= (b_1 - u_{12}x_2 - u_{13}x_3
\end{aligned}
$$

For general n, we obtain

```
x(n) = b(n)/U(n,n);
for i=n-1:-1:1
    x(i) = (b(i) - U(i,i+1:n)*x(i+1:n))/U(i,i);
end
```

As in the lower triangular case, the computations can be arranged so that a column-oriented, saxpy update procedure is obtained:

```
    function x = UTriSol(U,b)
%
% Pre:
%   U    n-by-n nonsingular upper triangular matrix
%   b    n-by-1
%
% Post:
%   x    Lx = b

    n = length(b);
    x = zeros(n,1);
    for j=n:-1:2
        x(j) = b(j)/U(j,j);
        b(1:j-1) = b(1:j-1) - x(j)*U(1:j-1,j);
    end
    x(1) = b(1)/U(1,1);
```

This algorithm is called *backward substitution.*

6.1.3   Multiple Right-Hand Sides

In many applications we must solve a sequence of triangular linear systems where the matrix stays the same, but the right-hand sides vary. For example, if L is lower triangular and $B \in \mathbb{R}^{n \times r}$ is given, our task is to find $X \in \mathbb{R}^{n \times r}$ so that $LX = B$. Looking at the kth column of this matrix equation, we see that

$$LX(:,k) = B(:,k)$$

One way to solve for X is merely to apply the single right-hand side forward substitution algorithm r times:

```
X = zeros(n,r);
for k=1:r
   X(:,k) = LTriSol(L,B(:,k));
end
```

However, if we expand the call to LTriSol,

```
X = zeros(n,r);
for k=1:r
   for j=1:n-1
      X(j,k)      = B(j,k)/L(j,j);
      B(j+1:n,k) = B(j+1:n,k) - L(j+1:n,j)*X(j,k);
   end
   X(n,k) = B(n,k)/L(n,n);
end
```

and modify the order of computation, then we can vectorize "on k." To do this, note that in the preceding script we solve for $X(:,1)$, and then $X(:,1)$, and then $X(:,3)$, etc. Instead, we can solve for $X(1,:)$, and then $X(2,:)$, and then $X(3,:)$, etc. This amounts to reversing the order of the k- and j- loops:

```
X = zeros(n,r);
for j=1:n-1
  for k=1:r
      X(j,k)      = B(j,k)/L(j,j);
      B(j+1:n,k) = B(j+1:n,k) - L(j+1:n,j)*X(j,k);
   end
end
for k=1:r
   X(n,k) = B(n,k)/L(n,n);
end
```

Vectorizing the k-loops, we obtain

```
   function x = LTriSolM(L,B)
%
% Pre:
%   L    n-by-n nonsingular lower triangular matrix
%   B    n-by-r
%
% Post:
%   X    LX = B
```

```
[n,r] = size(B);
X = zeros(n,r);
for j=1:n-1
    X(j,1:r)     = B(j,1:r)/L(j,j);
    B(j+1:n,1:r) = B(j+1:n,1:r) - L(j+1:n,j)*X(j,1:r);
end
X(n,1:r) = B(n,1:r)/L(n,n);
```

In high-performance computing environments, maneuvers like this are often the key to efficient matrix computations.

As an example, we use LTriSolM to compute the inverse of the n-by-n *Forsythe* matrix $F_n = (f_{ij})$ defined as follows:

$$f_{ij} = \begin{cases} 0 & \text{if } i < j \\ 1 & \text{if } i = j \\ -1 & \text{if } i > j \end{cases}$$

This is accomplished by solving $F_n X = I_n$. For example,

$$\begin{bmatrix} 1 & 0 & 0 & 0 & 0 \\ -1 & 1 & 0 & 0 & 0 \\ -1 & -1 & 1 & 0 & 0 \\ -1 & -1 & -1 & 1 & 0 \\ -1 & -1 & -1 & -1 & 1 \end{bmatrix} X = \begin{bmatrix} 1 & 0 & 0 & 0 & 0 \\ 0 & 1 & 0 & 0 & 0 \\ 0 & 0 & 1 & 0 & 0 \\ 0 & 0 & 0 & 1 & 0 \\ 0 & 0 & 0 & 0 & 1 \end{bmatrix}$$

Running the script file

```
% Script File: ShowTri
%
% Inverse of the 5-by-5 Forsythe Matrix.

n = 5;
L = eye(n,n) - tril(ones(n,n),-1)
X = LTriSolM(L,eye(n,n))
```

we find

$$X = \begin{bmatrix} 1 & 0 & 0 & 0 & 0 \\ 1 & 1 & 0 & 0 & 0 \\ 2 & 1 & 1 & 0 & 0 \\ 4 & 2 & 1 & 1 & 0 \\ 8 & 4 & 2 & 1 & 1 \end{bmatrix}$$

Problems

P6.1.1 Modify LTriSol so that if $\ell_{nnn} = 0$, then it returns a vector x with $x_n = -1$ so that $Lx = 0$.

P6.1.2 Develop a vectorized method for solving the upper triangular multiple right-hand-side problem.

P6.1.3 Suppose $T \in \mathbf{R}^{m \times m}$ and $S \in \mathbf{R}^{n \times n}$ are given upper triangular matrices and that $B \in \mathbf{R}^{m \times n}$. Write a MATLAB function X = Sylvester(S,T,B) that solves the matrix equation $SX + XT = B$ for X. Note that if we compare kth columns in this equation, we obtain

$$SX(:,k) + \sum_{j=1}^{k} T(j,k)X(:,j) = B(:,k)$$

That is,

$$(S - T(k,k)I)X(:,k) = B(:,k) - \sum_{j=1}^{k-1} T(j,k)X(:,j)$$

By using this equation for $k = 1:n$, we can solve for $X(:,1), \ldots, X(:,n)$ in turn. Moreover, the matrix $S - T(k,k)I$ is upper triangular so that we can apply UTriSol. Assume that no diagonal entry of S is a diagonal entry of T.

P6.1.4 Repeat the previous problem, assuming that S and T are both lower triangular.

P6.1.5 Note that by solving the multiple right-hand-side problem $TX = B$ with $B = I$, then the solution is the inverse of T. Write a MATLAB function X = UTriInv(U) that computes the inverse of a nonsingular upper triangular matrix. Be sure to exploit any special patterns that arise because of B's special nature.

P6.1.6 As a function of n, i, and j, give an expression for the (i, j) entry of the inverse of the n-by-n Forsythe matrix.

6.2 Banded Problems

Before we embark on the development of Gaussian elimination for general linear systems, we take time out to look at the linear equation problem in two special cases where the matrix of coefficients already has a large number of zeros.

In the spline interpolation problem of §3.3, we have to solve a tridiagonal linear system whose matrix of coefficients looks like this:

$$A = \begin{bmatrix} \times & \times & 0 & 0 & 0 & 0 \\ \times & \times & \times & 0 & 0 & 0 \\ 0 & \times & \times & \times & 0 & 0 \\ 0 & 0 & \times & \times & \times & 0 \\ 0 & 0 & 0 & \times & \times & \times \\ 0 & 0 & 0 & 0 & \times & \times \end{bmatrix}$$

In such a system, each unknown x_i is coupled to at most two of its "neighbors." For example, the fourth equation relates x_4 to x_3 and x_5. This kind of local coupling among the unknowns occurs in a surprising number of applications. It is a happy circumstance because the matrix comes to us with zeros in many of the places that would ordinarily be zeroed by the elimination process.

Upper Hessenberg systems provide a second family of specialized problems that are useful to consider. An *upper Hessenberg* matrix has lower bandwidth 1. For example,

$$A = \begin{bmatrix} \times & \times & \times & \times & \times & \times \\ \times & \times & \times & \times & \times & \times \\ 0 & \times & \times & \times & \times & \times \\ 0 & 0 & \times & \times & \times & \times \\ 0 & 0 & 0 & \times & \times & \times \\ 0 & 0 & 0 & 0 & \times & \times \end{bmatrix}$$

Upper Hessenberg linear systems arise in many applications where the eigenvalues and eigenvectors of a matrix are required.

For these two specially structured linear equation problems, we set out to show how the matrix of coefficients can be factored into a product $A = LU$, where L is lower triangular and U is upper triangular. If such a factorization is available, then the solution to $Ax = b$ follows from a pair of triangular system solves:

$$\left. \begin{array}{l} Ly = b \\ \\ Ux = y \end{array} \right\} \;\; \Rightarrow \;\; Ax = (LU)x = L(Ux) = Ly = b$$

In terms of the functions LTriSol and UTriSol,

```
y = LTriSol(L,b);
x = UTriSol(U,y);
```

For tridiagonal and Hessenberg systems, the computation of L and U is easier to explain than for general matrices.

6.2.1   Tridiagonal Systems

Consider the following 5-by-5 tridiagonal linear system:

$$\begin{bmatrix} d_1 & f_1 & 0 & 0 & 0 \\ e_2 & d_2 & f_2 & 0 & 0 \\ 0 & e_3 & d_3 & f_3 & 0 \\ 0 & 0 & e_4 & d_4 & f_4 \\ 0 & 0 & 0 & e_5 & d_5 \end{bmatrix} \begin{bmatrix} x_1 \\ x_2 \\ x_3 \\ x_4 \\ x_5 \end{bmatrix} = \begin{bmatrix} b_1 \\ b_2 \\ b_3 \\ b_4 \\ b_5 \end{bmatrix}$$

One way to derive the LU factorization for a tridiagonal A is to equate entries in the following equation:

$$\begin{bmatrix} d_1 & f_1 & 0 & 0 & 0 \\ e_2 & d_2 & f_2 & 0 & 0 \\ 0 & e_3 & d_3 & f_3 & 0 \\ 0 & 0 & e_4 & d_4 & f_4 \\ 0 & 0 & 0 & e_5 & d_5 \end{bmatrix} = \begin{bmatrix} 1 & 0 & 0 & 0 & 0 \\ \ell_2 & 1 & 0 & 0 & 0 \\ 0 & \ell_3 & 1 & 0 & 0 \\ 0 & 0 & \ell_4 & 1 & 0 \\ 0 & 0 & 0 & \ell_5 & 1 \end{bmatrix} \begin{bmatrix} u_1 & f_1 & 0 & 0 & 0 \\ 0 & u_2 & f_2 & 0 & 0 \\ 0 & 0 & u_3 & f_3 & 0 \\ 0 & 0 & 0 & u_4 & f_4 \\ 0 & 0 & 0 & 0 & u_5 \end{bmatrix}$$

Doing this, we find

$$
\begin{array}{lll}
(1,1): & d_1 = u_1 & \Rightarrow \quad u_1 = d_1 \\
(2,1): & e_2 = \ell_2 u_1 & \Rightarrow \quad \ell_2 = e_2/u_1 \\
(2,2): & d_2 = \ell_2 f_1 + u_2 & \Rightarrow \quad u_2 = d_2 - \ell_2 f_1 \\
(3,2): & e_3 = \ell_3 u_2 & \Rightarrow \quad \ell_3 = e_3/u_2 \\
(3,3): & d_3 = \ell_3 f_2 + u_3 & \Rightarrow \quad u_3 = d_3 - \ell_3 f_2 \\
(4,3): & e_4 = \ell_4 u_3 & \Rightarrow \quad \ell_4 = e_4/u_3 \\
(4,4): & d_4 = \ell_4 f_3 + u_4 & \Rightarrow \quad u_4 = d_4 - \ell_4 f_3 \\
(5,4): & e_5 = \ell_5 u_4 & \Rightarrow \quad \ell_5 = e_5/u_4 \\
(5,5): & d_5 = \ell_5 f_4 + u_5 & \Rightarrow \quad u_5 = d_5 - \ell_5 f_4
\end{array}
$$

In general, for $i = 2{:}n$ we have

$$
\begin{array}{llll}
(i, i-1){:} & e_i = \ell_i u_{i-1} & \Rightarrow & \ell_i = e_i / u_{i-1} \\
(i, i){:} & d_i = \ell_i f_{i-1} + u_i & \Rightarrow & u_i = d_i - \ell_i f_{i-1}
\end{array}
$$

which leads to the following procedure:

```
function [l,u] = TriDiLU(d,e,f)
%
% Pre:
%   d,e,f    n-vectors that define a tridiagonal matrix
%            A = diag(e(2:n),-1) + diag(d) + diag(f(1:n-1),1).
%            A has an LU factorization.
% Post:
%   l        n-vector with the property that
%            L = eye + diag(l(2:n),-1)
%   u        n-vector with the property that
%            U = diag(u) + diag(f(1:n-1),1).

n = length(d);
l = zeros(n,1);
u = zeros(n,1);
u(1) = d(1);
for i=2:n
   l(i) = e(i)/u(i-1);
   u(i) = d(i) - l(i)*f(i-1);
end
```

This process requires $3n$ flops to carry out and is defined as long as $u_1, \ldots, u_{n-1}$ are nonzero.

As mentioned previously, to solve $Ax = b$ we must solve

$$
Ly \;=\;
\begin{bmatrix}
1 & 0 & 0 & 0 & 0 \\
\ell_2 & 1 & 0 & 0 & 0 \\
0 & \ell_3 & 1 & 0 & 0 \\
0 & 0 & \ell_4 & 1 & 0 \\
0 & 0 & 0 & \ell_5 & 1
\end{bmatrix}
\begin{bmatrix}
y_1 \\ y_2 \\ y_3 \\ y_4 \\ y_5
\end{bmatrix}
=
\begin{bmatrix}
b_1 \\ b_2 \\ b_3 \\ b_4 \\ b_5
\end{bmatrix}
= b
$$

for y, and

$$
Ux \;=\;
\begin{bmatrix}
u_1 & f_1 & 0 & 0 & 0 \\
0 & u_2 & f_2 & 0 & 0 \\
0 & 0 & u_3 & f_3 & 0 \\
0 & 0 & 0 & u_4 & f_4 \\
0 & 0 & 0 & 0 & u_5
\end{bmatrix}
\begin{bmatrix}
x_1 \\ x_2 \\ x_3 \\ x_4 \\ x_5
\end{bmatrix}
=
\begin{bmatrix}
y_1 \\ y_2 \\ y_3 \\ y_4 \\ y_5
\end{bmatrix}
= y
$$

for x. These *bidiagonal* systems can be solved very simply. Looking at $Ly = b$, we see by comparing components that

$$
\begin{array}{llll}
y_1 = b_1 & \Rightarrow & y_1 = b_1 \\
\ell_2 y_1 + y_2 = b_2 & \Rightarrow & y_2 = b_2 - \ell_2 y_1 \\
\ell_3 y_2 + y_3 = b_3 & \Rightarrow & y_3 = b_3 - \ell_3 y_2 \\
\ell_4 y_3 + y_4 = b_4 & \Rightarrow & y_4 = b_4 - \ell_4 y_3 \\
\ell_5 y_4 + y_5 = b_5 & \Rightarrow & y_5 = b_5 - \ell_5 y_4
\end{array}
$$

From this we conclude

```
function x = LBidiSol(l,b)
%
% Pre:
%    l   n-vector that defines the lower bidiagonal matrix
%        L = I + diag(l(2:n),-1).
%    b   n-vector
%
% Post:
%    x   n-vector that solves Lx = b

   n = length(b);
   x = zeros(n,1);
   x(1) = b(1);
   for i=2:n
       x(i) = b(i) - l(i)*x(i-1);
   end
```

This requires $2n$ flops. Likewise, the upper bidiagonal system $Ux = y$ can be solved as follows:

```
function x = UBidiSol(u,f,b)
%
% Pre:
%   u,f  n-vector that defineS the upper bidiagonal matrix
%        U = diag(u) + diag(f(1:n-1),1)
%    b   n-vector
%
% Post:
%    x   n-vector that solves Ux = b

   n = length(b);
   x = zeros(n,1);
   x(n) = b(n)/u(n);
   for i=n-1:-1:1
       x(i) = (b(i) - f(i)*x(i+1))/u(i);
   end
```

Summarizing the overall solution process, the fragment

```
[l,u] = TriDiLU(d,e,f);
y = LBiDiSol(l,b);
x = UBiDiSol(u,f,y);
```

solves the tridiagonal system $Ax = b$, assuming that d, e, and f house the diagonal, subdiagonal, and superdiagonal of A. The following table indicates the amount of arithmetic required:

Operation	Procedure	Flops
$A = LU$	TriDiLU	$3n$
$Ly = b$	LBiDiSol	$2n$
$Ux = y$	UBiDiS1	$3n$

Run the script file ShowTriD, which illustrates some of the key ideas behind tridiagonal system solving.

6.2.2   Hessenberg Systems

We derived the algorithm for tridiagonal LU by equating coefficients in $A = LU$. We could use this same strategy for Hessenberg LU. However, in anticipation of the general LU computation, we proceed in "elimination terms."

Presented with a Hessenberg system

$$
\begin{bmatrix}
\times & \times & \times & \times & \times & \times \\
\times & \times & \times & \times & \times & \times \\
0 & \times & \times & \times & \times & \times \\
0 & 0 & \times & \times & \times & \times \\
0 & 0 & 0 & \times & \times & \times \\
0 & 0 & 0 & 0 & \times & \times
\end{bmatrix}
\begin{bmatrix}
\times \\ \times \\ \times \\ \times \\ \times \\ \times
\end{bmatrix}
=
\begin{bmatrix}
\times \\ \times \\ \times \\ \times \\ \times \\ \times
\end{bmatrix}
$$

we notice that we can get the first unknown to "drop out" of the second equation by multiplying the first equation by a_{21}/a_{11} and subtracting from the second equation. This transforms the system to

$$
\begin{bmatrix}
\times & \times & \times & \times & \times & \times \\
0 & \times & \times & \times & \times & \times \\
0 & \times & \times & \times & \times & \times \\
0 & 0 & \times & \times & \times & \times \\
0 & 0 & 0 & \times & \times & \times \\
0 & 0 & 0 & 0 & \times & \times
\end{bmatrix}
\begin{bmatrix}
\times \\ \times \\ \times \\ \times \\ \times \\ \times
\end{bmatrix}
=
\begin{bmatrix}
\times \\ \times \\ \times \\ \times \\ \times \\ \times
\end{bmatrix}
$$

Then we notice that we can eliminate the second unknown from equation 3 by multiplying the (new) second equation by a_{32}/a_{22} and subtracting from the third equation:

$$
\begin{bmatrix}
\times & \times & \times & \times & \times & \times \\
0 & \times & \times & \times & \times & \times \\
0 & 0 & \times & \times & \times & \times \\
0 & 0 & \times & \times & \times & \times \\
0 & 0 & 0 & \times & \times & \times \\
0 & 0 & 0 & 0 & \times & \times
\end{bmatrix}
\begin{bmatrix}
\times \\ \times \\ \times \\ \times \\ \times \\ \times
\end{bmatrix}
=
\begin{bmatrix}
\times \\ \times \\ \times \\ \times \\ \times \\ \times
\end{bmatrix}
$$

The pattern emerges regarding what happens to A during this process:

```
for k=1:n-1
   Set v(k) = A(k+1,k)/A(k,k).
   Update row k+1 by subtracting from it, v(k) times row k.
end
```

The modification of row $k + 1$ involves the following computations:

```
for j=1:n
   A(k+1,j) = A(k+1,j) - v(k)*A(k,j)
end
```

Note that since rows k and $k+1$ have zeros in their first $k-1$ positions, it makes sense to modify the loop range to j=k:n. Incorporating this change and vectorizing, we obtain

```
function [v,U] = HessLU(A)
%
% Pre:
%   A        n-by-n upper Hessenberg
% Post:
%   v        column n-vector
%   U        n-by-n upper triangular U
%            H = LU where L = I + diag(v(2:n),-1)
%
   [n,n] = size(A);
   v = zeros(n,1);
   for k=1:n-1
      v(k+1) = A(k+1,k)/A(k,k);
      A(k+1,k:n) = A(k+1,k:n) - v(k+1)*A(k,k:n);
   end
   U = triu(A);
```

It can be shown that this procedure requires n^2 flops. The connection between the vector v of multipliers and the lower triangular matrix L needs to be explained.

In the $n = 6$ case, steps 1 through 5 in HessLU involve the premultiplication of the matrix A by the matrices

$$M_1 = \begin{bmatrix} 1 & 0 & 0 & 0 & 0 & 0 \\ -v_2 & 1 & 0 & 0 & & \\ 0 & 0 & 1 & 0 & 0 & 0 \\ 0 & 0 & 0 & 1 & 0 & 0 \\ 0 & 0 & 0 & 0 & 1 & 0 \\ 0 & 0 & 0 & 0 & 0 & 1 \end{bmatrix}, \ldots, M_5 = \begin{bmatrix} 1 & 0 & 0 & 0 & 0 & 0 \\ 0 & 1 & 0 & 0 & 0 & 0 \\ 0 & 0 & 1 & 0 & 0 & 0 \\ 0 & 0 & 0 & 1 & 0 & 0 \\ 0 & 0 & 0 & 0 & 1 & 0 \\ 0 & 0 & 0 & 0 & -v_6 & 1 \end{bmatrix}$$

respectively. Run the script file ShowHessLU for an illustration of the reduction. From this we conclude that HessLU basically finds *multiplier matrices* $M_1, \ldots, M_{n-1}$ so that

$$M_{n-1} \cdots M_1 A = U$$

is upper triangular. The multiplier matrices are nonsingular, and it is easy to verify (for example)
that

$$
M_3^{-1} = \begin{bmatrix} 1 & 0 & 0 & 0 & 0 & 0 \\ 0 & 1 & 0 & 0 & 0 & 0 \\ 0 & 0 & 1 & 0 & 0 & 0 \\ 0 & 0 & -v_4 & 1 & 0 & 0 \\ 0 & 0 & 0 & 0 & 1 & 0 \\ 0 & 0 & 0 & 0 & 0 & 1 \end{bmatrix}^{-1} = \begin{bmatrix} 1 & 0 & 0 & 0 & 0 & 0 \\ 0 & 1 & 0 & 0 & 0 & 0 \\ 0 & 0 & 1 & 0 & 0 & 0 \\ 0 & 0 & v_4 & 1 & 0 & 0 \\ 0 & 0 & 0 & 0 & 1 & 0 \\ 0 & 0 & 0 & 0 & 0 & 1 \end{bmatrix}
$$

Thus

$$A = LU$$

where

$$L = M_1^{-1} \cdots M_{n-1}^{-1}$$

The product of the M_i^{-1} is lower triangular, and it can be shown that

$$
M_1^{-1} M_2^{-1} M_3^{-1} M_4^{-1} M_5^{-1} = \begin{bmatrix} 1 & 0 & 0 & 0 & 0 & 0 \\ v_2 & 1 & 0 & 0 & 0 & 0 \\ 0 & v_3 & 1 & 0 & 0 & 0 \\ 0 & 0 & v_4 & 1 & 0 & 0 \\ 0 & 0 & 0 & v_5 & 1 & 0 \\ 0 & 0 & 0 & 0 & v_6 & 1 \end{bmatrix}
$$

is a lower bidiagonal matrix. Thus if we define

```
function x = UTriSol(U,y)
% Assume that U is upper triangular and n-by-n.  Assume that y is n-by-1.
% Solves Ux = y.
```

then the script

```
[v,U] = HessLU(A);
y = LBiDiSol(v,b);
x = UTriSol(U,y);
```

solves the upper Hessenberg system $Hx = b$. The flop picture is this:

Operation	Procedure	Flops
$A = LU$	HessLU	n^2
$Ly = b$	LBiDiSol	$2n$
$Ux = y$	UTriSol	n^2

Thus Hessenberg system solving is a $2n^2$ flop computation.

Problems

P6.2.1 Write a MATLAB function x = HessTrans(A,b) that solves the linear system $A^T x = b$, where A is upper
Hessenberg. Hint: If $A = LU$, then $A^T = U^T L^T$. In the Hessenberg case, U^T is lower triangular and L^T is upper
bidiagonal.

P6.2.2 Incorporate the tridiagonal system solving codes into `CubicSpline` of Chapter 3. Quantify through benchmarks the improvement in efficiency. Notice also the reduction in memory requirements.

P6.2.3 Suppose $S \in \mathbf{R}^{m \times m}$ and $T \in \mathbf{R}^{n \times n}$ are given matrices with S upper Hessenberg and T upper triangular. Assume that $B \in \mathbf{R}^{m \times n}$. Write a MATLAB function $X = \mathtt{SylvesterH(H,T,B)}$ that solves the matrix equation $HX + XT = B$ for X.

P6.2.4 Assume that F is n-by-n upper triangular, G is n-by-n upper bidiagonal, and B is n-by-n and full. [Recall that $G = (g_{ij})$ is upper bidiagonal if $g_{ij} = 0$ whenever $i > j$ or $j > i+1$.] Assume that no diagonal entry of F is a diagonal entry of G. This problem is about finding an n-by-n matrix X so that so that $FX - XG = B$. (1) Show how $X(:1)$ can be computed by solving an upper triangular system with right hand side $B(:, 1)$. (2) For $k \geq 2$, show how $X(:, k)$ solves an upper triangular system with a righthand side that involves $B(:, k)$ and $X(:, k-1)$. (3) Give a complete MATLAB script for computing X. You may use `UTriSol(A,b)`.

6.3 Full Problems

We are now ready to develop a general linear equation solver. Again, the goal is to find a lower triangular L and an upper triangular U such that $A = LU$.

6.3.1   The $n = 3$ Case

The method of *Gaussian elimination* proceeds by systematically removing unknowns from equations. The core calculation is the multiplication of an equation by a scalar and its subtraction from another equation. For example, if we are given the system

$$
\begin{array}{rcrcrcr}
2x_1 & - & x_2 & + & 3x_3 & = & 13 \\
-4x_1 & + & 6x_2 & - & 5x_3 & = & -28 \\
6x_1 & + & 13x_2 & + & 16x_3 & = & 37
\end{array}
\tag{6.3.1}
$$

then we start by multiplying the first equation by $-4/2 = -2$ and subtracting it from the second equation. This removes x_1 from the second equation. Likewise we can remove x_1 from the third equation by subtracting from it $6/2 = 3$ times the first equation. With these two reductions we obtain

$$
\begin{array}{rcrcrcr}
2x_1 & - & x_2 & + & 3x_3 & = & 13 \\
& & 4x_2 & + & x_3 & = & -2 \\
& & 16x_2 & + & 7x_3 & = & -2
\end{array}
$$

We then multiply the (new) second equation by $16/4 = 4$ and subtract it from the (new) third equation, obtaining

$$
\begin{array}{rcrcrcr}
2x_1 & - & x_2 & + & 3x_3 & = & 13 \\
& & 4x_2 & + & x_3 & = & -2 \\
& & & & 3x_3 & = & 6
\end{array}
\tag{6.3.2}
$$

Thus the elimination transforms the given square system into an equivalent upper triangular system that has the same solution. The solution of triangular systems is discussed in §6.1. In our 3-by-3 example we proceed as follows:

$$
\begin{array}{rclcr}
x_3 & = & 6/3 & = & 2 \\
x_2 & = & (-2 - x_3)/4 & = & -1 \\
x_1 & = & (3 - 3x_3 + x_2)/2 & = & 3
\end{array}
$$

This description of Gaussian elimination can be succinctly described in matrix terms. In particular, the process finds a lower triangular matrix L and an upper triangular matrix U so $A = LU$. In the preceding example, we have

$$A = \begin{bmatrix} 2 & -1 & 3 \\ -4 & 6 & -5 \\ 6 & 13 & 16 \end{bmatrix} = \begin{bmatrix} 1 & 0 & 0 \\ -2 & 1 & 0 \\ 3 & 4 & 1 \end{bmatrix} \begin{bmatrix} 2 & -1 & 3 \\ 0 & 4 & 1 \\ 0 & 0 & 3 \end{bmatrix} \equiv LU$$

Notice that the subdiagonal entries in L are made up of the multipliers that arise during the elimination process. The diagonal elements of L are all equal to one. Lower triangular matrices with this property are called *unit lower triangular*.

In matrix computations, the language of "matrix factorizations" has assumed a role of great importance. It enables one to reason about algorithms at a high level, which in turn facilitates generalization and implementation on advanced machines. Thus we regard Gaussian elimination as a procedure for computing the LU factorization of a matrix. Once this factorization is obtained, then as we have discussed, the solution to $Ax = b$ requires a pair of triangular system solves: $Ly = b$, $Ux = y$. There are practical reasons why it is important to decouple the right-hand side from the elimination process. But the curious reader will note that the transformed right-hand side in (6.3.2) is the solution to the lower triangular system

$$\begin{bmatrix} 1 & 0 & 0 \\ -2 & 1 & -0 \\ 3 & 4 & 1 \end{bmatrix} \begin{bmatrix} y_1 \\ y_2 \\ y_3 \end{bmatrix} = \begin{bmatrix} 13 \\ -28 \\ 37 \end{bmatrix}$$

After this build-up for the LU factorization, it is disturbing to note that the elimination process on which it is based can break down on some very simple examples. For example, if we modify (6.3.1) by changing the $(1,1)$ coefficient from 2 to 0,

$$\begin{array}{rcrcrcr} & & x_2 & + & 3x_3 & = & 13 \\ -4x_1 & + & 6x_2 & - & 5x_3 & = & -28 \\ 6x_1 & + & 13x_2 & + & 16x_3 & = & 37 \end{array}$$

then the elimination process defined previously cannot get off the ground because we cannot use the first equation to get rid of x_1 in the second and third equations. A simple fix is proposed in §6.2.4. Until then, we assume that the matrices under discussion submit quietly to the LU factorization process without any numerical difficulty.

6.3.2   General n

We now turn our attention to the LU factorization of a general matrix. In looking at the system

$$\begin{bmatrix} \times & \times & \times & \times & \times & \times \\ \times & \times & \times & \times & \times & \times \\ \times & \times & \times & \times & \times & \times \\ \times & \times & \times & \times & \times & \times \\ \times & \times & \times & \times & \times & \times \\ \times & \times & \times & \times & \times & \times \end{bmatrix} \begin{bmatrix} \times \\ \times \\ \times \\ \times \\ \times \\ \times \end{bmatrix} = \begin{bmatrix} \times \\ \times \\ \times \\ \times \\ \times \\ \times \end{bmatrix}$$

we see that we can eliminate the unknown x_1 from equation i by subtracting from it a_{i1}/a_{11} times equation 1. If we do this for $i = 2{:}6$, then the given linear system transforms to

$$
\begin{bmatrix}
\times & \times & \times & \times & \times & \times \\
0 & \times & \times & \times & \times & \times \\
0 & \times & \times & \times & \times & \times \\
0 & \times & \times & \times & \times & \times \\
0 & \times & \times & \times & \times & \times \\
0 & \times & \times & \times & \times & \times
\end{bmatrix}
\begin{bmatrix}
\times \\ \times \\ \times \\ \times \\ \times \\ \times
\end{bmatrix}
=
\begin{bmatrix}
\times \\ \times \\ \times \\ \times \\ \times \\ \times
\end{bmatrix} .
$$

To remove x_2 from equation i, we scale (the new) 2nd equation by a_{i2}/a_{22} and subtract from row i. Doing this for $i = 3{:}6$ gives

$$
\begin{bmatrix}
\times & \times & \times & \times & \times & \times \\
0 & \times & \times & \times & \times & \times \\
0 & 0 & \times & \times & \times & \times \\
0 & 0 & \times & \times & \times & \times \\
0 & 0 & \times & \times & \times & \times \\
0 & 0 & \times & \times & \times & \times
\end{bmatrix}
\cdot
\begin{bmatrix}
\times \\ \times \\ \times \\ \times \\ \times \\ \times
\end{bmatrix}
=
\begin{bmatrix}
\times \\ \times \\ \times \\ \times \\ \times \\ \times
\end{bmatrix} .
$$

The pattern should now be clear regarding the operations that must be performed on A:

```
for k=1:n-1
    Compute the multipliers required to eliminate x(k) from equations
        k+1 through n and store in v(k+1:n).
    Update equations k+1 through n.
end
```

The multipliers required in the kth step are specified as follows:

```
for i=k+1:n
    v(i) = A(i,k)/A(k,k);
end
```

That is, `v(k+1:n) = A(k+1:n,k)/A(k,k)`. The act of multiplying row k by $v(i)$ and subtracting from row i can be implemented with `A(i,k:n) = A(i,k:n) - v(i)*A(k,k:n)`. The column range begins at k because the first $k-1$ entries in both rows k and i are zero. Incorporating these ideas, we get the following procedure for upper triangularizing A:

```
for k=1:n-1
    v(k+1:n) = A(k+1:n,k)/A(k,k);
    for i=k+1:n
        A(i,k+1:n) = A(i,k+1:n) - v(i)*A(k,k+1:n);
    end
end
U = triu(A)
```

(6.3.3)

The iloop oversees a collection of row-oriented saxpy operations. For example, if $n = 6$ and $k = 3$, then the three row saxpys

$$
\begin{aligned}
A(4, 4{:}6) &\leftarrow A(4, 4{:}6) - v(4)A(3, 4{:}6) \\
A(5, 4{:}6) &\leftarrow A(5, 4{:}6) - v(5)A(3, 4{:}6) \\
A(6, 4{:}6) &\leftarrow A(6, 4{:}6) - v(6)A(3, 4{:}6)
\end{aligned}
$$

are equivalent to

$$
A(4{:}6, 4{:}6) \;=\;
\begin{bmatrix}
A(4, 4{:}6) \\
A(5, 4{:}6) \\
A(6, 4{:}6)
\end{bmatrix}
\leftarrow
\begin{bmatrix}
A(4, 4{:}6) \\
A(5, 4{:}6) \\
A(6, 4{:}6)
\end{bmatrix}
-
\begin{bmatrix}
v_4 \\
v_5 \\
v_6
\end{bmatrix}
A(3, 4{:}6)
$$

Thus the i-loop in (6.3.3) can be replaced by a single outer product update:

```
for k=1:n-1
   v(k+1:n) = A(k+1:n,k)/A(k,k)
   A(k+1:n,k+1:n) = A(k+1:n,k+1:n) - v(k+1:n)*A(k,k+1:n);      (6.3.4)
end
U = triu(A);
```

So much for the production of U. To compute L, we proceed as in the Hessenberg case and show that it is made up of the multipliers. In particular, during the kth pass through the loop in (6.3.2), the current A matrix is premultiplied by a multiplier matrix of the form

$$
M_3 =
\begin{bmatrix}
1 & 0 & 0 & 0 & 0 & 0 \\
0 & 1 & 0 & 0 & 0 & 0 \\
0 & 0 & 1 & 0 & 0 & 0 \\
0 & 0 & -v_4 & 1 & 0 & 0 \\
0 & 0 & -v_5 & 0 & 1 & 0 \\
0 & 0 & -v_6 & 0 & 0 & 1
\end{bmatrix}
\qquad (n = 6, k = 3)
$$

After $n - 1$ steps

$$
M_{n-1} \cdots M_2 M_1 A = U
$$

is upper triangular and so

$$
A = \left(M_1^{-1} M_2^{-1} \cdots M_{n-1}^{-1} \right) U
$$

The inverse of a multiplier matrix has a particularly simple form. For example,

$$
M_3^{-1} =
\begin{bmatrix}
1 & 0 & 0 & 0 & 0 & 0 \\
0 & 1 & 0 & 0 & 0 & 0 \\
0 & 0 & 1 & 0 & 0 & 0 \\
0 & 0 & v_4 & 1 & 0 & 0 \\
0 & 0 & v_5 & 0 & 1 & 0 \\
0 & 0 & v_6 & 0 & 0 & 1
\end{bmatrix}
$$

Moreover, $L = M_1^{-1} M_2^{-1} \cdots M_{n-1}^{-1}$ is lower triangular with the property that $L(:,k)$ is the kth column of M_k^{-1}. Thus in the $n = 6$ case we have

$$L = M_1^{-1} M_2^{-1} M_3^{-1} M_4^{-1} M_5^{-1} = \begin{bmatrix} 1 & 0 & 0 & 0 & 0 & 0 \\ v_2^{(1)} & 1 & 0 & 0 & 0 & 0 \\ v_3^{(1)} & v_3^{(2)} & 1 & 0 & 0 & 0 \\ v_4^{(1)} & v_4^{(2)} & v_4^{(3)} & 1 & 0 & 0 \\ v_5^{(1)} & v_5^{(2)} & v_5^{(3)} & v_5^{(4)} & 1 & 0 \\ v_6^{(1)} & v_6^{(2)} & v_6^{(3)} & v_6^{(4)} & v_6^{(5)} & 1 \end{bmatrix}$$

where the superscripts are used to indicate the step associated with the multiplier. This suggests that the multipliers can be stored in the locations that they are designed to zero. For example, $v_5^{(2)}$ is computed during the second step in order to zero a_{52} and it can be stored in location $(5, 2)$. This leads to the following implementation of Gaussian elimination:

```
    function [L,U] = GE(A);
%
% Pre:
%    A        n-by-n
%
% Post:
%    L        n-by-n lower triangular.
%    U        n-by-n upper triangular.
%             A = LU

    [n,n] = size(A);
    for k=1:n-1
        A(k+1:n,k) = A(k+1:n,k)/A(k,k);
        A(k+1:n,k+1:n) = A(k+1:n,k+1:n) - A(k+1:n,k)*A(k,k+1:n);
    end
    L = eye(n,n) + tril(A,-1);
    U = triu(A);
```

It can be shown that this calculation requires $2n^3/3$ flops. The script ShowGE steps through this factorization process for the matrix

$$A = \begin{bmatrix} 17 & 24 & 1 & 8 & 15 \\ 23 & 5 & 7 & 14 & 16 \\ 4 & 6 & 13 & 20 & 22 \\ 10 & 12 & 19 & 21 & 3 \\ 11 & 18 & 25 & 2 & 9 \end{bmatrix}$$

and terminates with

$$
L = \begin{bmatrix}
1.0000 & 0 & 0 & 0 & 0 \\
1.3529 & 1.0000 & 0 & 0 & 0 \\
0.2353 & -0.0128 & 1.0000 & 0 & 0 \\
0.5882 & 0.0771 & 1.4003 & 1.0000 & 0 \\
0.6471 & -0.0899 & 1.9366 & 4.0578 & 1.0000
\end{bmatrix}
$$

and

$$
U = \begin{bmatrix}
17.0000 & 24.0000 & 1.0000 & 8.0000 & 15.0000 \\
0 & -27.4706 & 5.6471 & 3.1765 & -4.2941 \\
0 & 0 & 12.8373 & 18.1585 & 18.4154 \\
0 & 0 & 0 & -9.3786 & -31.2802 \\
0 & 0 & 0 & 0 & 90.1734
\end{bmatrix}
$$

Together with LTriSol and UTriSol, GE can be used to solve a linear system $Ax = b$:

```
[L,U] = GE(A);
y = LTriSol(L,b);
x = UTriSol(U,y)
```

But this assumes that no zero divides arise during the execution of GE.

6.3.3   Stability

The time has come to address the issue of breakdown in the Gaussian elimination process. We start with the grim fact that a matrix need not have an LU factorization. To see this, equate coefficients in

$$
\begin{bmatrix} 0 & 1 \\ 1 & 1 \end{bmatrix} = \begin{bmatrix} 1 & 0 \\ \ell_{21} & 1 \end{bmatrix} \begin{bmatrix} u_{11} & u_{12} \\ 0 & u_{22} \end{bmatrix}
$$

Equality in the $(1,1)$ position implies that $u_{11} = 0$. But then it is impossible to have agreement in the $(2,1)$ position since we must have $\ell_{21} u_{11} = 1$. Note that there is nothing "abnormal" about an $Ax = b$ problem in which $a_{11} = 0$. For example,

$$
\begin{bmatrix} 0 & 1 \\ 1 & 1 \end{bmatrix} \begin{bmatrix} x_1 \\ x_2 \end{bmatrix} = \begin{bmatrix} 1 \\ 2 \end{bmatrix}
$$

has solution $x = [1\ 1]^T$. It looks like corrective measures are needed to handle the undefined multiplier situation.

But problems also arise if the multipliers are large:

$$
A = \begin{bmatrix} \delta & 1 \\ 1 & 1 \end{bmatrix} = \begin{bmatrix} 1 & 0 \\ 1/\delta & 1 \end{bmatrix} \begin{bmatrix} \delta & 1 \\ 0 & 1 - \frac{1}{\delta} \end{bmatrix} = LU \qquad (\delta \neq 0)
$$

The following script solves

$$
Ax = \begin{bmatrix} \delta & 1 \\ 1 & 1 \end{bmatrix} \begin{bmatrix} x_1 \\ x_2 \end{bmatrix} = \begin{bmatrix} 1 + \delta \\ 2 \end{bmatrix} = b
$$

by computing $A = LU$ and then solving $Ly = b$ and $Ux = y$ in the usual fashion:

```
% Script File: NoPivot
%
% Examines solution to
%
%         [ delta 1 ; 1 1] [x1;x2] = [1+delta;2]
%
% for a sequence of diminishing delta values.
%
   clc
   home
   disp(' Delta              x(1)                        x(2)   ' )
   disp('-----------------------------------------------------')
   for delta = logspace(-2,-18,9)
      A = [delta 1; 1 1];
      b = [1+delta; 2];
      L = [ 1 0; A(2,1)/A(1,1) 1];
      U = [ A(1,1) A(1,2) ; 0 A(2,2)-L(2,1)*A(1,2)];
      y(1) = b(1);
      y(2) = b(2) - L(2,1)*y(1);
      x(2) = y(2)/U(2,2);
      x(1) = (y(1) - U(1,2)*x(2))/U(1,1);
      disp(sprintf(' %5.0e   %20.15f %20.15f',delta,x(1),x(2)))
   end
```

Here are the results:

Delta	x(1)	x(2)
1e-02	1.000000000000001	1.000000000000000
1e-04	0.999999999999890	1.000000000000000
1e-06	1.000000000028756	1.000000000000000
1e-08	0.999999993922529	1.000000000000000
1e-10	1.000000082740371	1.000000000000000
1e-12	0.999866855977416	1.000000000000000
1e-14	0.999200722162641	1.000000000000000
1e-16	2.220446049250313	1.000000000000000
1e-18	0.000000000000000	1.000000000000000

(You might want to deduce why $\hat{x}_2$ is exact.) A simple way to avoid this degradation is to introduce *row interchanges*. For our preceding example, this means we apply Gaussian elimination to compute the LU factorization of A with its rows reversed:

$$\begin{bmatrix} \delta & 1 \\ 1 & 1 \end{bmatrix} \begin{bmatrix} x_1 \\ x_2 \end{bmatrix} = \begin{bmatrix} 2 \\ 1+\delta \end{bmatrix}$$

A full precision answer is obtained.

6.3.4   Pivoting

To anticipate the row interchange process in the general case, we consider the third step in the $n = 6$ case. At the beginning of the step we face a partially reduced A that has the following form:

$$A = \begin{bmatrix} \times & \times & \times & \times & \times & \times \\ 0 & \times & \times & \times & \times & \times \\ 0 & 0 & a_{33} & \times & \times & \times \\ 0 & 0 & a_{43} & \times & \times & \times \\ 0 & 0 & a_{53} & \times & \times & \times \\ 0 & 0 & a_{63} & \times & \times & \times \end{bmatrix}$$

(Note that the displayed a_{ij} are not the original a_{ij}; they have been updated twice.) Ordinarily, we would use multipliers a_{43}/a_{33}, a_{53}/a_{33}, and a_{63}/a_{33} to zero entries $(4,3)$, $(5,3)$, and $(6,3)$ respectively. *Wouldn't it be nice if $|a_{33}|$ was the largest entry in $A(3{:}6,3)$?* That would ensure that all the multipliers are less than or equal to 1. This suggests that at the beginning of the kth step we swap row k and row q where it is assumed that a_{qk} has the largest absolute value of any entry in $A(k{:}n,k)$. When we emerge from this process we will have in hand the LU factorization of a row permuted version of A:

```
    function [L,U,piv] = GEpiv(A);
%
% Pre:
%   A        n-by-n
% Post:
%   L        n-by-n unit lower triangular with |L(i,j)|<=1.
%   U        n-by-n upper triangular
%   piv      integer n-vector that is a permutation of 1:n.
%            A(piv,:) = LU

    [n,n] = size(A);
    piv = 1:n;
    for k=1:n-1
       [maxv,r] = max(abs(A(k:n,k)));
       q = r+k-1;
       piv([k q]) = piv([q k]);
       A([k q],:) = A([q k],:);
       if A(k,k) ~= 0
          A(k+1:n,k) = A(k+1:n,k)/A(k,k);
          A(k+1:n,k+1:n) = A(k+1:n,k+1:n) - A(k+1:n,k)*A(k,k+1:n);
       end
    end
    L = eye(n,n) + tril(A,-1);
    U = triu(A);
```

A number of details need to be discussed. Applied to a vector, the MATLAB max function returns the largest entry and its index. Thus the preceding max computation assigns the largest of the numbers $|a_{kk}|, \ldots, |a_{nk}|$ to maxv and its index to r. However, the rth index of the length $n - k + 1$

vector $A(k{:}n, n)$ identifies an entry from row $r + k - 1$ of A. Thus to pick up the right row index, we need the adjustment q = r+k-1. With q so defined, rows q and k are swapped. Note that by swapping *all* of these two rows, some of the earlier multipliers are swapped. This ensures that the multipliers used in the elimination of unknowns from a given equation "stay with" that equation as it is reindexed.

The last issue to address concerns the recording of the interchanges. An integer vector piv(1:n) is used. It is initialized to 1:n. Every time rows are swapped in A, the corresponding entries in piv are swapped. Upon termination, piv(k) is the index of the equation that is now the kth equation in the permuted system. The integer vector piv is a representation of a *permutation matrix* P. A permutation matrix is obtained by reordering the rows of the identity matrix. If piv = [3 1 5 4 2], then it represents the permutation

$$
P = \begin{bmatrix}
0 & 0 & 1 & 0 & 0 \\
1 & 0 & 0 & 0 & 0 \\
0 & 0 & 0 & 0 & 1 \\
0 & 0 & 0 & 1 & 0 \\
0 & 1 & 0 & 0 & 0
\end{bmatrix}
$$

Observe that

$$
P \begin{bmatrix} b_1 \\ b_2 \\ b_3 \\ b_4 \\ b_5 \end{bmatrix} = \begin{bmatrix} b_3 \\ b_1 \\ b_5 \\ b_4 \\ b_2 \end{bmatrix} = b(piv)
$$

Likewise, $PA = A(piv, :)$. Thus GEpiv computes a factorization of the form $PA = LU$. To solve a linear system $Ax = b$ using GEpiv, we notice that x also satisfies $(PA)x = (Pb)$. Thus if $PA = LU$, $Ly = Pb$, and $Ux = y$, then $Ax = b$. Using the piv representation, the three-step process takes the following form:

```
[L,U,piv] = GEpiv(A);
y = LTriSol(L,b(piv));
x = UTriSol(U,y);
```

Here is the situation with respect to flops:

Operation	Procedure	Flops
$PA = LU$	GEpiv	$2n^3/3$
$Ly = b$	LTriSol	n^2
$Ux = y$	UTriSol	n^2

Thus the factorization dominates the overall computation for large n. Moreover, the pivoting amounts to an $O(n^2)$ overhead and so the stabilization purchased by the row swapping does not seriously affect efficiency.

The script `ShowGEpiv` steps through the factorization applied to the matrix.

$$A = \begin{bmatrix} 17 & 24 & 1 & 8 & 15 \\ 23 & 5 & 7 & 14 & 16 \\ 4 & 6 & 13 & 20 & 22 \\ 10 & 12 & 19 & 21 & 3 \\ 11 & 18 & 25 & 2 & 9 \end{bmatrix}$$

producing

$$L = \begin{bmatrix} 1.0000 & 0 & 0 & 0 & 0 \\ 0.7391 & 1.0000 & 0 & 0 & 0 \\ 0.1739 & 0.2527 & 1.0000 & 0 & 0 \\ 0.4348 & 0.4839 & 0.7231 & 1.0000 & 0 \\ 0.4783 & 0.7687 & 0.5164 & 0.9231 & 1.0000 \end{bmatrix}$$

$$U = \begin{bmatrix} 23.0000 & 5.0000 & 7.0000 & 14.0000 & 16.0000 \\ 0 & 20.3043 & -4.1739 & -2.3478 & 3.1739 \\ 0 & 0 & 24.8608 & -2.8908 & -1.0921 \\ 0 & 0 & 0 & 19.6512 & 18.9793 \\ 0 & 0 & 0 & 0 & -22.2222 \end{bmatrix}$$

and

$$piv = \begin{bmatrix} 2 & 1 & 5 & 3 & 4 \end{bmatrix}$$

Notice that entries in L are all ≤ 1 in absolute value.

6.3.5   The LU Mentality

It is important to interpret formulas that involve the inverse of a matrix in terms of the LU factorization. Consider the problem of computing

$$\alpha = c^T A^{-1} d$$

where $A \in \mathbb{R}^{n \times n}$ is nonsingular and $c, d \in \mathbb{R}^n$. Note that α is a scalar because it is the consequence of a dot product between the vectors c and $A^{-1}d$. At first glance, it looks like we actually need to compute the inverse of A. But the preferred approach is to recognize that $A^{-1}d$ is the solution to a linear system $Ax = d$ and so

```
[L,U,piv] = GEpiv(A);
y = LTriSol(L,d(piv));
x = UTriSol(U,y);
alpha = c'*x;
```

This is more efficient by a factor of 2 because the explicit formation of A^{-1} requires about $4n^3/3$ flops.

But a more dramatic payoff for "thinking LU" arises when several linear systems must be solved that involve the same matrix of coefficients. For example, suppose $v^{(0)}$ is a given n-vector and that for $j = 1{:}k$ we want to compute the solution $v^{(j)}$ to the system $Av = v^{(j-1)}$. For example, if $k = 4$ we need to solve the systems

$$
\begin{aligned}
Av^{(1)} &= v^{(0)} \\
Av^{(2)} &= v^{(1)} \\
Av^{(3)} &= v^{(2)} \\
Av^{(4)} &= v^{(3)}
\end{aligned}
$$

If we agree to assemble these solutions column by column in a matrix V, we obtain

```
b = v0;
V = zeros(n,k);
for j=1:k
    [L,U,piv] = GEpiv(A);
    y = LTriSol(L,b(piv));
    V(:,k) = UTriSol(U,y);
    b = V(:,k);
end
```

This approach requires $k(2n^3/3)$ flops since there are k applications of GEpiv. On the other hand, why repeat all the Gaussian elimination operations on A every time through the loop, since they are independent of k? A much more efficient approach is to factor A once and then "live off factors" as the linear systems come in the door:

```
b = v0;
V = zeros(n,k);
[L,U,piv] = GEpiv(A);
for j=1:k
    y = LTriSol(L,b(piv));
    V(:,k) = UTriSol(U,y);
    b = V(:,k);
end
```

This implementation involves $(2n^3/3 + O(kn^2))$ flops.

6.3.6 The MATLAB Linear System Tools

The built-in function LU can also be used to compute the $PA = LU$ factorization. A call of the form

```
[L,U,P] = LU(A)
```

returns the lower triangular factor in L, the upper triangular factor in U, and an explicit representation of the permutation matrix in P. The command piv = P*(1:n)' assigns to piv the same integer vector representation of P that is used in GEpiv.

We mention that x = A\b produces the same x as

```
[L,U,P] = LU(A);
y = LTriSol(L,P*b);
x = UTriSol(U,y);
```

Problems

P6.3.1 Use GEpiv, LTriSol, and UTriSol so compute $X \in \mathbf{R}^{n \times p}$ so that $AX = B$, where $A \in \mathbf{R}^{n \times n}$ is nonsingular and $B \in \mathbf{R}^{n \times p}$.

P6.3.2 Go into ShowGEpiv and modify the interchange strategy so that in the kth step, no interchange is performed if the current $|A(k, k)|$ is greater than or equal to largest value in $\alpha |A(k{:}n, k)|$, where $0 < \alpha \le 1$. If this is not the case, the largest value in $A(k{:}n, k)$ should be swapped into the (k, k) position. Try different values of α and observe the effect.

P6.3.3 How could GEpiv, LTriSol, and UTriSol be used to compute the (i, i) entry of A^{-1}?

P6.3.4 Assume that A, B, and C are given n-by-n matrices and that A and C are nonsingular. Assume that g and h are given n-by-1 vectors. Write a MATLAB fragment that computes n-vectors y and z so that both of the following equations hold:

$$
\begin{aligned}
Ay + Bz &= g \\
Cz &= h
\end{aligned}
$$

You may use the \ operator. Efficiency matters.

P6.3.5 Assume that A, B, and C are nonsingular n-by-n matrices and that f is an n-by-1 vector. Write an efficient MATLAB fragment that computes a vector x so that $ABCx = f$.

6.4   Analysis

We now turn our attention to the quality of the computed solution produced by Gaussian elimination with partial pivoting.

6.4.1   Residual Versus Error

Consider the following innocuous linear system:

$$
\begin{bmatrix} .780 & .563 \\ .913 & .659 \end{bmatrix} \begin{bmatrix} x_1 \\ x_2 \end{bmatrix} = \begin{bmatrix} .217 \\ .254 \end{bmatrix}
$$

Suppose we apply two different methods and get two different solutions:

$$
x^{(1)} = \begin{bmatrix} .341 \\ -.087 \end{bmatrix} \qquad x^{(2)} = \begin{bmatrix} .999 \\ -1.00 \end{bmatrix}
$$

Which is preferred? An obvious way to compare the two solutions is to compute the associated *residuals*:

$$
b - Ax^{(1)} = \begin{bmatrix} .0000001 \\ 0 \end{bmatrix} \qquad b - Ax^{(2)} = \begin{bmatrix} .001343 \\ .001572 \end{bmatrix}
$$

On the basis of residuals, it is clear that $x^{(1)}$ is preferred. However, it is easy to verify that the exact solution is given by

$$
x^{(exact)} = \begin{bmatrix} 1 \\ -1 \end{bmatrix}
$$

and this creates a dilemma. We see that $x^{(1)}$ renders a small residual while $x^{(2)}$ is much more accurate.

Reasoning in the face of such a dichotomy requires an understanding about the context in which the linear system arises. We may be in a situation where how well Ax predicts b is paramount. In this case small residuals are important. In other settings, accuracy is critical and the focus is on nearness to the true solution.

It is clear from the discussion that (1) the notion of a "good" solution can be ambiguous and (2) the intelligent appraisal of algorithms requires a sharper understanding of the mathematics behind the $Ax = b$ problem.

6.4.2   Problem Sensitivity and Nearness

In the preceding 2-by-2 problem, the matrix A is very close to singular. Indeed,

$$\tilde{A} = \begin{bmatrix} .780 & .563001095\ldots \\ .913 & .659 \end{bmatrix}$$

is exactly singular. Thus an $O(10^{-6})$ perturbation of the data renders our problem insoluble. Our intuition tells us that difficulties should arise if our given $Ax = b$ problem is "near" to a singular $Ax = b$ problem. In that case we suspect that small changes in the problem data (i.e., A and b) will induce relatively large changes in the solution. It is clear that we need to quantify such notions as "nearness to singularity" and "problem sensitivity."

We remark that these issues have *nothing* to do with the underlying algorithms. They are mathematical concepts associated with the $Ax = b$ problem. However, these concepts do clarify what we can expect from an algorithm in light of rounding errors.

To appreciate this point consider the hypothetical situation in which there is *no* roundoff during the entire solution process except when A and b are stored. This means from Theorem 5 that the computed solution $\hat{x}$ satisfies the perturbed linear system

$$(A + E)\hat{x} = b + f \tag{6.4.1}$$

where $\| E \|_1 \le \text{eps} \| A \|_1$ and $\| f \|_1 \le \text{eps} \| b \|_1$. Two fundamental questions arise: How can we guarantee that $A + E$ is nonsingular, and how close is $\hat{x}$ to the true solution x? The answer to both of these questions involves the quantity

$$\kappa_1(A) = \| A \|_1 \| A^{-1} \|_1$$

which is called the *condition number* of A (in the 1-norm). It can be shown that

- $\kappa_1(A) \ge 1$.

- $\kappa_1(\alpha A) = \kappa_1(A)$.

- $\kappa_1(A)$ is large if A is close to singular.

The last point is affirmed by the preceding 2-by-2 example.

$$A = \begin{bmatrix} .780 & .563 \\ .913 & .659 \end{bmatrix} \Rightarrow A^{-1} = 10^6 \begin{bmatrix} .659 & -.563 \\ -.913 & .780 \end{bmatrix} \Rightarrow \kappa_1(A) \approx 10^6$$

The following theorem uses the condition number to describe properties of the stored linear system (6.4.1).

Theorem 6 *Suppose $A \in \mathbb{R}^{n \times n}$ is nonsingular and that $Ax = b$, where $x, b \in \mathbb{R}^n$. If*

$$\text{eps } \kappa_1(A) < 1$$

then the stored linear system $\text{fl}(A)\hat{x} = \text{fl}(b)$ *is nonsingular and*

$$\frac{\| \hat{x} - x \|_1}{\| x \|_1} \leq \text{eps } \kappa_1(A) \left(1 + \frac{\| \hat{x} \|_1}{\| x \|_1} \right)$$

Proof. From Theorem 5 we know that $\text{fl}(A) = A + E$ and $\text{fl}(b) = b + f$, where $\| E \|_1 \leq \text{eps } \| A \|_1$ and $\| f \|_1 \leq \text{eps } \| b \|_1$. If $A + E$ is singular, then there exists a nonzero vector z so $(A + E)z = 0$. This implies that $z = -A^{-1}Ez$ and so

$$\| z \|_1 = \| A^{-1}Ez \|_1 \leq \| A^{-1} \|_1 \| A \|_1 \| z \|_1$$

But this contradicts the assumption that $\text{eps} \cdot \kappa_1(A) < 1$ and shows that the system $(A + E)\hat{x} = b + f$ is nonsingular.

Since $Ax = b$, it follows that $A(\hat{x} - x) = f - E\hat{x}$ and so $\hat{x} - x = A^{-1}f - A^{-1}E\hat{x}$. Thus

$$\| \hat{x} - x \| \leq \| A^{-1} \|_1 \| f \|_1 + \| A^{-1} \|_1 \| E \|_1 \| \hat{x} \|_1$$

$$\leq \text{eps } \| A^{-1} \|_1 \| b \|_1 + \text{eps } \| A \|_1 \| A^{-1} \|_1 \| \hat{x} \|_1$$

The theorem is established by dividing both sides by $\| x \|_1$ and observing that

$$\frac{\| b \|_1}{\| x \|_1} = \frac{\| Ax \|_1}{\| x \|_1} \leq \frac{\| A \|_1 \| x \|_1}{\| x \|_1} = \| A \|_1 \quad \square.$$

There are a number of things to say about this result. To begin with, the factor $\tau = \text{eps} \cdot \kappa_1(A)$ has a key roll to play. If this quantity is close to 1, then it is appropriate to think of A as *numerically singular*. If A is not numerically singular, then it is possible to show that the quotient $\| \hat{x} \|_1 / \| x \|_1$ contributes little to the upper bound. The main contribution of the theorem is thus to say that the 1-norm relative error in $\hat{x}$ is essentially bounded by τ.

In our discussion of $Ax = b$ sensitivity, we have been using the 1-norm. Condition numbers can also be defined in terms of the other norms mentioned in §5.2.4. The 2-norm condition number can be computed, at some cost, using the function cond. A cheap estimate of $\kappa_1(A)$ can be obtained with condest.

6.4.3 Backward Stability

It turns out that Gaussian elimination with pivoting produces a computed solution $\hat{x}$ that satisfies

$$(A + E)\hat{x} = b$$

where

$$\| E \| \approx \text{eps} \| A \|.$$

(It does not really matter which of the preceding norms we use, so the subscript on the norm symbol has been deleted.) This says that the Gaussian elimination solution is essentially as good as the "ideal" solution that we discussed earlier (i.e., $\hat{x}$ solves a nearby problem exactly).

Two important heuristics follow from this:

$$\| A\hat{x} - b \| \approx \mathbf{eps} \ \| A \| \ \| \hat{x} \| \tag{6.4.3}$$

$$\frac{\| \hat{x} - x \|}{\| x \|} \approx \mathbf{eps} \ \kappa(A) \tag{6.4.4}$$

The first essentially says that the algorithm produces small residuals compared to the size of A and $\hat{x}$. The second heuristic says that if the unit roundoff and condition satisfy $\mathbf{eps} \approx 10^{-t}$, and $\kappa_1(A) \approx 10^p$, then (roughly) $\tilde{x}$ has approximately $t - p$ correct digits.

The function GE2 can be used to illustrate these results. It solves a given 2-by-2 linear system in simulated three-digit arithmetic. Here is a sample result:

```
Stored A    =    .981x10^0    .726x10^0
                 .529x10^0    .384x10^0

Computed L =    .100x10^1    .000x10^0
                 .539x10^0    .100x10^1

Computed U =    .981x10^0    .726x10^0
                 .000x10^0   -.700x10^-2

Exact b          =   .255x10^0
                     .145x10^0

Exact Solution   =   .100x10^1
                    -.100x10^1

Computed Solution =  .110x10^1
                    -.114x10^1

cond(A) = 2.608e+02

Computed solution solves (A+E)*x = b where

         E =        0.001552     -0.001608
                    0.000377     -0.000391
```

Note that in this environment, the unit roundoff $\mathbf{eps}$ is approximately 10^{-3}. The script file ShowGE2 can be used to examine further examples.

The Pascal matrices provide another interesting source of test problems. Pascal(n) returns the n-by-n Pascal matrix, which has integer entries and is increasingly ill conditioned as n grows:

n	Condition of P_n
4	6.9e+02
8	2.0e+07
12	8.7e+11
16	4.2e+16

The script file CondEgs examines what happens when Gaussian elimination with pivoting is applied to a sequence of Pascal linear systems that are set up to have the vector of all ones as solution. Here is the output for $n = 12$:

```
cond(pascal(12)) = 8.7639e+11
True solution is vector of ones.


x =
        0.99999998079317
        1.00000022061699
        0.99999887633133
        1.00000337593939
        0.99999331477863
        1.00000919709849
        0.99999100757082
        1.00000625939438
        0.99999695671209
        1.00000098506337
        0.99999980884814
        1.00000001685318


Relative error = 4.7636e-06
Predicted value = EPS*cond(A) = 1.9460e-04
```

Notice that $-\log_{10}(\text{eps}\kappa(P_n))$ provides a reasonable upper bound for the number of correct significant digits in the computed solution.

Another nice feature of the Pascal matrices is that they have determinant 1 for any n. The preceding discussion confirms the *irrelevance* of the determinant in matters of linear system sensitivity and accuracy. The determinant usually figures quite heavily in any introduction to linear algebra. For example,

$$\det(A) = 0 \iff A \text{ singular}$$

It is natural to think that

$$\det(A) \approx 0 \iff A \text{ approximately singular}$$

However, this is not the case. The Pascal example shows that nearly singular matrices can have determinant 1. Diagonal matrices of the form $D = \alpha I_n$ have unit condition number but determinant α^n, further weakening the correlation between determinant size and condition.

Problems

P6.4.1 Gaussian elimination with pivoting is used to solve a 2-by-2 system $Ax = b$ on a computer with machine precision 10^{-16}. It is known that $\| A \|_1 \| A^{-1} \|_1 \approx 10^{10}$ and that the exact solution is given by

$$x = \left[\begin{array}{c} 1.234567890123456 \\ .0000123456789012 \end{array} \right]$$

Underline the digits in x_1 and x_2 that can probably agree with the corresponding digits in the computed solution. Explain the heuristic assumptions used to answer the question.

P6.4.2 It is known that the components of the exact solution to a linear system $Ax = b$ range from 10^{-1} to 10^3 and that $cond(A)$ is about 10^4. What must the machine precision be in order to ensure that the smallest component of

```
x = A\b
```

has at least five significant digits of accuracy? No proof is necessary, just a reasonable heuristic argument.

M-Files and References

Script Files

ShowTri	Uses LTriSolM to get inverse of Forsythe matrix.
ShowTriD	Illustrates Tridiagonal system solving.
ShowHessLU	Illustrates the Hessenberg LU factorization.
ShowGE	Illustrates GE applied to a 5-by-5 system.
ShowGEpiv	Illustrates GEpiv applied to a 5-by-5 system.
ShowGE2	Applies GE2 to three different examples.
CondEgs	Accuracy of some ill-conditioned Pascal linear systems.

Function Files

LTriSol	Solves lower triangular system $Lx = b$.
UTriSol	Solves upper triangular system $Ux = b$.
LTriSolM	Solves multiple right-hand-side lower triangular systems.
TriDiLU	LU factorization of a tridiagonal matrix.
LBiDiSol	Solves lower bidiagonal systems.
UBiDiSol	Solves upper bidiagonal systems.
HessLU	Hessenberg LU factorization.
GE	General LU factorization without pivoting.
GEpiv	General LU factorization with pivoting.
GE2	Illustrates 2-by-2 GE in three-digit arithmetic.

References

G.E. Forsythe and C. Moler (1967). *Computer Solution of Linear Algebraic Systems,* Prentice-Hall, Englewood Cliffs, NJ.

W.W. Hager (1988). *Applied Numerical Linear Algebra,* Prentice-Hall, Englewood Cliffs, NJ.

G.W. Stewart (1973). *Introduction to Matrix Computations,* Academic Press, New York.

G.H. Golub and C.F. Van Loan (1996). *Matrix Computations, Third Edition,* Johns Hopkins University Press, Baltimore, MD.

Chapter 7

The QR and Cholesky Factorizations

§**7.1** Least Squares Fitting

§**7.2** The QR Factorization

§**7.3** The Cholesky Factorization

§**7.4** High-Performance Cholesky

The solution of overdetermined systems of linear equations is central to computational science. If there are more equations than unknowns in $Ax = b$, then we must lower our aim and be content to make Ax close to b. Least squares fitting results when the 2-norm of $Ax - b$ is used to quantify success. In §7.1 we introduce the least squares problem and solve a simple fitting problem using built-in MATLAB features.

In §7.2 we present the QR factorization and show how it can be used to solve the least squares problem. Orthogonal rotation matrices are at the heart of the method and represent a new class of transformations that can be used to introduce zeros into a matrix.

The solution of systems of linear equations with symmetric positive definite coefficient matrices is discussed in §7.3 and a special "symmetric version" of the LU factorization called the Cholesky factorization is introduced. Several different implementations are presented that stress the importance of being able to think at the matrix-vector level. In the last section we look at two Cholesky implementations that have appeal in advanced computing environments.

7.1 Least Squares Fitting

It is not surprising that a square nonsingular system $Ax = b$ has a unique solution: there are exactly the same number of unknowns as equations. On the other hand, if we have more equations than unknowns, then it may not be possible to find an x so that $Ax = b$. Consider the 3-by-2 case:

$$
\begin{bmatrix} a_{11} & a_{12} \\ a_{21} & a_{22} \\ a_{31} & a_{32} \end{bmatrix} \begin{bmatrix} x_1 \\ x_2 \end{bmatrix} = \begin{bmatrix} b_1 \\ b_2 \\ b_3 \end{bmatrix}
$$

For this *overdetermined* $Ax = b$ problem to have a solution, it is necessary for b to be in the span of A's two columns. This is not a forgone conclusion since this span is a proper subspace of $\mathbb{R}^3$. For example, if we try to find x_1 and x_2 so that

$$
\begin{bmatrix} 1 & 2 \\ 3 & 4 \\ 5 & 6 \end{bmatrix} \begin{bmatrix} x_1 \\ x_2 \end{bmatrix} = \begin{bmatrix} 0 \\ 1 \\ 1 \end{bmatrix}
$$

then the first equation says that $x_1 = -2x_2$. Substituting this into the second equation implies $1 = 3x_1 + 4x_2 = 3(-2x_2) + 4x_2 = -2x_2$, while substituting it into equation 3 says that $1 = 5x_1 + 6x_2 = 5(-2x_2) + 6x_2 = -4x_2$. Since the requirements $x_2 = -1/2$ and $x_2 = -1/4$ conflict, the system has no solution.

So with more equations than unknowns we need to adjust our aims. Instead of trying to "reach" b with Ax, we try to get as close as possible. Vector norms can be used to quantify the degree of success. If we work with the 2-norm, then we obtain this formulation:

Given $A \in \mathbb{R}^{m \times n}$ and $b \in \mathbb{R}^m$, find $x \in \mathbb{R}^n$ to minimize $\| Ax - b \|_2$

This is the *least squares* (LS) problem, apt terminology because the 2-norm involves a sum of squares:

$$
\| Ax - b \|_2 = \sqrt{ \sum_{i=1}^{m} \left(A(i,:)x - b(i) \right)^2 }
$$

The goal is to minimize the discrepancies in each equation:

$$
\left(A(i,:)x - b(i) \right)^2 = \left(a_{i1}x_1 + \cdots + a_{in}x_n - b_i \right)^2
$$

From the column point of view, the goal is to find a linear combination of A's columns that gets as close as possible to b in the 2-norm sense.

7.1.1   Setting Up Least Squares Problems

LS fitting problems often arise when scientists attempt to fit a model to experimentally obtained data. Suppose a biologist conjectures that plant height h is a function of four soil nutrient concentrations a_1, a_2, a_3, and a_4:

$$
h = a_1 x_1 + a_2 x_2 + a_3 x_3 + a_4 x_4.
$$

This is a *linear model*, and x_1, x_2, x_3, and x_4 are *model parameters* whose value must be determined. To that end, the biologist performs m (a large number) experiments. The ith experiment consists of establishing the four nutrient values a_{i1}, a_{i2}, a_{i3} and a_{i4} in the soil, planting the seed, and observing the resulting height h_i. If the model is perfect, then

$$
h_i = a_{i1}x_1 + a_{i2}x_2 + a_{i3}x_3 + a_{i4}x_4
$$

for $i = 1{:}m$. That is,

$$\begin{bmatrix} a_{11} & a_{12} & a_{13} & a_{14} \\ a_{21} & a_{22} & a_{23} & a_{24} \\ a_{31} & a_{32} & a_{33} & a_{34} \\ a_{41} & a_{42} & a_{43} & a_{44} \\ \vdots & \vdots & \vdots & \vdots \\ a_{m1} & a_{m2} & a_{m3} & a_{m4} \end{bmatrix} \begin{bmatrix} x_1 \\ x_2 \\ x_3 \\ x_4 \end{bmatrix} = \begin{bmatrix} h_1 \\ h_2 \\ h_3 \\ h_4 \\ \vdots \\ h_m \end{bmatrix}$$

Of course, the model will not be perfect, making it impossible to find such an x. The aims of the biologist are lowered and the minimizer of $\| Ax - h \|_2$ is sought. If the minimum sum of squares is small, then the biologist has reason to be happy with the chosen model. Otherwise, additional factors may be brought into play (e.g., a fifth nutrient or the amount of sunlight). The linear model may be exchanged for a nonlinear one with twice the number of parameters: $h = c_1 e^{-x_1 a_1} + \cdots + c_4 e^{-x_1 a_4}$. The treatment of such problems is briefly discussed in the next chapter.

LS fitting also arises in the approximation of known functions. Suppose the designer of a built-in square root function needs to develop an approximation to the function $f(x) = \sqrt{x}$ on the interval $[.25, 1]$. A linear approximation of the form $\ell(x) = \alpha + \beta x$ is sought. [Think of $\ell(x)$ as a two-parameter model with parameters α and β.] We could set this function to be just the linear interpolant of f at two well-chosen points. Alternatively, if a partition

$$.25 = x_1 < \cdots < x_m = 1$$

is given, then the parameters α and β can be chosen so that the quantity

$$\phi_m(\alpha, \beta) = \sum_{i=1}^{m} \left[(\alpha + \beta x_i) - \sqrt{x_i} \right]^2$$

is minimized. Note that if we set $f_i = \sqrt{x_i}$, then in the language of matrices, vectors, and norms we have

$$\phi_m(\alpha, \beta) = \left\| \begin{bmatrix} 1 & x_1 \\ 1 & x_2 \\ \vdots & \vdots \\ 1 & x_m \end{bmatrix} \begin{bmatrix} \alpha \\ \beta \end{bmatrix} - \begin{bmatrix} f_1 \\ f_2 \\ \vdots \\ f_m \end{bmatrix} \right\|_2^2$$

Thus an m-by-2 least squares problem needs to be solved in order to resolve α and β.

It is important to recognize that any norm could be used to quantify the error in the fit of a model. However, the 2-norm is particularly important for two reasons: (1) In many experimental settings the fitting errors are normally distributed. The underlying statistics can then be used to make a rigorous case for 2-norm minimization. (2) The mathematics of LS fitting is rich and supportive of interesting and powerful algorithms. We examine some of those methods in the following.

7.1.2 MATLAB's Least Squares Tools

The backslash operator can be used to solve the LS problem in MATLAB once it is cast in the matrix/vector terms (i.e., $\min \| Ax - b \|_2$). Here is a script that solves the square root fitting problem mentioned in the previous subsection:

```
% Script File: ShowLSFit
%
% Displays the LS fits to the function f(x) = sqrt(x) on [.25,1]

    close all
    z = linspace(.25,1);
    fz = sqrt(z);
    for m = [2 100 ]
        x = linspace(.25,1,m)';
        A = [ones(m,1) x];
        b = sqrt(x);
        xLS = A\b;
        alpha = xLS(1);
        beta  = xLS(2);
        figure
        plot(z,fz,z,alpha+beta*z)
        title(sprintf('m = %2.0f,   alpha = %10.6f,   beta = %10.6f',m,alpha,beta))
    end
```

The two fits are displayed in Fig 7.1. Note that if $m = 2$, then we just obtain the interpolant to the square root function at $x = .25$ and $x = 1.00$. For large m it follows from the rectangle rule that

$$\frac{.75}{m} \sum_{i=1}^{m} [(\alpha + \beta x_i) - \sqrt{x_i}]^2 \approx \int_{.25}^{1} [(\alpha + \beta x) - \sqrt{x}]^2 dx \equiv \phi_\infty(\alpha, \beta)$$

where $x_i = .25 + .75(i-1)/(m-1)$ for $i = 1{:}m$. Thus as $m \to \infty$ the minimizer of $\phi_m(\alpha, \beta)$ converges to the minimizer of $\phi_\infty(\alpha, \beta)$. From the equations

$$\frac{\partial \phi_\infty}{\partial \alpha} = 0 \qquad \frac{\partial \phi_\infty}{\partial \beta} = 0$$

we are led to a 2-by-2 linear system that specifies the α_* and β_* that minimize $\phi_\infty(\alpha, \beta)$:

$$\begin{bmatrix} \frac{3}{4} & \frac{15}{32} \\ \frac{15}{32} & \frac{21}{64} \end{bmatrix} \begin{bmatrix} \alpha_* \\ \beta_* \end{bmatrix} = \begin{bmatrix} \frac{7}{12} \\ \frac{31}{80} \end{bmatrix}$$

The solution is given by

$$\begin{bmatrix} \alpha_* \\ \beta_* \end{bmatrix} = \begin{bmatrix} 0.370370370 \\ 0.651851851 \end{bmatrix}$$

In general, if we try to fit data points $(x_1, f_1), \ldots, (x_m, f_m)$ in the least squares sense with a polynomial of degree d, then an m-by-$(d+1)$ least squares problem arises. The MATLAB function Polyfit can be used to solve this problem. (See also Polyval.)

Problems

P7.1.1 Consider the problem

$$\min_{x \in \mathbb{R}} \left\| \begin{bmatrix} 1 \\ 1 \\ 1 \end{bmatrix} x - \begin{bmatrix} b_1 \\ b_2 \\ b_3 \end{bmatrix} \right\|_p$$

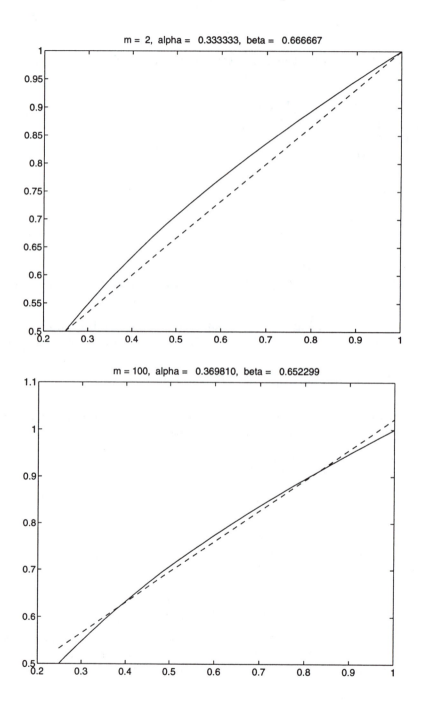

FIGURE 7.1 The LS Fitting of a Line to $\sqrt{x}$.

where $p = 1, 2$, or ∞. Suppose $b_1 \geq b_2 \geq b_3 \geq 0$. Show that

$$
\begin{array}{rcl}
p = 1 & \Rightarrow & x_{opt} = b_2 \\
p = 2 & \Rightarrow & x_{opt} = (b_1 + b_2 + b_3)/3 \\
p = \infty & \Rightarrow & x_{opt} = (b_1 + b_3)/2
\end{array}
$$

P7.1.2 Complete the following function:

```
function [a,b,c] = LSFit(L,R,fname,m)
% Assume that fname is a string that names an available function f(x)
% that is defined on [L,R]. Returns in a,b, and c values so that if
% q(x) = ax^2 + bx + c, then
% [q(x(1)) - f(x(1))]^2 + ... + [q(x(m)) - f(x(m))]^2
% is minimized.  Here, x(i) = L + h(i-1) where h = (R-L)/(m-1) and m >=3.
```

Efficiency matters.

7.2 The QR Factorization

In the linear equation setting, the Gaussian elimination approach converts the given linear system $Ax = b$ into an equivalent, easy-to-solve linear system $(MA)x = (Mb)$. The transition from the given system to the transformed system is subject to a strict rule: All row operations applied to A must also be applied to b. We seek a comparable strategy in the least squares setting. Our goal is to produce an m-by-m matrix Q so the given least squares problem

$$
\min_{x \in \mathbf{R}^n} \| Ax - b \|_2
$$

is equivalent to a transformed problem

$$
\min_{x \in \mathbf{R}^n} \| (Q^T A)x - (Q^T b) \|_2
$$

where, by design, $Q^T A$ is "simple."

A family of matrices known as *orthogonal matrices* can be used for this purpose. A matrix $Q \in \mathbb{R}^{m \times m}$ is orthogonal if $Q^T = Q^{-1}$, or, equivalently, if $QQ^T = Q^T Q = I$. Here is a 2-by-2 example:

$$
Q = \left[\begin{array}{cc} \cos(\theta) & \sin(\theta) \\ -\sin(\theta) & \cos(\theta) \end{array} \right]
$$

For this particular Q, it is easy to show that Qx is obtained by rotating x clockwise by θ radians.

The key property of orthogonal matrices that makes them useful in the least squares context is that they preserve 2-norm. Indeed, if $Q \in \mathbb{R}^{m \times m}$ is orthogonal and $r \in \mathbb{R}^m$, then

$$
\| Q^T r \|_2^2 = (Q^T r)^T (Q^T r) = (r^T Q)(Q^T r) = r^T (QQ^T)r = r^T I_m r = r^T r = \| r \|_2^2
$$

If Q is orthogonal, then any x that minimizes $\| Ax - b \|_2$ also minimizes $\| (Q^T A)x - (Q^T b) \|_2$ since

$$
\| (Q^T A)x - (Q^T b) \|_2 = \| Q^T (Ax - b) \|_2 = \| Ax - b \|_2
$$

Our plan is to apply a sequence of orthogonal transformations to A that reduce it to upper triangular form. This will render an equivalent, easy-to-solve problem. For example,

$$\min_{x \in \mathbf{R}^2} \| Ax - b \|_2 = \min_{x \in \mathbf{R}^2} \| (Q^T A)x - (Q^T b) \|_2 = \min_{x \in \mathbf{R}^2} \left\| \begin{bmatrix} r_{11} & r_{12} \\ 0 & r_{22} \\ 0 & 0 \\ 0 & 0 \end{bmatrix} \begin{bmatrix} x_1 \\ x_2 \end{bmatrix} - \begin{bmatrix} c_1 \\ c_2 \\ c_3 \\ c_4 \end{bmatrix} \right\|_2,$$

Here, $Q^T b = c$ is the transformed right-hand side. Note that no matter how x_1 and x_2 are chosen, the sum of squares must be at least $c_3^2 + c_4^2$. Thus we "write off" components 3 and 4 and focus on the reduction in size of components 1 and 2. It follows that the optimum choice for x is that which solves the 2-by-2 upper triangular system

$$\begin{bmatrix} r_{11} & r_{12} \\ 0 & r_{22} \end{bmatrix} \begin{bmatrix} x_1 \\ x_2 \end{bmatrix} = \begin{bmatrix} c_1 \\ c_2 \end{bmatrix}.$$

Call this solution x_{LS} and note that in this case

$$\min_{x \in \mathbf{R}^2} \| Ax - b \|_2 = \| Ax_{LS} - b \|_2 = \sqrt{c_3^2 + c_4^2}.$$

The columns of an orthogonal matrix define an *orthonormal basis*. From the equation $Q^T Q = I$ we see that the inner product of $Q(:,j)$ with any other column is zero. The columns of Q are mutually orthogonal. Moreover, since $Q(:,j)^T Q(:,j) = 1$, each column has unit 2-norm, explaining the "normal" in "orthonormal."

Finding an orthogonal $Q \in \mathbb{R}^{m \times m}$ and an upper triangular $R \in \mathbb{R}^{m \times n}$ so that $A = QR$ is the QR factorization problem. It amounts to finding an orthonormal basis for the subspace defined by the columns of A. To see this, note that the jth column of the equation $A = QR$ says that

$$A(:,j) = Q(:,1) * R(1,j) + Q(:,2) * R(2,j) + \cdots + Q(:,j) * R(j,j)$$

Thus *any* column of A is in the span of $\{Q(:,1), \ldots, Q(:,n)\}$. In the example

$$\begin{bmatrix} 1 & -8 \\ 2 & -1 \\ 2 & 14 \end{bmatrix} = \begin{bmatrix} 1/3 & -2/3 & -2/3 \\ 2/3 & -1/3 & 2/3 \\ 2/3 & 2/3 & -1/3 \end{bmatrix} \begin{bmatrix} 3 & 6 \\ 0 & 15 \\ 0 & 0 \end{bmatrix}$$

we see that

$$\begin{bmatrix} 1 \\ 2 \\ 2 \end{bmatrix} = 3 \begin{bmatrix} 1/3 \\ 2/3 \\ 2/3 \end{bmatrix} \qquad \begin{bmatrix} -8 \\ -1 \\ 14 \end{bmatrix} = 6 \begin{bmatrix} 1/3 \\ 2/3 \\ 2/3 \end{bmatrix} + 15 \begin{bmatrix} -2/3 \\ -1/3 \\ 2/3 \end{bmatrix}$$

The observation that the QR factorization can help us solve the least squares problem tells us once again that finding the "right" basis is often the key to solving a linear algebra problem. Recall from Chapter 2 the attractiveness of the Newton basis for the polynomial interpolation problem.

7.2.1 Rotations

The Q in the QR factorization can be computed using a special family of orthogonal matrices that are called *rotations*. A 2-by-2 *rotation* is an orthogonal matrix of the form

$$G = \begin{bmatrix} c & s \\ -s & c \end{bmatrix} \qquad c = \cos(\theta), \ s = \sin(\theta)$$

If $x = [\, x_1 \ x_2 \,]^T$, then it is possible to choose (c, s) so that if $y = Gx$, then $y_2 = 0$. Indeed, from the requirement that

$$y_2 = -sx_1 + cx_2 = 0$$

and the stipulation that $c^2 + s^2 = 1$ we merely set

$$c = x_1/\sqrt{x_1^2 + x_2^2} \qquad s = x_2/\sqrt{x_1^2 + x_2^2}$$

However, a preferred algorithm for computing the (c, s) pair is the following:

```
function [c,s] = Rotate(x1,x2);
% Pre:
%   x1,x2    scalars
% Post:
%   c,s      c^2+s^2=1 so -s*x1 + c*x2 = 0.

  if x2==0
     c = 1; s = 0;
  else
     if abs(x2)>=abs(x1)
        cotangent = x1/x2; s = 1/sqrt(1+cotangent^2); c = s*cotangent;
     else
        tangent = x2/x1;   c = 1/sqrt(1+tangent^2);   s = c*tangent;
     end
  end
```

In this alternative, we guard against the squaring of arbitrarily large numbers and thereby circumvent the problem of *overflow*. Note that the sine and cosine are computed *without* computing the underlying rotation angle θ. No inverse trigonometric functions are involved in the implementation.

Introducing zeros into a vector by rotation extends to higher dimensions. Suppose $m = 4$ and define

$$G(1,3,\theta) = = \begin{bmatrix} c & 0 & s & 0 \\ 0 & 1 & 0 & 0 \\ -s & 0 & c & 0 \\ 0 & 0 & 0 & 1 \end{bmatrix} \qquad c = \cos(\theta), \ s = \sin(\theta)$$

This is a rotation in the $(1,3)$ plane. It is easy to check that $G(1,3,\theta)$ is orthogonal. Note that

$$G(1,3,\theta) \begin{bmatrix} x_1 \\ x_2 \\ x_3 \\ x_4 \end{bmatrix} = \begin{bmatrix} cx_1 + sx_3 \\ x_2 \\ -sx_1 + cx_3 \\ x_4 \end{bmatrix}$$

and we can determine the cosine-sine pair so that the third component is zeroed as follows:

```
[c,s] = rotate(x(1),x(3))
```

For general m, rotations in the (i, k) plane look like this:

$$
G(i,k,\theta) = \begin{bmatrix}
1 & \cdots & & 0 & \cdots & 0 & \cdots & 0 \\
\vdots & \ddots & & \vdots & & \vdots & & \vdots \\
0 & \cdots & & c & \cdots & s & \cdots & 0 \\
\vdots & & & \vdots & \ddots & \vdots & & \vdots \\
0 & \cdots & & -s & \cdots & c & \cdots & 0 \\
\vdots & & & \vdots & & \vdots & \ddots & \vdots \\
0 & \cdots & & 0 & \cdots & 0 & \cdots & 1
\end{bmatrix} \begin{matrix} \\ \\ i \\ \\ k \\ \\ \ \end{matrix}
$$
$$ i \qquad\quad k$$

where $c = \cos(\theta)$ and $s = \sin(\theta)$ for some θ. Premultiplication of a matrix by $G(i, k, \theta)$ may be organized as follows assuming that c and s house the cosine and sine:

```
A([i k],:)  = [c s ; -s c]*A([i k],:)
```

The integer vector [i k] is used in this context to extract rows i and k from the matrix A. Note that premultiplication by $G(i, k, \theta)$ effects only rows i and k.

7.2.2 Reduction to Upper Triangular Form

We now show how a sequence of row rotations can be used to upper triangularize a rectangular matrix. The 4-by-3 case illustrates the general idea:

$$
\begin{bmatrix} \times & \times & \times \\ \times & \times & \times \\ \times & \times & \times \\ \times & \times & \times \end{bmatrix} \xrightarrow{(3,4)}
\begin{bmatrix} \times & \times & \times \\ \times & \times & \times \\ \times & \times & \times \\ 0 & \times & \times \end{bmatrix} \xrightarrow{(2,3)}
\begin{bmatrix} \times & \times & \times \\ \times & \times & \times \\ 0 & \times & \times \\ 0 & \times & \times \end{bmatrix} \xrightarrow{(1,2)}
$$

$$
\begin{bmatrix} \times & \times & \times \\ 0 & \times & \times \\ 0 & \times & \times \\ 0 & \times & \times \end{bmatrix} \xrightarrow{(3,4)}
\begin{bmatrix} \times & \times & \times \\ 0 & \times & \times \\ 0 & \times & \times \\ 0 & 0 & \times \end{bmatrix} \xrightarrow{(2,3)}
\begin{bmatrix} \times & \times & \times \\ 0 & \times & \times \\ 0 & 0 & \times \\ 0 & 0 & \times \end{bmatrix} \xrightarrow{(3,4)}
\begin{bmatrix} \times & \times & \times \\ 0 & \times & \times \\ 0 & 0 & \times \\ 0 & 0 & 0 \end{bmatrix}
$$

The index pairs over the arrows indicate the rows that are being rotated. Notice that the zeroing proceeds column-by-column and that within each column, the the entries are zeroed from the bottom row on up to the subdiagonal entry. In the m-by-n case we have

```
for j=1:n
   for i=m:-1:j+1
      %Zero A(i,j)
      [c,s] = Rotate(A(i-1,j),A(i,j));
      A(i-1:i,:)  = [c s ; -s c]*A(i-1:i,:);
   end
end
```

Note that when working to zero the subdiagonal elements in column j, columns 1 through $j-1$ are already zero. It follows that the A update is better written as

```
A(i-1:i,j:n) = [c s ; -s c]*A(i-1:i,j:n);
```

The algorithm produces a sequence of rotations $G_1, G_2, \ldots, G_t$ with the property that

$$G_t \cdots G_1 A = R \quad \text{(upper triangular)}$$

Thus if we define

$$Q^T = G_t \cdots G_1$$

then $Q^T A = R$ and

$$A = I \cdot A = (QQ^T)A = Q(Q^T A) = QR$$

The Q matrix can be built up by accumulating the rotations as they are produced:

$$
\begin{aligned}
Q &\leftarrow I \\
Q &\leftarrow QG_1^T \\
Q &\leftarrow QG_2^T \\
&\vdots
\end{aligned}
$$

Packaging all these ideas, we get

```
      function [Q,R] = QRrot(A)
%
% Pre:
%    A    m-byn with m>=n
%
% Post:
%    Q  is an m-by-m orthogonal matrix and R is an m-by-n upper triangular
%    such that  A = QR.

      [m,n] = size(A);
      Q = eye(m,m);
      for j=1:n
         for i=m-1:-1:j
            %Zero A(i,j)
            [c,s] = Rotate(A(i,j),A(i+1,j));
            A(i:i+1,j:n) = [c s; -s c]*A(i:i+1,j:n);
            Q(:,i:i+1) = Q(:,i:i+1)*[c s; -s c]';
         end
      end
      R = triu(A);
```

A flop count reveals that this algorithm requires about $(9m - 4n)n^2$ flops. If $n = m$, then this is about seven times as costly as the LU factorization.

7.2.3   The Least Squares Solution

Once we have the QR factorization of A, then the given least squares problem of minimizing $\| Ax - b \|_2$ transforms as follows:

$$\| Ax - b \|_2 \;=\; \| Q^T(Ax - b) \|_2 \;=\; \| (Q^T A)x - (Q^T b) \|_2 \;=\; \| Rx - c \|_2$$

$$=\; \left\| \left[\begin{array}{c} A_1 \\ 0 \end{array} \right] x - \left[\begin{array}{c} c_1 \\ c_2 \end{array} \right] \right\|_2 \;=\; \left\| \left(\begin{array}{c} A_1 x - c_1 \\ -c_2 \end{array} \right) \right\|_2$$

Here $A_1 = Q(:,1{:}n)^T A$ is made up of the first n rows of $Q^T A$, $c_1 = Q(:,1{:}n)^T b$, is made up of the top n components of $Q^T b$, and $c_2 = Q(:,n+1{:}m)^T b$ is made up of the bottom $m - n$ components of $Q^T b$. Notice that no matter how x is chosen,

$$\| Ax - b \|_2 \geq \| c_2 \|_2$$

The lower bound can be achieved if we solve the upper triangular system $A_1 x = c_1$. It follows that the least squares solution x_{LS} and the norm of the *residual* $\| Ax_{LS} - b \|_2$ are prescribed by

```
      function [xLS,res] = LSq(A,b)
%
%  Pre:
%     A        m-by-n and rank(A) = n.
%     b        m-by-1.
%  Post:
%     xLS      n-by-1, minimizes the 2-norm of Ax-b.
%     res      norm(A*xLS-b)
%
      [m,n] = size(A);
      for j=1:n
         for i=m:-1:j+1
            %Zero A(i,j)
            [c,s] = Rotate(A(i-1,j),A(i,j));
            A(i-1:i,j:n) = [c s; -s c]*A(i-1:i,j:n);
            b(i-1:i) = [c s; -s c]*b(i-1:i);
         end
      end
      xLS = UTriSol(A(1:n,1:n),b(1:n));
      if m==n
         res = 0;
      else
         res = norm(b(n+1:m));
      end
```

In this implementation, Q is not explicitly formed as in LSq. Instead, the rotations that "make up" Q are applied to b as they are generated. This algorithm requires about $3mn^2 - 2n^3$ flops. Note that if $m = n$, then the solution to the square linear system $Ax = b$ is produced at a cost of n^3 flops, 50% more than what is required by Gaussian elimination.

7.2.4 Solution Sensitivity

As in the linear equation problem, it is important to appreciate the sensitivity of the LS solution to perturbations in the data. The notion of condition extends to rectangular matrices. We cannot use the definition $\kappa^2(A) = \| A \| \| A^{-1} \|$ anymore since A, being rectangular, does not have an inverse. (It is natural to use the 2-norm in the LS problem.) Instead we use an equivalent formulation that makes sense for both square and rectangular matrices:

$$\frac{1}{\kappa_2(A)} = \min_{\text{rank}(A+E) < n} \frac{\| E \|_2}{\| A \|_2}$$

Roughly speaking, the inverse of the condition is the distance to the nearest rank deficient matrix. If the columns of A are nearly dependent, then its condition number is large.

It can be shown with reasonable assumptions that if $\tilde{x}_{LS}$ solves the perturbed LS problem

$$\min \| (A + \Delta A)x - (b + \Delta b) \|_2 \qquad \| \Delta A \|_2 \approx \text{eps} \| A \|_2, \ \| \Delta b \|_2 \approx \text{eps} \| b \|_2 \qquad (7.2.1)$$

then

$$\frac{\| \tilde{x}_{LS} - x_{LS} \|}{\| x_{LS} \|} \approx \text{eps} \left(\kappa_2(A) + \rho_{LS}\kappa_2(A)^2 \right)$$

where $\rho_{LS} = \| Ax_{LS} - b \|_2$. If b is in the span of A's columns, then this is essentially the same result obtained for the linear equation problem: $O(\text{eps})$ errors in the data show up as $O(\kappa_2(A)\text{eps})$ errors in the solution. However, if the minimum residual is nonzero, then the *square* of the condition number is involved, and this can be much larger than $\kappa_2(A)$. The script file ShowLSq illustrates this point. Here are the results for a pair of randomly generated, 10-by-4 LS problems:

```
m = 10, n =  4, cond(A) = 1.000e+07
```

Zero residual problem:

Exact Solution	Computed Solution
1.0000000000000000	1.0000000003318523
1.0000000000000000	1.0000000003900065
1.0000000000000000	0.9999999992968892
1.0000000000000000	1.0000000001162717

Nonzero residual problem:

Exact Solution	Computed Solution
1.0000000000000000	1.0010816095684492
1.0000000000000000	1.0012711526858316
1.0000000000000000	0.9977083453465591
1.0000000000000000	1.0003789656343001

Both problems are set up so that the exact solution is the vector of all ones. Notice that the errors are greater in the nonzero residual example, reflecting the effect of the $\kappa_2(A)^2$ factor.

The computed solution obtained by LSq satisfies a nearby LS problem in the sense of (7.2.1). The method is therefore stable. But as we learned in the previous chapter and as the preceding example shows, stability does not guarantee accuracy.

Problems

P7.2.1 Suppose $b \in \mathbb{R}^n$ and $H \in \mathbb{R}^{n \times n}$ is upper upper Hessenberg ($h_{ij} = 0$, $i > j + 1$). Show how to construct rotations $G_1, \ldots, G_{n-1}$ so that $G_{n-1}^T \cdots G_1^T H = R$ is upper triangular. (Hint: G_k should zero $h_{k+1,k}$.) Write a MATLAB function x = HessQRSolve(H,b) that uses this reduction to solve the linear system $Hx = b$.

P7.2.2 Given

$$A = \left[\begin{array}{cc} w & x \\ y & z \end{array} \right]$$

show how to construct a rotation

$$Q = \left[\begin{array}{cc} c & s \\ -s & c \end{array} \right]$$

so that the $(1,2)$ and $(2,1)$ entries in $Q^T A$ are the same.

P7.2.3 Suppose $A \in \mathbb{R}^{n \times n}$. Show how to premultiply A by a sequence of rotations so that A is transformed to *lower* triangular form.

P7.2.4 Modify QRrot so that it efficiently handles the case when A is upper Hessenberg.

P7.2.5 Write a MATLAB fragment that prints the a and b that minimize

$$\phi(a,b) = \sum_{i=1}^m [(a + bx_i) - \sqrt{x_i}]^2 \qquad x = .25{:}h{:}1, \ h = .75/(m-1)$$

for $m = 2, 3, 4, 5, 6, 7, 8, 9, 10, 20, 40, 80, 200$. Use LS.

P7.2.6 Write a MATLAB function Rotate1 that is just like Rotate except that it zeros the top component of a 2-vector.

7.3 The Cholesky Factorization

A matrix $A \in \mathbb{R}^{n \times n}$ is *symmetric* if $A = A^T$ and *positive definite* if $x^T A x > 0$ for all nonzero $x \in \mathbb{R}^n$. Symmetric positive definite (SPD) matrices are the single most important class of specially structured matrices that arise in applications. Here are some important facts associated with such matrices:

1. If $A = (a_{ij})$ is SPD, then its diagonal entries are positive and for all i and j,

$$|a_{ij}| \leq (a_{ii} + a_{jj})/2$$

This says that the largest entry in an SPD matrix is on the diagonal and more qualitatively, that SPD matrices have more "mass" on the diagonal than off the diagonal.

2. If $A \in \mathbb{R}^{n \times n}$ and

$$|a_{ii}| > \sum_{\substack{j=1 \\ j \neq i}}^{n} |a_{ij}| \qquad i = 1{:}n$$

then A is *strictly diagonally dominant*. If A is also symmetric with positive diagonal entries, then it is SPD.

3. If $C \in \mathbb{R}^{m \times n}$ has independent columns and $A = C^T C$, then A is SPD. This fact paves the way to the *method of normal equations* for the full-rank LS problem $\min \| Ax - b \|_2$. It turns out that x_{LS} solves the SPD system $A^T A x = A^T b$. (See §7.3.5.)

4. If A is SPD, then it is possible to find a lower triangular matrix G so that

$$A = GG^T$$

This is called the *Cholesky factorization*. Once it is obtained the solution to $Ax = b$ can be determined via forward and back substitution: $Gy = b$, $G^T x = y$. It is an attractive solution procedure, because it involves half the number of flops required by Gaussian elimination and there is no need to pivot for stability.

Our primary goal is to develop a sequence of Cholesky implementations that feature different "kernel operations" just as we did with matrix multiplication in §5.2.3. The design of a high-performance linear equation solver for an advanced computer hinges on being able to formulate the algorithm in terms of linear algebra that is "friendly" to the underlying architecture. Our Cholesky presentation is an occasion to elevate our matrix-vector skills in this direction.

7.3.1 Positive Definiteness

We start with an application that builds intuition for positive definiteness and leads to a particularly simple Cholesky problem. Positive definite matrices frequently arise when differential equations are discretized. Suppose $p(x)$, $q(x)$, and $r(x)$ are known functions on an interval $[a, b]$ and that we wish to find an unknown function $u(x)$ that satisfies

$$- D\left[p(x)Du(x)\right] + q(x)u(x) = r(x) \qquad a \le x \le b \qquad (7.3.1)$$

with $u(a) = u(b) = 0$. Here, D denotes differentiation with respect to x. This is an example of a *two-point boundary value problem*, and there are several possible solution frameworks. We illustrate the method of *finite differences*, a technique that discretizes the derivatives and culminates in a system of linear equations.

Let n be a positive integer and set $h = (b - a)/(n - 1)$. Define

$$
\begin{array}{rcll}
x_i &=& a + (i - 1)h & i = 1{:}n \\
p_i &=& p(x_i + h/2) & i = 2{:}n \\
q_i &=& q(x_i) & i = 1{:}n \\
r_i &=& r(x_i) & i = 1{:}n
\end{array}
$$

and let u_i designate an approximation to $u(x_i)$ whose value we seek. Set $u_1 = u_n = 0$ since $u(x)$ is zero at the endpoints. From our experience with divided differences,

$$Du(x_i + h/2) \approx \frac{u_{i+1} - u_i}{h}$$

$$Du(x_i - h/2) \approx \frac{u_i - u_{i-1}}{h}$$

and so we obtain the following approximation:

$$D\left[p(x)Du(x)\right]\big|_{x=x_i} \approx \frac{p_iDu(x_i + h/2) - p_{i-1}Du(x_i - h/2)}{h} \approx \frac{p_i\dfrac{u_{i+1} - u_i}{h} - p_{i-1}\dfrac{u_i - u_{i-1}}{h}}{h}$$

If we set $x = x_i$ in (7.3.1) and substitute this discretized version of the $D[p(x)Du(x)]$ term, then we get

$$\frac{1}{h^2}\left(-p_{i-1}u_{i-1} + (p_{i-1} + p_i)u_i - p_iu_{i+1}\right) + q_iu_i = r_i \qquad i = 2{:}n-1$$

Again recalling that $u_1 = u_n = 0$, we see that this defines a system of $n - 2$ equations in the $n - 2$ unknowns $u_2, \ldots, u_{n-1}$. For example, if $n = 6$, then we obtain

$$\begin{bmatrix} p_1 + p_2 + h^2q_2 & -p_2 & 0 & 0 \\ -p_2 & p_2 + p_3 + h^2q_3 & -p_3 & 0 \\ 0 & -p_3 & p_3 + p_4 + h^2q_4 & -p_4 \\ 0 & 0 & -p_4 & p_4 + p_5 + h^2q_5 \end{bmatrix} \begin{bmatrix} u_2 \\ u_3 \\ u_4 \\ u_5 \end{bmatrix} = \begin{bmatrix} h^2r_2 \\ h^2r_3 \\ h^2r_4 \\ h^2r_5 \end{bmatrix}$$

This is a symmetric tridiagonal system. If the functions $p(x)$ and $q(x)$ are positive on $[a, b]$, then the matrix T has strict diagonal dominance and is positive definite. The following theorem offers a rigorous proof of this fact.

Theorem 7 *If*

$$T = \begin{bmatrix} d_1 & e_2 & 0 & \cdots & 0 \\ e_2 & d_2 & \ddots & & 0 \\ 0 & \ddots & \ddots & \ddots & \vdots \\ \vdots & & \ddots & d_{n-1} & e_{n-1} \\ 0 & 0 & \cdots & e_{n-1} & d_n \end{bmatrix}$$

has the property that

$$d_i > \begin{cases} |e_2| & \text{if } i = 1 \\ |e_i| + |e_{i+1}| & \text{if } 2 \leq i \leq n - 1 \\ |e_n| & \text{if } i = n \end{cases}$$

then T is positive definite.

Proof. If $n = 5$, then

$$x^T T x = \begin{bmatrix} x_1 & x_2 & x_3 & x_4 & x_5 \end{bmatrix} \begin{bmatrix} d_1 & e_2 & 0 & 0 & 0 \\ e_2 & d_2 & e_3 & 0 & 0 \\ 0 & e_3 & d_3 & e_4 & 0 \\ 0 & 0 & e_4 & d_4 & e_5 \\ 0 & 0 & 0 & e_5 & d_5 \end{bmatrix} \begin{bmatrix} x_1 \\ x_2 \\ x_3 \\ x_4 \\ x_5 \end{bmatrix}$$

$$= \begin{bmatrix} x_1 & x_2 & x_3 & x_4 & x_5 \end{bmatrix} \begin{bmatrix} d_1 x_1 + e_2 x_2 \\ e_2 x_1 + d_2 x_2 + e_3 x_3 \\ e_3 x_2 + d_3 x_3 + e_4 x_4 \\ e_4 x_3 + d_4 x_4 + e_5 x_5 \\ e_5 x_4 + d_5 x_5 \end{bmatrix}$$

$$= (d_1 x_1^2 + d_2 x_2^2 + d_3 x_3^2 + d_4 x_4^2 + d_5 x_5^2) + 2(e_2 x_1 x_2 + e_3 x_2 x_3 + e_4 x_3 x_4 + e_5 x_4 x_5)$$

In general, we find that

$$x^T T x = \sum_{i=1}^{n} d_i x_i^2 + 2 \sum_{i=2}^{n} e_i x_{i-1} x_i$$

Our goal is to show that if $x \neq 0$, then this quantity is strictly positive. The first summation is obviously positive. The challenge is to show that the second summation cannot be too negative. Since $0 \leq (x_{i-1} - x_i)^2 = x_{i-1}^2 - 2x_{i-1}x_i + x_i^2$, it follows that $2|x_{i-1}||x_i| \leq x_{i-1}^2 + x_i^2$. Therefore,

$$x^T T x \geq \sum_{i=1}^{n} d_i x_i^2 - 2 \sum_{i=2}^{n} |e_i||x_{i-1}||x_i|$$

$$\geq \sum_{i=1}^{n} d_i x_i^2 - \sum_{i=2}^{n} |e_i|(x_{i-1}^2 + x_i^2)$$

$$= (d_1 - |e_2|) x_1^2 + \sum_{i=2}^{n-1} (d_i - |e_i| - |e_{i+1}|) x_i^2 + (d_n - |e_n|) x_n^2$$

The theorem follows because in this last expression, every quantity in parentheses is positive and at least one of the x_i is nonzero. $\square$

Some proofs of positive definiteness are straightforward like this, and others are more difficult. Once positive definiteness is established, then we have a license to compute the Cholesky factorization. We are ready to show how this is done.

7.3.2   Tridiagonal Cholesky

It turns out that a symmetric tridiagonal positive definite matrix

$$
T = \begin{bmatrix}
d_1 & e_2 & 0 & 0 & 0 \\
e_2 & d_2 & e_3 & 0 & 0 \\
0 & e_3 & d_3 & e_4 & 0 \\
0 & 0 & e_4 & d_4 & e_5 \\
0 & 0 & 0 & e_5 & d_5
\end{bmatrix} \qquad (n = 5)
$$

has a Cholesky factorization $T = GG^T$ with a lower bidiagonal G:

$$
G = \begin{bmatrix}
g_1 & 0 & 0 & 0 & 0 \\
h_2 & g_2 & 0 & 0 & 0 \\
0 & h_3 & g_3 & 0 & 0 \\
0 & 0 & h_4 & g_4 & 0 \\
0 & 0 & 0 & h_5 & g_5
\end{bmatrix}
$$

To see this, we equate coefficients

$$
\begin{array}{llll}
(1,1): & d_1 = g_1^2 & \Rightarrow & g_1 = \sqrt{d_1} \\
(2,1): & e_2 = h_2 g_1 & \Rightarrow & h_2 = e_2/g_1 \\
(2,2): & d_2 = h_2^2 + g_2^2 & \Rightarrow & g_2 = \sqrt{d_2 - h_2^2} \\
(3,2): & e_3 = h_3 g_2 & \Rightarrow & h_3 = e_3/g_2 \\
(3,3): & d_3 = h_3^2 + g_3^2 & \Rightarrow & g_3 = \sqrt{d_3 - h_3^2}
\end{array}
$$

$$\vdots$$

and conclude that

$$
\begin{array}{lll}
g_i & = & \sqrt{d_i - h_i^2} \qquad (h_1 \equiv 0) \\
h_i & = & e_i/g_{i-1} \qquad\quad (i > 1)
\end{array}
$$

We therefore obtain

```
    function [g,h] = CholTrid(d,e)
%
% Pre:  d and e are n-vectors with the property that
%       A = diag(d) + diag(e(2:n),-1) + diag(e(2:n),1) is positive definite.
%
% Post: g,h are n-vectors with the property that if G is the lower bidiagonal
%       matrix G = diag(g) + diag(h(2:n),-1), then A = GG^T.
%
    n = length(d); g = zeros(n,1); h = zeros(n,1);
    g(1) = sqrt(d(1));
    for i=2:n
        h(i) = e(i)/g(i-1);
        g(i) = sqrt(d(i) - h(i)^2);
    end
```

It is clear that this algorithm requires $O(n)$ flops. Since G is lower bidiagonal, the solution of $Ax = b$ via $Gy = b$ and $G^T x = y$ proceeds as follows:

```
      function x = CholTridSol(g,h,b)
%
% Pre:  g, h, and b are n-vectors and g has no zero entries.
% Post: x is a column n-vector that solves GG'x = b where
%          g(1:n) and h(2:n) are the diagonal and subdiagonal of G.

      n = length(g); y = zeros(n,1); x = zeros(n,1);

% Solve Gy = b
      y(1) = b(1)/g(1);
      for k=2:n
         y(k) = (b(k) - h(k)*y(k-1))/g(k);
      end

% Solve G'x = y
      x(n) = y(n)/g(n);
      for k=n-1:-1:1
         x(k) = (y(k) - h(k+1)*x(k+1))/g(k);
      end
```

See §6.2 for a discussion of bidiagonal system solvers. Overall we see that the complete solution to a tridiagonal SPD system involves $O(n)$ flops. The following script uses these functions to solve a very simple two-point BVP:

```
% Script File: ShowBVP
%
% Illustrates the numerical solution to
%         -D^2 u + u(x) = 2xsin(x) - 2cos(x)    u(0) = u(pi) = 0.
% Exact solution = u(x) = xsin(x).

      n =  100;
      x =  linspace(0,pi,n)';
      hx = pi/(n-1);
      d =  2*ones(n-2,1) + hx^2;
      e =  -ones(n-2,1);
      [g,h] = CholTrid(d,e);
      b = hx^2*( 2*x(2:n-1).*sin(x(2:n-1)) - 2*cos(x(2:n-1)));
      umid  = CholTridSol(g,h,b);
      u = [0;umid;0];
      plot(x,u,x,x.*sin(x))
      err = norm(u - x.*sin(x),'inf');
      title('Solution to   -D^2 u + u = 2xsin(x) - 2cos(x), u(0)=u(pi)=0')
      xlabel(sprintf(' n = %3.0f    norm(u - xsin(x),''inf'') = %10.6f',n,err))
```

The results are displayed in Fig 7.2.

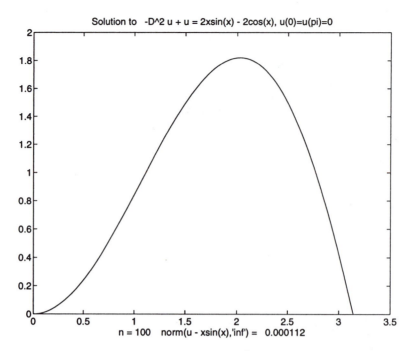

FIGURE 7.2 Numerical solution to $-D^2 u + u = 2x \sin(x) - 2 \cos(x)$

7.3.3   Five Implementations for Full Matrices

Here is an example of a 3-by-3 Cholesky factorization:

$$
\begin{bmatrix} 4 & -10 & 2 \\ -10 & 34 & 17 \\ 2 & -17 & 18 \end{bmatrix} = \begin{bmatrix} 2 & 0 & 0 \\ -5 & 3 & 0 \\ 1 & -4 & 1 \end{bmatrix} \begin{bmatrix} 2 & -5 & 1 \\ 0 & 3 & -4 \\ 0 & 0 & 1 \end{bmatrix}
$$

An algorithm for computing the entries in the Cholesky factor can be derived by equating entries in the equation $A = GG^T$:

$$
\begin{bmatrix} a_{11} & a_{12} & a_{13} \\ a_{21} & a_{22} & a_{23} \\ a_{31} & a_{32} & a_{33} \end{bmatrix} = \begin{bmatrix} g_{11} & 0 & 0 \\ g_{21} & g_{22} & 0 \\ g_{31} & g_{32} & g_{33} \end{bmatrix} \begin{bmatrix} g_{11} & g_{21} & g_{31} \\ 0 & g_{22} & g_{32} \\ 0 & 0 & g_{33} \end{bmatrix}
$$

Since

$$a_{11} = g_{11}^2$$

$$a_{21} = g_{11} g_{21} \qquad a_{22} = g_{21}^2 + g_{22}^2$$

$$a_{31} = g_{11} g_{31} \qquad a_{32} = g_{21} g_{31} + g_{22} g_{32} \qquad a_{33} = g_{31}^2 + g_{32}^2 + g_{33}^2$$

we obtain

$$g_{11} = \sqrt{a_{11}}$$

$$g_{21} = a_{21}/g_{11}$$

$$g_{22} = \sqrt{a_{22} - g_{21}^2}$$

$$g_{31} = a_{31}/g_{11}$$

$$g_{32} = (a_{32} - g_{21}g_{31})/g_{22}$$

$$g_{33} = \sqrt{a_{33} - g_{31}^2 - g_{32}^2}$$

The algorithm would break down if any of the g_{ii} are zero or complex. But the property of being positive definite guarantees that neither of these things happen.

To derive the Cholesky algorithm for general n, we repeat the preceding methodology and compare entries in the equation $A = GG^T$. If $i \geq j$, then

$$a_{ij} = \sum_{k=1}^{j} g_{ik}g_{jk} \quad \Rightarrow \quad g_{ij}g_{jj} = a_{ij} - \sum_{k=1}^{j-1} g_{ik}g_{jk} \equiv s_{ij}$$

and so

$$g_{ij} = \begin{cases} \sqrt{s_{jj}} & i = j \\ \\ s_{ij}/g_{jj} & i > j \end{cases}$$

If we compute the lower triangular matrix G row by row as we did in the 3-by-3 example, then we obtain the following implementation:

```
function G = CholScalar(A);
%
% Pre:  A is a symmetric and positive definite matrix.
% Post: G is lower triangular so A = G*G'.

[n,n] = size(A);
G = zeros(n,n);
for i=1:n
   % Compute G(i,1:i)
   for j=1:i
      s = A(j,i);
      for k=1:j-1
         s = s - G(j,k)*G(i,k);
      end
      if j<i
         G(i,j) = s/G(j,j);
      else
         G(i,i) = sqrt(s);
      end
   end
end
```

An assessment of the work involved reveals that this implementation of Cholesky requires $n^3/3$ flops. This is half of the work required by Gaussian elimination, to be expected since the problem involves half of the data.

Notice that the k-loop in CholScalar oversees an inner product between subrows of G. With this observation we obtain the following dot product implementation:

```
function G = CholDot(A);
%
% Pre:  A is a symmetric and positive definite matrix.
% Post: G is lower triangular and A = G*G'.

[n,n] = size(A);
G = zeros(n,n);
for i=1:n
    % Compute G(i,1:i)
    for j=1:i
        if j==1
            s = A(j,i);
        else
            s = A(j,i) - G(j,1:j-1)*G(i,1:j-1)';
        end
        if j<i
            G(i,j) = s/G(j,j);
        else
            G(i,i) = sqrt(s);
        end
    end
end
```

An inner product is an example of a *level-1* linear algebra operation. Level-1 operations involve $O(n)$ work and $O(n)$ data. Inner products, vector scaling, vector addition, and saxpys are level-1 operations. Notice that CholDot is row oriented because the ith and jth rows of G are accessed during the inner product.

A column-oriented version that features the saxpy operation can be derived by comparing the j-th columns in the equation $A = GG^T$:

$$A(:,j) = \sum_{k=1}^{j} G(:,k)G(j,k)$$

This can be solved for $G(:,j)$:

$$G(:,j)G(j,j) = A(:,j) - \sum_{k=1}^{j-1} G(:,k)G(j,k) \equiv s$$

But since G is lower triangular, we need only focus on the nonzero portion of this vector:

$$G(j{:}n, j)G(j, j) = A(j{:}n, j) - \sum_{k=1}^{j-1} G(j{:}n, k)G(j, k) \equiv s(j{:}n)$$

Since $G(j, j) = \sqrt{s(j)}$, it follows that

$$G(j{:}n, j) = s(j{:}n)/\sqrt{s(j)}$$

This leads to the following implementation:

```
    function G = CholSax(A);
%
% Pre:  A is a symmetric and positive definite matrix.
% Post: G is lower triangular so A = G*G'.

    [n,n] = size(A);
    G = zeros(n,n);
    s = zeros(n,1);
    for j=1:n
       s(j:n) = A(j:n,j);
       for k=1:j-1
          s(j:n) = s(j:n) - G(j:n,k)*G(j,k);
       end
       G(j:n,j) = s(j:n)/sqrt(s(j));
    end
```

Notice that as i ranges from j to n, the kth column of G is accessed in the inner loop. From the flop point of view, CholSax is identical to CholDot.

An update of the form

$$\text{Vector} \leftarrow \text{Vector} + \text{Matrix} \times \text{Vector}$$

is called a *gaxpy* operation. Notice that the k-loop in CholSax oversees the gaxpy operation

$$s(j{:}n) \quad \leftarrow \quad s(j{:}n) - \begin{bmatrix} G(j, 1{:}j-1) \\ G(j+1, 1{:}j-1) \\ \vdots \\ G(n, 1{:}j-1) \end{bmatrix} \begin{bmatrix} G(j, 1) \\ G(j, 2) \\ \vdots \\ G(j, j-1) \end{bmatrix}$$

$$= \quad s(j{:}n) - G(j{:}n, 1{:}j-1)G(j, 1{:}j-1)^T$$

Substituting this observation into CholSax gives

```
    function G = CholGax(A);
%
% Pre:  A is a symmetric and positive definite matrix.
% Post: G is lower triangular so A = G*G'.

    [n,n] = size(A);
    G = zeros(n,n);
    s = zeros(n,1);
    for j=1:n
       if j==1
          s(j:n) = A(j:n,j);
       else
          s(j:n) = A(j:n,j) - G(j:n,1:j-1)*G(j,1:j-1)';
       end
       G(j:n,j)  = s(j:n)/sqrt(s(j));
    end
```

The gaxpy operation is a *level-2* operation. Level-2 operations are characterized by quadratic work and quadratic data. [E.g., for an m-by-n gaxpy operation, $O(mn)$ data is involved and $O(mn)$ flops are required].

Finally, we developed a recursive implementation. Suppose that $n > 1$ and that

$$A = \begin{bmatrix} B & v \\ v & \alpha \end{bmatrix}$$

is SPD where $B = A(1{:}n-1, 1{:}n-1)$, $v = A(1{:}n-1, n)$, and $\alpha = A(n, n)$. It is easy to show that B is also SPD. If

$$G = \begin{bmatrix} G_1 & 0 \\ w & \beta \end{bmatrix}$$

is partitioned the same way and we equate blocks in the equation

$$\begin{bmatrix} B & v \\ v^T & \alpha \end{bmatrix} = \begin{bmatrix} G_1 & 0 \\ w^T & \beta \end{bmatrix} \begin{bmatrix} G_1 & 0 \\ w^T & \beta \end{bmatrix}^T = \begin{bmatrix} G_1 & 0 \\ w^T & \beta \end{bmatrix} \begin{bmatrix} G_1^T & w \\ 0 & \beta \end{bmatrix}$$

then we find that $B = G_1 G_1^T$, $v = G_1 w$, and $\alpha = \beta^2 + w^T w$. This says that if G is to be the Cholesky factor of A, then it can be synthesized by (1) computing the Cholesky factor G_1 of B, (2) solving the lower triangular system $G_1 w = v$ for w, and (3) computing $\beta = \sqrt{\alpha - w^T w}$. The square root is guaranteed to render a real number because

$$\begin{aligned}
0 &< \begin{bmatrix} -B^{-1}v \\ 1 \end{bmatrix}^T \begin{bmatrix} B & v \\ v & \alpha \end{bmatrix} \begin{bmatrix} -B^{-1}v \\ 1 \end{bmatrix} \\
&= \alpha - v^T B^{-1} v \\
&= \alpha - v^T (G_1 G_1^T)^{-1} v \\
&= \alpha - (G_1^{-1}v)^T (G_1^{-1}v) \\
&= \alpha - w^T w
\end{aligned}$$

This is the basis for the following recursive implementation:

```
      function G = CholRecur(A);
%
% Pre:  A is a symmetric and positive definite matrix.
% Post: G is lower triangular so A = G*G',

      [n,n] = size(A);
      if n==1
         G = sqrt(A);
      else
         G(1:n-1,1:n-1) = CholRecur(A(1:n-1,1:n-1));
         G(n,1:n-1)     = LTriSol(G(1:n-1,1:n-1),A(1:n-1,n))';
         G(n,n)         = sqrt(A(n,n) - G(n,1:n-1)*G(n,1:n-1)');
      end
```

If this function is applied to the matrix

$$
A = \begin{bmatrix} 1 & 0 & 0 & 0 \\ 2 & 3 & 0 & 0 \\ 4 & 5 & 6 & 0 \\ 7 & 8 & 9 & 10 \end{bmatrix} \begin{bmatrix} 1 & 0 & 0 & 0 \\ 2 & 3 & 0 & 0 \\ 4 & 5 & 6 & 0 \\ 7 & 8 & 9 & 10 \end{bmatrix}^T
$$

and the semicolon is deleted from the recursive call command, then the following sequence of matrices is displayed:

$$
\begin{bmatrix} 1 \end{bmatrix} \rightarrow \begin{bmatrix} 1 & 0 \\ 2 & 3 \end{bmatrix} \rightarrow \begin{bmatrix} 1 & 0 & 0 \\ 2 & 3 & 0 \\ 4 & 5 & 6 \end{bmatrix} \rightarrow \begin{bmatrix} 1 & 0 & 0 & 0 \\ 2 & 3 & 0 & 0 \\ 4 & 5 & 6 & 0 \\ 7 & 8 & 9 & 10 \end{bmatrix}
$$

From this we infer that CholRecur computes the Cholesky factor G row by row, just like CholScalar.

7.3.4   Efficiency, Stability, and Accuracy

The script file CholBench benchmarks the five implementations. Here are the results from a sample run:

```
            n = 32

            Algorithm     Time     Flops
            ------------------------------
            CholScalar    1.000    11986
            CholDot       0.375    12915
            CholSax       0.319    11522
            CholGax       0.058    12017
            CholRecur     0.503    13224
```

The times reported are relative to the time required by CholScalar. The concern is not so much about the actual values in the table but the fact that different implementations of the same matrix algorithm can have different levels of performance.

In Gaussian elimination we had to worry about large multipliers. This is not an issue in the Cholesky factorization because the equation

$$a_{ii} = \sum_{j=1}^{i} g_{ij}^2$$

implies $|g_{ij}| \leq \sqrt{a_{ii}}$. Thus no entry in G can be larger than the square root of A's largest diagonal entry. This does *not* imply that all symmetric positive definite systems are well conditioned. The computed solution is prone to the same kind of error that we saw in Gaussian elimination. In particular, if $\hat{x}$ is the computed vector produced by

```
G = CholScalar(A);
y = LTriSol(G,b);
x = UTriSol(G',y);
```

then

$$(A + E)\hat{x} = b \qquad \| E \| \approx \text{eps} \| A \|$$

and

$$\frac{\| \hat{x} - x \|}{\| x \|} \approx \text{eps } \kappa(A)$$

The script file CholErr can be used to examine these heuristics. It solves a series of SPD systems that involve the *Hilbert* matrices. These matrices can be generated with the built-in MATLAB function hilb(n). Their condition number increases steeply with n and their exact inverse can be obtained by calling invhilb. This enables us to compute the true solution and therefore the exact relative error. Here are the results:

n	cond(A)	relerr
2	1.928e+01	1.063e-16
3	5.241e+02	1.231e-15
4	1.551e+04	1.393e-13
5	4.766e+05	1.293e-12
6	1.495e+07	9.436e-11
7	4.754e+08	3.401e-09
8	1.526e+10	1.090e-08
9	4.932e+11	3.590e-06
10	1.603e+13	1.081e-04
11	5.216e+14	2.804e-03
12	1.668e+16	5.733e-02

Relative error deteriorates with increasing condition number in the expected way. CholErr uses CholScalar, but the results would be no different were any of the other implementations used.

7.3.5   The MATLAB CHOL Function and the LS Problem

When applied to an SPD matrix A, the MATLAB function Chol(A) returns an *upper* triangular matrix R so that $A = R^T R$. Thus R is the transpose of the matrix G that we have been calling the Cholesky factor.

The MATLAB Cholesky style highlights an important connection between the Cholesky and QR factorizations. If $A \in \mathbb{R}^{m \times n}$ has full column rank and $A = QR$ is its QR factorization, then $\tilde{R} = R(1{:}n, 1{:}n)$ defines the MATLAB Cholesky factor of $A^T A$:

$$A^T A = (QR)^T (QR) = (R^T Q^T)(QR) = R^T (Q^T Q)R = R^T R = \tilde{R}^T \tilde{R}$$

Note that

$$A = QR = Q(:, 1{:}n)\tilde{R}$$

and so the LS minimizer of $\| Ax - b \|_2$ is given by

$$x_{LS} = \tilde{R}^{-1} Q(:, 1{:}n)^T b = \tilde{R}^{-1}(A\tilde{R}^{-1})^T b = (\tilde{R}^T \tilde{R})^{-1} A^T b = (A^T A)^{-1} A^T b.$$

Thus x_{LS} is the solution to the SPD system

$$A^T A x = A^T b$$

Solving the LS problem by computing the Cholesky factorization of $A^T A$ is called the method of normal equations. The method works well on some problems but in general is not as numerically sound as the QR method outlined in the previous section.

Problems

P7.3.1 Complete the following function:

```
    function uvals = TwoPtBVP(n,a,b,pname,qname,rname)
%
% Pre:
%    a,b       reals with a<b.
%    n         integer >= 3
%    pname     string that names a positive function defined on [a,b].
%    qname     string that names a positive function defined on [a,b].
%    rname     string that names a function defined on [a,b].
%
% Post:
%    uvals     column n-vector with the property that uvals(i) approximates
%              the solution to the 2-point boundary value problem
%
%                 -D[p(x)Du(x)] + q(x)u(x) = r(x)     a<=x<=b
%                  u(a) = u(b) = 0
%
%              at x = a+(i-1)h where h = (b-a)/(n-2)
```

Use it to solve

$$-D[(2 + x \cos(20\pi x) Du(x)] + (20 \sin(1/(.1 + x)^2))u(x) = 1000(x - .5)^3$$

with $u(0) = u(1) = 0$. Set $n = 400$ and plot the solution.

P7.3.2 The derivation of an outer product Cholesky implementation is based on the following triplet of events:

- Compute G(1,1) = sqrt(A(1,1)).

- Scale to get the rest of G's first column: `G(2:n,1) = A(2:n,1)/G(1,1)`.

- Repeat on the reduced problem `A(2:n,2:n)-G(2:n,1)*G(2:n,1)'`.

Using this framework, write a recursive implementation of CholOuter. To handle the base case, observe that if $n = 1$, then the Cholesky factor is just the square root of $A = A(1,1)$.

P7.3.3 A positive definite matrix of the form

$$
A = \begin{bmatrix}
d_1 & e_2 & f_3 & 0 & 0 & 0 & 0 & 0 \\
e_2 & d_2 & e_3 & f_4 & 0 & 0 & 0 & 0 \\
f_3 & e_3 & d_3 & e_4 & f_5 & 0 & 0 & 0 \\
0 & f_4 & e_4 & d_4 & e_5 & f_6 & 0 & 0 \\
0 & 0 & f_5 & e_5 & d_5 & e_6 & f_7 & 0 \\
0 & 0 & 0 & f_6 & e_6 & d_6 & e_7 & f_8 \\
0 & 0 & 0 & 0 & f_7 & e_7 & d_7 & e_8 \\
0 & 0 & 0 & 0 & 0 & f_8 & e_8 & d_8
\end{bmatrix}
$$

has a factorization $A = GG^T$, where

$$
G = \begin{bmatrix}
g_1 & 0 & 0 & 0 & 0 & 0 & 0 & 0 \\
h_2 & g_2 & 0 & 0 & 0 & 0 & 0 & 0 \\
p_3 & h_3 & g_3 & 0 & 0 & 0 & 0 & 0 \\
0 & p_4 & h_4 & g_4 & 0 & 0 & 0 & 0 \\
0 & 0 & p_5 & h_5 & g_5 & 0 & 0 & 0 \\
0 & 0 & 0 & p_6 & h_6 & g_6 & 0 & 0 \\
0 & 0 & 0 & 0 & p_7 & h_7 & g_7 & 0 \\
0 & 0 & 0 & 0 & 0 & p_8 & h_8 & g_8
\end{bmatrix}
$$

By comparing coefficients in the equation $A = GG^T$, develop a MATLAB function `[g,h,p] = Chol5(d,e,f)` that computes the vectors `g(1:n)`, `h(2:n)`, and `p(3:n)` from `d(1:n)`, `e(2:n)`, and `f(3:n)`. Likewise, develop triangular system solvers `LTriSol5(g,h,p,b)` and `UTriSol5(g,h,p,b)` that can be used to solve systems of the form $Gx = b$ and $G^Tx = b$. To establish the correctness of your functions, use them to solve a 10-by-10 system $Ax = b$ where $d_i = 100 + i$, $e_i = 10 + i$, $f_i = i$ with $b = 100 * ones(10,1)$. Produce a plot that shows how long it takes your functions to solve an n-by-n system for $10 \leq n \leq 100$. Print a table showing the number of flops required to solve $Ax = b$ for $n = 10, 20, 40, 80, 160$.

7.4   High-Performance Cholesky

We discuss two implementations of the Cholesky factorization that shed light on what it is like to design a linear equation solver that runs fast in an advanced computing environment. The block Cholesky algorithm shows how to make a linear algebra computation rich in matrix-matrix multiplication. This has the effect of reducing the amount of data motion. The shared memory implementation shows how the work involved in a matrix factorization can be subdivided into roughly equal parts that can be simultaneously computed by separate processors.

7.4.1   Level-3 Implementation

Level-3 operations involve a quadratic amount of data and a cubic amount of work. Matrix multiplication is the leading example of a level-3 operation. In the n-by-n, $C = AB$ case, there are $2n^2$ input values and $2n^3$ flops to perform. Unlike level-1 and level-2 operations, the ratio of work to data grows with problem size in the level-3 setting. This makes it possible to bury data motion overheads in a computation that is rich in level-3 operations. Algorithms with this property usually run very fast on advanced architectures.

As an exercise in "level-3 thinking" we show how the Cholesky computation can be arranged so that all but a small fraction of the arithmetic occurs in the context of matrix multiplication. The key is to partition the A and G matrices into blocks (submatrices) and to organize the calculations at that level. For simplicity, we develop a block version of CholScalar. Assume for that $n = pm$ and partition A and G into p-by-p blocks as follows:

$$A = \begin{bmatrix} A_{11} & \cdots & A_{1m} \\ \vdots & & \vdots \\ A_{m1} & \cdots & A_{mm} \end{bmatrix} \qquad G = \begin{bmatrix} G_{11} & \cdots & 0 \\ \vdots & \ddots & \vdots \\ G_{m1} & \cdots & G_{mm} \end{bmatrix}$$

This means that we are regarding A and G as m-by-m matrices with p-by-p blocks. For example, here is a partitioning of a 12-by-12 symmetric positive definite matrix into 3-by-3 blocks:

$$A = \left[\begin{array}{ccc|ccc|ccc|ccc}
34 & 1 & 14 & 17 & 12 & 9 & 6 & 17 & 5 & 9 & 12 & 8 \\
1 & 38 & 10 & 11 & 10 & 9 & 17 & 11 & 8 & 7 & 16 & 10 \\
14 & 10 & 45 & 10 & 2 & 8 & 11 & 9 & 9 & 18 & 6 & 11 \\
\hline
17 & 11 & 10 & 43 & 6 & 16 & 17 & 6 & 6 & 9 & 7 & 8 \\
12 & 10 & 2 & 6 & 48 & 10 & 2 & 14 & 11 & 7 & 6 & 19 \\
9 & 9 & 8 & 16 & 10 & 40 & 4 & 9 & 17 & 12 & 14 & 15 \\
\hline
6 & 17 & 11 & 17 & 2 & 4 & 44 & 17 & 7 & 9 & 14 & 11 \\
17 & 11 & 9 & 6 & 14 & 9 & 17 & 38 & 14 & 4 & 6 & 15 \\
5 & 8 & 9 & 6 & 11 & 17 & 7 & 14 & 40 & 12 & 14 & 10 \\
\hline
9 & 7 & 18 & 9 & 7 & 12 & 9 & 4 & 12 & 30 & 8 & 2 \\
12 & 16 & 6 & 7 & 6 & 14 & 14 & 6 & 14 & 8 & 38 & 11 \\
8 & 10 & 11 & 8 & 19 & 15 & 11 & 15 & 10 & 2 & 11 & 35
\end{array}\right]$$

Note that the (i, j) block is the transpose of the (j, i) block. For example,

$$A_{23} = \begin{bmatrix} 17 & 6 & 6 \\ 2 & 14 & 11 \\ 4 & 9 & 17 \end{bmatrix} = \begin{bmatrix} 17 & 2 & 4 \\ 6 & 14 & 9 \\ 6 & 11 & 17 \end{bmatrix}^T = A_{32}^T$$

Comparing (i, j) blocks in the equation $A = GG^T$ with $i \geq j$ gives

$$A_{ij} = \sum_{k=1}^{j} G_{ik} G_{jk}^T$$

and so

$$G_{ij} G_{jj}^T = A_{ij} - \sum_{k=1}^{j-1} G_{ik} G_{jk}^T \equiv S_{ij}$$

Corresponding to the derivation of CholScalar, we obtain the following framework for computing the G_{ij}:

```
for i=1:m
   for j=1:i
      Compute S_ij
      if i<j
         G_ij is the solution of XG_jj^T = S_ij
      else
         G_ii is the Cholesky factor of S_ii.
      end
   end
end
```

A number of observations need to be made in order to turn this into MATLAB code.

To begin with, since the blocks are p-by-p, we see that the ivec = ((i-1)*p+1):(i*p) and jvec = ((j-1)*p+1):(j*p) assign the row and column indices associated with the (i,j) block. In the preceding example, A(7:9,4:6) contains A_{23}. It follows that if $1 < j \le i \le n$, then

```
ivec = ((i-1)*p+1):(i*p);
jvec = ((j-1)*p+1):(j*p);
S = A(ivec,jvec);
for k=1:j-1
   kvec = ((k-1)*p+1):(k*p);
   S = S + G(ivec,kvec)*G(jvec,kvec)';
end
```

assigns the p-by-p matrix

$$S_{ij} = A_{ij} - \sum_{k=1}^{} j - 1 G_{ik} G_{jk}^T$$

to S. If $i = j$, then the $j,j)$ block of G can be computed with a call to any of the Cholesky procedures developed thus far. For example,

```
G(ivec,ivec) = CholScalar(S)
```

If $j < i$, then the equation $G_{ij} G_{jj}^T = S$ requires repeated triangular system solves since

$$G_{jj} G_{ij}^T = S^T.$$

In particular, for $q = 1:p$,

$$G_{jj}(q\text{th column of } G_{ij}^T) = (q\text{th column of } S^T)$$

These are lower triangular system problems involving the p-by-p lower triangular matrix G_{jj}. Noting that the qth column of a matrix transpose is the qth row of the matrix, we obtain

```
for q=1:p
   G(q+(i-1)*p,jvec)= (LTriSol(G(jvec,jvec),S(q,:)'))';
end
```

Putting it all together, we obtain

```
    function G = CholBlock(A,p);
%
% Pre:  A is an n-by-n symmetric and positive definite matrix
%       and p is an integer that divides n.
% Post: G is lower triangular and A = G*G'.

    [n,n] = size(A);
    m = n/p;
    G = zeros(n,n);
    for i=1:m
        ivec = ((i-1)*p+1):(i*p);
        for j=1:i
            jvec = ((j-1)*p+1):(j*p);
            %Find the (i,j) block of G, i.e., G(ivec,jvec).
            S = A(ivec,jvec);
            for k=1:j-1
                kvec = ((k-1)*p+1):(k*p);
                S = S - G(ivec,kvec)*G(jvec,kvec)';
            end
            if j<i
                for q=1:p
                G(q+(i-1)*p,jvec)= (LTriSol(G(jvec,jvec),S(q,:)'))';
                end
            else
                G(ivec,ivec) = CholScalar(S);
            end
        end
    end
```

This algorithm involves $n^3/3$ flops, the same number of flops as any of the §7.1 methods. The only flops that are *not* level-3 flops are those associated with the computation of the p-by-p Cholesky factors $G_{11}, \ldots, G_{mm}$. It follows that the fraction of flops that are level-3 flops is given by

$$L_3 = \frac{(n^3/3) - (mp^3/3)}{n^3/3} = 1 - \frac{1}{m^2}$$

A tacit assumption in all this is that the block size p is large enough that true level-3 performance is extracted during the computation of S. Intelligent block size determination is a function of algorithm and architecture and typically involves careful experimentation. The script file ShowCholBlock benchmarks CholBlock on an $n = 48$ problem for various block sizes p. Here are the results with the unit block size time normalized to 1.000:

Block Size	Time
1	1.000
2	0.292
3	0.174
4	0.141
6	0.120
8	0.117
12	0.124
16	0.138
24	0.179
48	0.415

Notice that the optimum block size is around $\sqrt{n}$.

7.4.2   A Shared Memory Implementation

We now turn our attention to the implementation of the Cholesky factorization in a shared memory environment. This framework for parallel computation is introduced in §4.5. Assume at the start that $A \in \mathbb{R}^{n \times n}$ is housed in shared memory and that we have p processors to apply to the problem, $p << n$. Our goal is to write the node program for Proc(1),...,Proc(p).

We identify the computation of $G(j{:}n, j)$ as the jth task. Analogous to the development of CholSax, we compare jth columns in the equation $A = GG^T$,

$$A(:,j) = \sum_{k=1}^{j} G(:,k)G(j,k)$$

and obtain the key result

$$G(j{:}n, j)G(j,j) = A(j{:}n, j) \; - \; \sum_{k=1}^{j-1} G(j{:}n, k)G(j,k) \; \equiv \; s(j{:}n)$$

Note that $G(j{:}n, j) = s(j{:}n)/\sqrt{s_j}$. From this we conclude that the processor in charge of computing $G(j{:}n, j)$ must oversee the update

$$A(j{:}n, j) \leftarrow A(j{:}n, j) \; - \; \sum_{k=1}^{j-1} G(j{:}n, k)G(j,k) \tag{7.4.1}$$

and the scaling

$$A(j{:}n, j) \leftarrow A(j{:}n, j)/\sqrt{A(j,j)} \tag{7.4.2}$$

This requires about $2n(n-j)$ flops. Since the cost of computing column j is a decreasing function of j, it does not makes sense for Proc(1) to handle tasks 1 to n/p, Proc(2) to handle tasks $1 + n/p$ through $2n/p$, etc. To achieve some measure of flop load balancing, we assign to Proc(μ) tasks $\mu{:}p{:}n$. This amounts to "dealing" out tasks as you would a deck of n cards. Here is a who-does-what table for the case $n = 22$, $p = 4$:

Processor	Tasks					
1	1	5	9	13	17	21
2	2	6	10	14	18	22
3	3	7	11	15	19	
4	4	8	12	16	20	

Thus $\text{Proc}(\mu)$ is charged with the production of $G(:,\mu{:}p{:}n)$.

Suppose $G(k{:}n, k)$ has just been computed and is contained in A(k:n,k) in shared memory. $\text{Proc}(\mu)$ can then carry out the updates

```
for j=k+1:n
   if j is one of the integers in mu:p:n
      A(k:n,j) = A(k:n,j) - A(k:n,k)A(j,k)
   end
end
```

In other words, for each of the remaining tasks that it must perform (i.e., the indices in mu:p:n that are greater than k), $\text{Proc}(\mu)$ incorporates the kth term in the corresponding update summation (7.4.1). Assuming that A is in shared memory and that all other variables are local, here is the node program for $\text{Proc}(\mu)$:

```
MyCols = mu:p:n;
for k=1:n
   if any(Mcols==k)
      % My turn to generate a G-column
      A(k:n,k) = A(k:n,k)/sqrt(A(k,k));
   end
   barrier
   % Update columns whose indices are greater than k and in mu:p:n.
   for j=k+1:n
      if any(MyCols==j)
         A(k:n,j) = A(k:n,j)-A(k:n,k)*A(j,k);
      end
   end
   barrier
end
```

The first if is executed by only one processor, the processor that "owns" column k. The first barrier is necessary to ensure that no processor uses the new G-column until it is safely stored in shared memory. Once that happens, the p processors share in the update of columns k+1 through n. The second barrier is necessary to guarantee that column k of A is ready at the start of the next step.

See Fig.7.3 for clarification about what goes on at each step in an $n = 22$, $p = 4$ example. The boxed integers indicate the index of the G-column being produced. The other indices name columns that the processor must update.

Problems

P7.4.1 Generalize CholBlock so that it can handle the case when the dimension n is not a multiple of the block dimension p. (This means that there may be some less-than-full-size blocks on the right and bottom edges of A.)

k	Proc(1)	Proc(2)	Proc(3)	Proc(4)
1	[1] ,5,9,13,17,21	2,6,10,14,18,22	3,7,11,15,19	4,8,12,16,20
2	5,9,13,17,21	[2] ,6,10,14,18	3,7,11,15,19	4,8,12,16,20
3	5,9,13,17,21	6,10,14,18,22	[3] ,7,11,15,19	4,8,12,16,20
4	5,9,13,17,21	6,10,14,18,22	7,11,15,19	[4] ,8,12,16,20
5	[5] , 9,13,17,21	6,10,14,18,22	7,11,15,19	8,12,16,20
6	9,13,17,21	[6] ,10,14,18,22	7,11,15,19	8,12,16,20
7	9,13,17,21	10,14,18,22	[7] ,11,15,19	8,12,16,20
8	9,13,17,21	10,14,18,22	11,15,19	[8] ,12,16,20
9	[9] ,13,17,21	10,14,18	11,15,19	12,16,20
10	13,17,21	[10] ,14,18,22	11,15,19	12,16,20
11	13,17,21	14,18,22	[11] ,15,19	12,16,20
12	13,17,21	14,18,22	15,19	[12] ,16,20
13	[13] ,17,21	14,18,22	15,19	16,20
14	17,21	[14],18,22	15,19	16,20
15	17,21	18,22	[15],19	16,20
16	17,21	18,22	19	[16] ,20
17	[17],21	18,22	19	20
18	21	[18],22	19	20
19	21	22	[19]	20
20	21	22		[20]
21	[21]	22		
22		[22]		

FIGURE 7.3 Distribution of tasks ($n = 22$, $p = 4$)

P7.4.2 Write a function f = FlopBalance(n,p) that returns a p-vector f with the property that f_μ is the number of flops that Proc(μ) must perform in shared memory implementation developed in the text.

P7.4.3 Assume that $n = mp$. Rewrite the node program with the assumption that Proc(μ) computes $G(:, (\mu - 1)m + 1{:}\mu m)$. As in the previous problem, analyze the distribution of flops.

M-Files and References

Script Files

ShowLSFit	Fits a line to the square root function.
ShowLSq	Sensitivity of LS solutions.
ShowQR	Illustrates QRrot.
ShowBVP	Solves a two-point boundary value problem.
CholBench	Benchmarks various Cholesky implementations.
CholErr	Sensitivity of symmetric positive definite systems.
ShowCholBlock	Explores block size selection in CholBlock.

Function Files

Rotate	Computes a rotation to zero bottom of 2-vector.
QRrot	QR factorization via rotations.
LSq	Least squares solution via QR factorization.
CholTrid	Cholesky factorization (tridiagonal version).
CholTridSol	Solves a factored tridiagonal system.
CholScalar	Cholesky factorization (scalar version).
CholDot	Cholesky factorization (dot product version).
CholSax	Cholesky factorization (saxpy version).
CholGax	Cholesky factorization (gaxpy version).
CholRecur	Cholesky factorization (recursive version).
CholBlock	Cholesky factorization (block version).

References

Å. Björck (1996). *Numerical Methods for Least Squares Problems,* SIAM Publications, Philadelphia, PA.

J. Dongarra, I. Duff, D. Sorensen, and H. van der Vorst (1990). *Solving Linear Systems on Vector and Shared Memory Computers,* SIAM Publications, Philadelphia, PA.

C.L. Lawson and R.J. Hanson (1996). *Solving Least Squares Problems*, SIAM Publications, Philadelphia, PA.

G.H. Golub and C.F. Van Loan (1996). *Matrix Computations, Third Edition,* Johns Hopkins University Press, Baltimore, MD.

Chapter 8

Nonlinear Equations and Optimization

§**8.1** Finding Roots

§**8.2** Minimizing a Function of One Variable

§**8.3** Minimizing Multivariate Functions

§**8.4** Solving Systems of Nonlinear Equations

In this chapter we consider several types of nonlinear problems. They differ in whether or not the solution sought is a vector or a scalar and whether or not the goal is to produce a root or a minimizer of some given function. The presentation is organized around a family of "orbit" problems. Suppose the vector-valued functions

$$
\mathcal{P}_1(t) = \left[\begin{array}{c} x_1(t) \\ y_1(t) \end{array} \right] \qquad \mathcal{P}_2(t) = \left[\begin{array}{c} x_2(t) \\ y_2(t) \end{array} \right]
$$

specify the location at time t of a pair of planets that go around the Sun. Assume that the orbits are elliptical and that the Sun is situated at $(0,0)$.

Question 1. At what times t are the planets and the Sun collinear? If $f(t)$ is the sine of the angle between $\mathcal{P}_1$ and $\mathcal{P}_2$, then this problem is equivalent to finding a zero of $f(t)$. Root-finding problems with a single unknown are covered in §8.1. We focus on the bisection and Newton methods, and the MATLAB zero-finder `fzero` is also presented.

Question 2. How close do the two *planets* get for $t \in [0, t_{max}]$? If $f(t)$ is the distance from $\mathcal{P}_1$ to $\mathcal{P}_2$, then this is a single-variable minimization problem. In §8.2 we develop the method of golden section search and discuss the
Matlab minimizer `fmin`. The role of graphics in building intuition about a search-for-a-min problem is highlighted.

Question 3. How close do the two *orbits* get? This is a minimization problem in two variables. For example,

$$
\min_{t_1, t_2} \| \mathcal{P}_1(t_1) - \mathcal{P}_2(t_2) \|_2
$$

The method of steepest descent and the MATLAB multivariable minimizer fmins are designed to solve problems of this variety. They are discussed in §8.3 along with the idea of a line search.

Question 4. Where (if at all) do the two orbits intersect? This is an example of a multivariable root-finding problem:

$$F(t_1, t_2) = \mathcal{P}_2(t_2) - \mathcal{P}_1(t_1) = \left[\begin{array}{c} x_2(t_2) - x_1(t_1) \\ y_2(t_2) - y_1(t_1) \end{array} \right] = 0$$

The Newton framework for systems of nonlinear equations is discussed in §8.4. Related topics include the use of finite differences to approximate the Jacobian and the Gauss-Newton method for the nonlinear least squares problem.

8.1 Finding Roots

Suppose a chemical reaction is taking place and the concentration of a particular ion at time t is given by $10e^{-3t} + 2e^{-5t}$. We want to know when this concentration is one-half of its value at $t = 0$. Since $f(0) = 12$, this is equivalent to finding a zero of the function

$$f(t) = 10e^{-3t} + 2e^{-5t} - 6$$

This particular problem has a unique root. In other zero-finding applications, the function involved may have several roots, some of which are required. For example, we may want to find the largest and smallest zeros (in absolute value) of the quintic polynomial $x^5 - 2x^4 + x^2 - x + 7$. In this section we discuss the bisection and Newton frameworks that can be applied to problems of this kind.

Algorithms in this area are *iterative* and proceed by producing a sequence of numbers that (it is hoped) converge to a root of interest. The implementation of any iterative technique requires that we deal with three major issues:

- Where do we start the iteration?

- Does the iteration converge, and how fast?

- How do we know when to terminate the iteration?

To build an appreciation for these issues, we start with a discussion about computing square roots.

8.1.1   The Square Root Problem

Suppose A is a positive real number and that we want to compute its square root. Geometrically, this task is equivalent to constructing a square whose area is A. Having an approximate square root x_c is equivalent to having an approximating rectangle with sides x_c and A/x_c. To make the approximating rectangle "more square," it is reasonable to replace x_c with

$$x_+ = \frac{1}{2}\left(x_c + \frac{A}{x_c}\right)$$

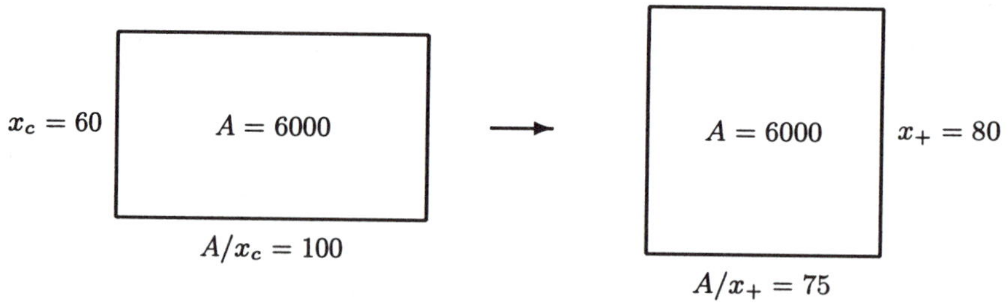

FIGURE 8.1. A better square root.

since the value sought is clearly in between x_c and A/x_c. (See Fig 8.1 from which we conclude that $\sqrt{6000}$ is in between $x_c = 60$ and $A/x_c = 100$.) The process can obviously be repeated and it appears to converge:

xc	6000/xc
60.0000000000000000	100.0000000000000000
80.0000000000000000	75.0000000000000000
77.5000000000000000	77.4193548387096797
77.4596774193548470	77.4596564289432479
77.4596669241490474	77.4596669241476263
77.4596669241483369	77.4596669241483369
77.4596669241483369	77.4596669241483369
:	:

So much for how to improve an approximate square root. How do we produce an initial guess? The "search space" for the square root problem can be significantly reduced by writing A in the form

$$A = m \times 4^e \qquad \frac{1}{4} \le m < 1$$

where e is an integer, for then

$$\sqrt{A} = \sqrt{m} \times 2^e$$

Thus an arbitrary square root problem reduces to the problem of computing the square root of a number in the range $[.25, 1]$.

A good initial guess for the reduced range problem can be obtained by linear interpolation with

$$L(m) = (1 + 2m)/3$$

This function interpolates $f(m) = \sqrt{m}$ at $m = .25$ and $m = 1$. (See Fig 8.2.) Moreover, it can

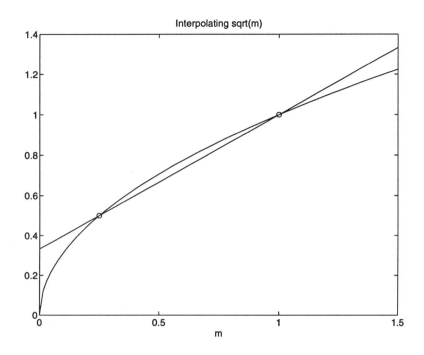

FIGURE 8.2 Approximating $\sqrt{m}$ with $L(m) = (1 + 2m/3)$

be shown that

$$|L(m) - \sqrt{m}| \leq .05 \tag{8.1.1}$$

for all m in the interval. It follows that $L(m)$ can serve as a reasonable starting value.

This brings us to the analysis of the iteration itself. We can use the equation

$$x_+ - \sqrt{A} = \frac{1}{2}\left(x_c + \frac{A}{x_c}\right) - \sqrt{A} = \frac{1}{2}\left(\frac{x_c - \sqrt{A}}{\sqrt{x_c}}\right)^2 \tag{8.1.2}$$

to bound the error after k steps. Switching to subscripts, let $x_0 = L(m)$ be the initial guess and let x_k be the kth iterate. If $e_k = x_k - \sqrt{A}$, then (8.1.2) says that

$$e_{k+1} = \frac{1}{2x_k}e_k^2$$

It can be shown that the iterates x_k are always in the interval $[.5, 1]$, and so $e_{k+1} \leq e_k^2$. Thus

$$e_4 \leq e_3^2 \leq e_2^4 \leq e_1^8 \leq e_0^{16} \leq (.05)^{16}$$

implying that four steps are adequate if the unit roundoff is in the vicinity of 10^{-16}. Combining the range reduction with this analysis, we obtain the following implementation:

```
function x = Mysqrt(A)
%
% Pre:  A  is non-negative.
% Post: x  the square root of A.
%
   if A==0
      x = 0;
   else
      m:=A;
      TwoPower = 1;
      while m>=1
         m = m/4;
         TwoPower = 2*TwoPower;
      end
      while m < .25
         m = m*4;
         TwoPower = TwoPower/2;
      end;
      x = (1+2*m)/3;
      for k=1:4
         x = (x + (m/x))/2;
      end
      x = x*TwoPower;
   end
```

Notice the special handling of the $A = 0$ case. The script ShowMySqrt plots the relative error of MySqrt across any specified interval. (See Fig 8.3.)

Problems

P8.1.1 Another idea for a starting value in MySqrt is to choose the parameters s and t so that the maximum value of $|s + t\alpha - \sqrt{\alpha}|$ is minimized on $[.25, 1]$. Implement this change in MySqrt and examine the error.

P8.1.2 Repeat the previous problem with s and t chosen to minimize

$$e(s, t) = \int_{.25}^{1} [s + tm - \sqrt{m}]^2 dm$$

P8.1.3 Prove (8.1.1).

P8.1.4 Vectorize MySqrt so that it can handle the case when A is a matrix of nonnegative numbers.

8.1.2   Bisection

If a continuous function $f(x)$ changes sign on an interval, then it must have at least one root in the interval. This fact can be used to produce a sequence of ever-smaller intervals that each bracket a root of $f(x)$. To see this, assume that $f(a)f(b) \leq 0$ and let $m = (a+b)/2$, the midpoint of $[a, b]$. It follows that either $f(a)f(m) \leq 0$ or $f(m)f(b) \leq 0$. In the former case we know that

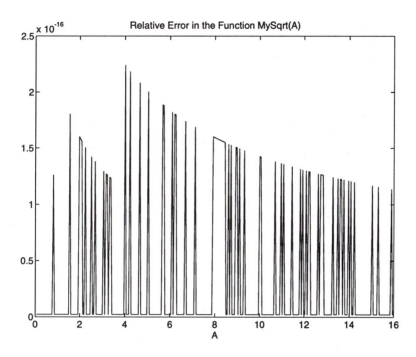

FIGURE 8.3 Relative error in `MySqrt`

there is a root in $[a, m]$ while the latter situation implies that there is a root in $[m, b]$. In either case, we are left with a half-sized bracketing interval. The halving process can continue until the current interval is shorter than a designated positive tolerance `delta`:

```
while abs(a-b) > delta
   if f(a)*f((a+b)/2) <= 0
      b = (a+b)/2;
   else
      a = (a+b)/2;
   end
end
root = (a+b)/2;
```

This is "version zero" of the method of *bisection*. It needs to be cleaned up in two ways. First, the `while`-loop may never terminate if `delta` is smaller than the spacing of the floating point numbers between a and b. To rectify this, we change the `while`-loop control to

```
while abs(a-b) > delta+eps*max(abs(a),abs(b))
```

This guarantees termination even if `delta` is too small. The second flaw in the preceding implementation is that it requires two function evaluations per iteration. However, with a little manipulation, it is possible to reduce this overhead to just one f-evaluation per step. Overall we obtain

```
    function root = Bisection(fname,a,b,delta)
%
% Pre:
%     fname    string that names a continuous function  f(x) of
%              a single variable.
%       a,b    define an interval upon which f is defined and f(a)f(b) <= 0
%     delta    non-negative real.
% Post:
%      root    the midpoint of an interval [alpha,beta]
%                with the property that f(alpha)f(beta)<=0 and
%                |beta-alpha| <= delta+eps*max(|alpha|,|beta|)
%
    fa = feval(fname,a);
    fb = feval(fname,b);
    while abs(a-b) > delta+eps*max(abs(a),abs(b))
       mid = (a+b)/2;
       fmid = feval(fName,mid);
       if fa*fmid<=0
          % There is a root in [a,mid].
          b  = mid;
          fb = fmid;
       else
          % There is a root in [mid,b].
          a  = mid;
          fa = fmid;
       end
    end
    root = (a+b)/2;
```

The script ShowBisect can be used to trace the progress of the iteration. It should be stressed that the time required by a root-finder like Bisection is proportional to the *number of function-evaluations.* The arithmetic that takes place outside of the f-evaluations is typically insignificant.

If $[a_k, b_k]$ is the k-th bracketing interval, then $|a_k - b_k| \le |a_0 - b_0|/2^k$ showing that convergence is guaranteed. Thinking of $x_k = (a_k + b_k)/2$ as the kth iterate, we see that there exists a root x_* for $f(x)$ with the property that

$$|t_k - t_*| \le \frac{|a_0 - b_0|}{2^{k+1}}$$

The error bound is halved at each step. This is an example of *linear convergence.* In general, a sequence $\{x_k\}$ converges to x_* linearly if there is a constant c in the interval $[0, 1)$ and an integer k_0 such that

$$|x_{k+1} - x_*| \le c|x_k - x_*|$$

for all $k \ge k_0$. To illustrate what linear convergence is like, we apply Bisection to $f(x) = \tan(x/4) - 1$ with initial interval $[a_0, b_0] = [2, 4]$:

k	a_k	b_k	$b_k - a_k$
0	2.00000000000000	4.00000000000000	2.00000000000000
1	3.00000000000000	4.00000000000000	1.00000000000000
2	3.00000000000000	3.50000000000000	0.50000000000000
3	3.00000000000000	3.25000000000000	0.25000000000000
4	3.12500000000000	3.25000000000000	0.12500000000000
5	3.12500000000000	3.18750000000000	0.06250000000000
$\vdots$	$\vdots$	$\vdots$	$\vdots$
43	3.14159265358967	3.14159265358990	0.00000000000023
44	3.14159265358978	3.14159265358990	0.00000000000011
45	3.14159265358978	3.14159265358984	0.00000000000006
46	3.14159265358978	3.14159265358981	0.00000000000003
47	3.14159265358978	3.14159265358980	0.00000000000001

Notice that a new digit of π is acquired every three or so iterations. This kind of uniform acquisition of significant digits is the hallmark of methods that converge linearly.

Problems

P8.1.5 What can go wrong in `Bisection` if the comparison `fa*fmid<=0` is changed to `fa*fmid<0`? Give an example.

P8.1.6 Consider the following recursive formulation of bisection:

```
function x = rBisection(fname,a,b,delta)
   if abs(a-b) <= delta
      x = (a+b)/2;
   else
      mid = (a+b)/2;
      if eval([ fname '(a)*' fname '(c) <=0])
         x = rBisection(fname,a,c,delta);
      else
         x = rBisection(fname,c,b,delta);
      end
   end
```

Modify this procedure so that it does not involve redundant function evaluations.

P8.1.7 Complete the following MATLAB function:

```
function r = MiddleRoot(a,b,c,d)
% Pre: The cubic equation ax^3 + bx^2 + cx + d = 0 has three real distinct roots.
%
% Post:  r is within 100*EPS of a true root.
```

Write a script that prints the middle root of all cubics of the form $x^3 + bx^2 + cx + d$ where b, c, and d are integers from the interval $[-2, 2]$ so that the resulting cubic has three distinct real roots. Note that a cubic $c(x)$ has three real distinct roots if $c'(x)$ has two real distinct roots r_1 and r_2 Note that the interval defined by r_1 and r_2 brackets the middle root of $c(x)$.

P8.1.8 Assume that the function $f(x)$ is positive on $[0, 1]$ and that

$$\int_0^1 f(x)dx = 1.$$

Write a complete, efficient MATLAB fragment that uses bisection to compute z so that

$$\left| \int_0^z f(x)dx - \frac{1}{2} \right| \leq 10^{-14}$$

Assume f is an available MATLAB function. Compute all integrals via the MATLAB function Quad(fname,a,b,tol) with tol = 10*EPS. [For your information, q = Quad('g',a,b,tol) assigns to q the integral of $g(x)$ from a to b with an error tolerance *tol*.]

P8.1.9 Let A be the 8-by-8 pentadiagonal matrix

$$A = \begin{bmatrix}
6 & -4 & 1 & 0 & 0 & 0 & 0 & 0 \\
-4 & 6 & -4 & 1 & 0 & 0 & 0 & 0 \\
1 & -4 & 6 & -4 & 1 & 0 & 0 & 0 \\
0 & 1 & -4 & 6 & -4 & 1 & 0 & 0 \\
0 & 0 & 1 & -4 & 6 & -4 & 1 & 0 \\
0 & 0 & 0 & 1 & -4 & 6 & -4 & 1 \\
0 & 0 & 0 & 0 & 1 & -4 & 6 & -4 \\
0 & 0 & 0 & 0 & 0 & 1 & -4 & 6
\end{bmatrix}$$

Let $b = (1:8)'$. To within 12 significant digits, determine $\alpha > 0$ so that the solution to $(A + \alpha I)x = ones(8,1)$ satisfies $x^T x = 1$. (Here, I is the 8×8 identity.)

8.1.3   The Newton Method Idea

Suppose we have the value of a function $f(x)$ and its derivative $f'(x)$ at $x = x_c$. The tangent line

$$L_c(x) = f(x_c) + (x - x_c)f'(x_c)$$

can be thought of as a linear model of the function at this point. (See Fig 8.4.) The zero x_+ of

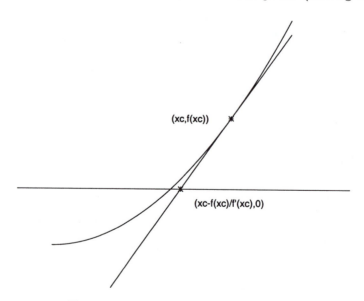

FIGURE 8.4 Modeling a nonlinear function

$L_c(x)$ is given by

$$x_+ = x_c - \frac{f(x_c)}{f'(x_c)}$$

If $L_c(x)$ is a good model over a sufficiently wide range of values, then x_+ should be a good approximation to a root of f. The repeated use of the x_+ update formula defines the *Newton framework*:

```
xc = input('Enter starting value:');
fc = feval(fname,xc);
fpc = feval(fpName,xc);
while input('Newton step?  (0=no, 1=yes)');
    xnew = xc - fc/fpc;
    xc = xnew;
    fc = feval(fName,xc);
    fpc = feval(fpName,xc);
end
```

This assumes that `fname` and `fpname` are strings that house respectively the name of a function and the name of its derivative (e.g., `fname = 'sin'`; `fpname = 'cos'`). As an illustration, we apply the Newton framework to the problem of finding a zero of the function $f(x) = \tan(x/4) - 1$:

k	x_k	$|x_k - \pi|$
0	1.00000000000000	2.14159265358979
1	3.79631404657234	0.65472139298255
2	3.25943543617547	0.11784278258568
3	3.14513155420752	0.00353890061772
4	3.14159578639006	0.00000313280027
5	3.14159265359225	0.00000000000245
6	3.14159265358979	0.00000000000000

This should be compared to the 47 bisection steps required to produce the same accuracy. Apparently, after a few iterations we reach a stage at which the error is approximately squared at each iteration. This characterizes methods that converge *quadratically*. For a sequence with this property, there exists an integer k_0 and a positive constant c such that

$$|x_{k+1} - x_*| \leq c|x_k - x_*|^2$$

for all $k \geq k_0$. Quadratic convergence is a much-sought-after commodity in the root-finding business.

However, unlike the plodding but guaranteed-to-converge method of bisection, big things can go wrong during the Newton iteration. There is no guarantee that x_+ is is closer to a root than x_c. After all, it is produced by a linear model and f may be very nonlinear. Moreover, a small value for $f'(x_*)$ works against rapid convergence. If the first derivative is small at the solution, then the tangent line on which the Newton step is based is nearly horizontal and it has trouble predicting x_*. Here is a table showing the number of iterations that are required to zero the function $f(x) = x^2 - a^2$ with starting value $t = 2$:

a	Iterations
10^0	5
10^{-4}	18
10^{-8}	31
10^{-12}	43
0	50

Note that $f'(x_*) = 2a$ and that the rate of convergence drifts from strongly quadratic to essentially linear as a decreases from 1 to 0.

But even worse things can happen because of small derivatives. If $f'(x_c)$ is small relative to $f(x_c)$, then the Newton step can transport us to a region far away from a root. The classic example of this is the function $f(x) = \arctan(x)$. Since $f'(x) = 1/(1+x^2)$, the Newton update has the form $x_+ = x_c - (1+x_c^2)\arctan(x_c)$. It can be shown that if $|x_c| > 1.3917$, then $|x_+| > |x_c|$. This implies that the iteration *diverges* for starting values outside $[-1.3917, 1.3917]$.

Explore the reliability of the Newton iteration by experimenting with the script ShowNewton. To guarantee convergence for any starting value in the vicinity of a root, you will discover that f cannot be too nonlinear and that the f' must be bounded away from zero. The following theorem makes this precise.

Theorem 8 *Suppose $f(x)$ and $f'(x)$ are defined on an interval $I = [x_* - \mu, x_* + \mu]$, where $f(x_*) = 0$, $\mu > 0$, and positive constants ρ and δ exist such that*

$$(1) \quad |f'(x)| \geq \rho \text{ for all } x \in I.$$
$$(2) \quad |f'(x) - f'(y)| \leq \delta|x - y| \text{ for all } x, y \in I.$$
$$(3) \quad \mu \leq \rho/\delta.$$

If $x_c \in I$, then $x_+ = x_c - f(x_c)/f'(x_c) \in I$ and

$$|x_+ - x_*| \leq \frac{\delta}{2\rho}|x_c - x_*|^2 \leq \frac{1}{2}|x_c - x_*|$$

Thus for any $x_c \in I$, a Newton step more than halves the distance to x_*, thereby guaranteeing convergence for any starting value in I. The rate of convergence is quadratic.

Proof. From the fundamental theorem of calculus

$$-f(x_c) = f(x_*) - f(x_c) = \int_{x_c}^{x_*} f'(z)dz$$

and so

$$-f(x_c) - (x_* - x_c)f'(x_c) = \int_{x_c}^{x_*} (f'(z) - f'(x_c))\, dz$$

Dividing both sides by $f'(x_c)$ (which we know is nonzero), we get

$$\left(x_c - \frac{f(x_c)}{f'(x_c)}\right) - x_* = \frac{1}{f'(x_c)}\int_{x_c}^{x_*} (f'(z) - f'(x_c))dz$$

Taking absolute values, using the definition of x_+, and invoking (1) and (2) we obtain

$$|x_+ - x_*| \leq \frac{1}{\rho}\int_{x_c}^{x_*} |f'(z) - f'(x_c)|\, dz \leq \frac{\delta}{\rho}\int_{x_c}^{x_*} |z - x_c|\, dz = \frac{\delta}{2\rho}|x_c - x_*|^2$$

Since $x_c \in I$, it follows that $|x_c - x_*| \leq \mu$, and so by (3) we have

$$\frac{\delta}{2\rho} |x_c - x_*|^2 \leq \frac{1}{2} \frac{\delta \mu}{\rho} |x_c - x_*| \leq \frac{1}{2} |x_c - x_*|$$

This completes the proof since it shows that $x_+ \in I$. □

The theorem shows that if (1) f' does not change sign in a neighborhood of x_*, (2) f is not too nonlinear, and (3) the Newton iteration starts close enough to x_*, then convergence is guaranteed and at a quadratic rate.

For all but the simplest functions, it may be impossible to verify that the conditions of the theorem apply. A general-purpose, Newton-based nonlinear equation solver requires careful packaging in order for it to be useful in practice. This includes the intelligent handling of termination. Unlike bisection, where the current interval length bounds the error, we have to rely on heuristics in order to conclude that an iterate is acceptable as a computed root. One possibility is to quit as soon as $|x_+ - x_c|$ is small enough. After all, if $\lim x_k = x_*$, then $\lim |x_{k+1} - x_k| = 0$. But the converse does not necessarily follow. If $f(x) = \tan(x)$ and $x_c = \pi/2 - 10^{-5}$, then $|x_+ - x_c| \approx 10^{-15}$ even though the nearest root is at $x = 0$. Alternatively, we could terminate as soon as $|f(x_c)|$ is small. But this does *not* mean that x_c is close to a true root. For example, if $f(x) = (x - 1)^{10}$, then $x_c = 1.1$ makes f very small even though it is relatively far away from the root $x_* = 1$.

8.1.4 Globalizing the Newton Iteration

The difficulties just discussed can be handled with success by combining the Newton and bisection ideas in a way that captures the best features of each framework. At the beginning of a step we assume that we have a bracketing interval $[a, b]$ and that x_c equals one of the endpoints. If

$$x_+ = x_c - \frac{f(x_c)}{f'(x_c)} \in [a, b]$$

then we proceed with either $[a, x_+]$ or $[x_+, b]$, whichever is bracketing. The new x_c equals x_+. If the Newton step carries us out of $[a, b]$, then we take a bisection step setting the new x_c to $(a + b)/2$. To check whether the Newton step stays in $[a, b]$, we use

```
    function ok = StepIsIn(x,fx,fpx,a,b)
%
% Pre:
%       x       current iterate.
%       fx      the value of f  at x.
%       fpx     the value of f' at x.
%       a,b     [a,b] is the interval of interest.
%
% Post:
%       ok      1 if the Newton step  x - fx/fpx is in [a,b].
%               0 if not.
%
```

```
      if fpx > 0
         ok = ((a-x)*fpx <= -fx) & (-fx <= (b-x)*fpx);
      elseif fpx < 0
         ok = ((a-x)*fpx >= -fx) & (-fx >= (b-x)*fpx);
      else
         ok = 0;
      end
```

To guarantee termination, we bring the iteration to a halt if any of the following three conditions hold:

- The length of the current bracketing interval is less than tolx, a specified nonnegative tolerance. This ensures that the computed root is within tolx of a true root. It does *not* guarantee that f is small. For example, if $f(x) = (x - 1)^{.1}$, then $x_c = 1.0000000001$ is very close to the root $x_* = 1$, but $f(x_c)$ is not particularly small.

- The absolute value of $f(x_c)$ is less than or equal to tolf, a specified nonnegative tolerance. This does *not* mean that x_c is close to a true root. For example, if $f(x) = (x - 1)^{10}$, then $x_c = 1.1$ makes f very small even though it is relatively far away from the root $x_* = 1$.

- The number of f-evaluations exceeds a specified positive integer nEvalsMax. This means that neither of the preceding two conditions is satisfied.

Putting it all together, we obtain

```
      function [x,fx,nEvals,aLast,bLast] =
                  GlobalNewton(fName,fpName,a,b,tolx,tolf,nEvalsMax)
%
% Pre:
%    fName       string that names a function f(x).
%    fpName      string that names the derivative function f'(x).
%    a,b         A root of f(x) is sought in the interval [a,b]
%                and f(a)*f(b)<=0.
%    tolx,tolf   Nonnegative termination criteria.
%    nEvalsMax   Maximum number of derivative evaluations.
%
% Post:
%    x           An approximate zero of f.
%    fx          The value of f at x.
%    nEvals      The number of derivative evaluations required.
%    aF,bF       The final bracketing interval is [aF,bF].
%
% Comments:
%    Iteration terminates as soon as x is with tolx of a true zero or
%    if |f(x)|<= tolf or after nEvalMax f-evaluations.
```

```
fa  = feval(fName,a);
fb  = feval(fName,b);
x   = (a+b)/2;
fx  = feval(fName,x);
fpx = feval(fpName,x);

nEvals = 1;
while (abs(a-b)>tolx) & (abs(fx)>tolf) & ((nEvals<nEvalsMax) | (nEvals==1))
    %[a,b] brackets a root and x = a or x = b.
    if StepIsIn(x,fx,fpx,a,b)
        %Take Newton Step
        x = x-fx/fpx;
    else
        %Take a Bisection Step:
        x = (a+b)/2;
    end
    fx  = feval(fName,x);
    fpx = feval(fpName,x);
    nEvals = nEvals+1;
    if fa*fx<=0
        % There is a root in [a,x]. Bring in right endpoint.
        b  = x;
        fb = fx;
    else
        % There is a root in [x,b]. Bring in left endpoint.
        a  = x;
        fa = fx;
    end
end
aF = a;
bF = b;
```

In a typical situation, a number of bisection steps are taken before the Newton iteration takes over. Here is what happens when we try to find a zero of $f(x) = \sin(x)$ with initial bracketing interval $[-7\pi/2, 15\pi + .1]$:

Step	a	x	b
<Start>	-10.995574287564276	-10.995574287564276	47.223889803846895
Bisection	-10.995574287564276	18.114157758141310	18.114157758141310
Bisection	-10.995574287564276	3.559291735288517	3.559291735288517
Newton	3.115476144648328	3.115476144648328	3.559291735288517
Newton	3.115476144648328	3.141598592990409	3.141598592990409
Newton	3.141592653589793	3.141592653589793	3.141598592990409

Problems

P8.1.10 The function $f(z) = z^4 - 1$ has four roots: $r_1 = 1$, $r_2 = i$, $r_3 = -1$, and $r_4 = -i$. The Newton iteration is defined for complex numbers, so if we repeatedly apply the update

$$z_+ = z_c - \frac{z_c^4 - 1}{4z_c^3} = \frac{1}{4}\left(3z_c - \frac{1}{z_c^3}\right)$$

the iterates converge to one of the four roots. (1) Determine the largest value of ρ so that if z_0 is within ρ of a root, then Newton iteration converges to the same root. (2) Write a function [steps,i] = WhichRoot(z0) that returns the index i of the root obtained by the Newton iteration when z0 is the starting value. The number of Newton steps required to get with ρ of r_i should be returned in steps. (3) We say that the line segment that connects z_1 and z_2 has property *two root* if WhichRoot(z1) and WhichRoot(z2) indicate convergence to two different roots. Write a function

```
function [w1,w2] = Close(z1,z2,delta)
Pre:
   z1,z2    complex scalars that define a two root segment
   delta    positive real scalar.
Post:
   w1,w2    complex scalars that define a two-root segment
            and satisfy |w1-w2| <= delta*|z1-z2|
```

8.1.5   Avoiding Derivatives

A problem with the Newton framework is that it requires software for both f and its derivative f'. In many instances, the latter is difficult or impossible to encode. A way around this is to approximate the required derivative with an appropriate divided difference:

$$f'(x_c) \approx \frac{f(x_c + \delta_c) - f(x_c)}{\delta_c}$$

This leads to the *finite difference Newton framework*:

```
fval = feval(fname,x);
Choose delta.
fpval = (feval(fname,x+delta) - fval)/delta']);
x = x - fval/fpval;
```

The choice of δ_c requires some care if quadratic-like convergence is to be ensured. A good heuristic would involve $|x_c|$, $|f(x_c)|$, and the machine precision. Note that a step requires two f-evaluations.

In the *secant method*, $f'(x_c)$ is approximated with the divided difference

$$f'(x_c) \approx \frac{f(x_c) - f(x_-)}{x_c - x_-}$$

where x_- is the iterate before x_c. A step in the secant iteration requires a single function evaluation:

```
fpc = (fc - f_)/(xc - x_);
xnew = xc - fc/fpc;
x_ = xc;  f_ = fc;
xc = xnew;  fc = feval(fName,xc);
```

The method requires two starting values. Typically, an initial guess $\tilde{x}$ and a slight perturbation $\tilde{x} + \delta$ are used.

If the iteration converges, then it is usually the case that

$$|x_{k+1} - x_*| \leq c|x_k - x_*|^r$$

where

$$r = \frac{1 + \sqrt{5}}{2} \approx 1.6$$

for all sufficiently large k. Applying the secant method to the function $f(x) = \tan(x/4) - 1$ with starting values $x_0 = 1$ and $x_1 = 2$, we get

k	x_k	$|x_k - \pi|$
0	1.00000000000000	2.14159265358979
1	2.00000000000000	1.14159265358979
2	3.55930926415136	0.41771661056157
3	3.02848476491863	0.11310788867116
4	3.12946888739926	0.01212376619053
5	3.14193188940880	0.00033923581901
6	3.14159162639551	0.00000102719428
7	3.14159265350268	0.00000000008712
8	3.14159265358979	0.00000000000000

Thus there is only a small increase in the number of iterations compared to the Newton method. Note that two starting values are required. A finite difference Newton step could be used to get the extra function value that is necessary to initiate the iteration.

It is important to recognize that the total number of function evaluations required by a root-finding framework depends on the rate of convergence of the underlying method *and* how long it takes for the iterates to get close enough for the local convergence theory to "take hold." This makes it impossible, for example, to assert that the Newton approach is quicker than the secant approach. There are lots of application-specific factors that enter the efficiency equation: quality of initial guess, the second derivative behavior of f, the relative cost of f-evaluations and f'-evaluations, etc.

Problems

P8.1.11 Write a script file that illustrates the finite difference Newton method framework. Experiment with the choice of the difference parameter δ_c.

P8.1.12 Write a ShowSecant script file that illustrates the secant method framework.

8.1.6 The MATLAB fzero Function

The MATLAB function fzero is a general-purpose root finder that does not require derivatives. A simple call involves only the name of the function and a starting value. For example, if

```
function y = decay(t);
y = 10*exp(-3*t) +2*exp(-2*t) -6
```

is available, then

```
root = fzero('decay',1)
```

assigns the value of the root .24620829278302 to root.

If all goes well, then fzero returns a computed root r that is within $|r|$EPS of a true root. A less stringent error tolerance can be used by specifying a third parameter. The call

```
root = fzero('decay',1,.001)
```

returns .24666930469558, which is correct to three significant digits.

The function to be passed to fzero must be a function of a single scalar variable. Thus fzero cannot be used to find a zero of

```
function y = cubic(a,b,c,d,x);
    y = ((a*x + b)*x + c)*x + d;
```

However, if a,b,c, and d are global variables, then we can apply fzero to

```
function y = cubic(x);
    global a b c d
    y = ((a*x + b)*x + c)*x + d;
```

As a more involved application of fzero, suppose the locations of planets M and E at time t are given by

$$
\begin{align}
x_M(t) &= -11.9084 + 57.9117\cos(2\pi t/87.97) \tag{8.1.3}\\
y_M(t) &= 56.6741\sin(2\pi t/87.97) \tag{8.1.4}
\end{align}
$$

and

$$
\begin{align}
x_E(t) &= -2.4987 + 149.6041\cos(2\pi t/365.25) \tag{8.1.5}\\
y_E(t) &= 149.5832\sin(2\pi t/365.25) \tag{8.1.6}
\end{align}
$$

These are crude impersonations of Mercury and Earth. Both orbits are elliptical with one focus, the Sun, situated at $(0,0)$. To an observer on E, M is in *conjunction* if it is located on the Sun-to-Earth line segment. Clearly there is a conjunction at $t = 0$. Our goal is to compute the time of the next 10 conjunctions and the spacing between them. To that end, we define the function

```
function s = SineMercEarth(t)

% Post: The sine of the Mercury-Sun-Earth angle.

  % Mercury location:
  xm = -11.9084 +  57.9117*cos(2*pi*t/87.97);
  ym =             56.6741*sin(2*pi*t/87.97);
  % Earth location:
  xe =  -2.4987 + 149.6041*cos(2*pi*t/365.25);
  ye =            149.5832*sin(2*pi*t/365.25);
  s = (xm.*ye - xe.*ym)./(sqrt(xm.^2 + ym.^2).*sqrt(xe.^2 + ye.^2));
```

and note that at the time of conjunction, this function is zero. Before we go after the next 10 conjunction times, we plot SineMercEarth to get a rough idea about the spacing of the conjunctions. (See Fig 8.5.) At first glance it looks like one occurs every 60 or so days. However,

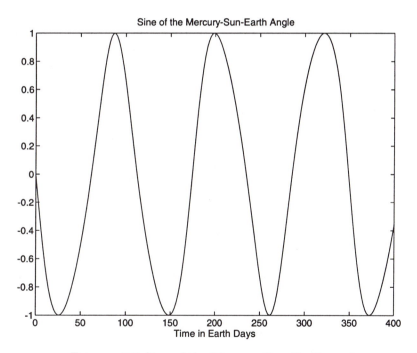

FIGURE 8.5 Sine of the Mercury-Sun-Earth angle

SineMercEarth is also zero whenever the two planets are aligned with the sun in the middle. (*M* is then said to be in *Opposition.*) It follows that the spacing between conjunctions is about 120 days and so it is reasonable to call fzero with starting value $120k$ in order to compute the time of the kth conjunction. Here is a script file that permits the input of an arbitrary spacing factor and then computes the conjunctions accordingly:

```
% Script File: FindConj
%
% Estimates spacing between Mercury-Earth Conjunctions

    clc
    GapEst = input('Spacing Estimate = ');
    clc
    disp(sprintf('Next ten conjunctions, Spacing Estimate = %8.3f',GapEst))
    disp(' ')
    t = zeros(11,1);
    disp('Conjunction    Time        Spacing')
    disp('----------------------------------')
    for k=1:10;
    t(k+1) = fzero('SineMercEarth',k*GapEst);
    disp(sprintf('    %2.0f       %8.3f    %8.3f',k,t(k+1),t(k+1)-t(k)))
    end
```

If we respond with a spacing factor equal to 115, then here is the output:

```
          Next ten conjunctions, Spacing Estimate =  115.000

          Conjunction    Time      Spacing
          ------------------------------------
                1       112.476     112.476
                2       234.682     122.206
                3       348.554     113.872
                4       459.986     111.433
                5       581.491     121.505
                6       697.052     115.561
                7       807.815     110.762
                8       928.020     120.206
                9      1045.440     117.420
               10      1155.908     110.467
```

Problems

P8.1.13 A call of the form root = fzero('f',x0) can be used to compute a zero of the function $f(x)$ near $x = x_0$. Write a MATLAB function z = cubeRoot(a) that uses fzero to compute the cube root of a. (You cannot just set z = z^(1/3).) Notice that if x is the cube root of a then $x^3 = a$. You may assume that $1 \leq a \leq 8$ and that $(6 + a)/7$ is a good initial guess for the required root. Hint: You will have to write a second function whose name is passed to fzero. Be sure to indicate clearly any global variables that are required by your solution.

P8.1.14 Determine the longest possible interval $[L, R]$ so that if GapEst is in this interval, then the 10 required conjunctions are computed in FindConj.

8.2 Minimizing a Function of a Single Variable

Suppose we want to compute the minimum Mercury-Earth separation over the next 1000 days given that their respective locations $[(x_M(t), y_M(t))$ and $(x_E(t), y_E(t))]$ are specified by (8.1.3) through (8.1.6). Thus we want to solve the problem

$$\min_{t \in S} \; f(t)$$

where

$$f(t) = \sqrt{(x_E(t) - x_M(t))^2 + (y_E(t) - y_M(t))^2}$$

and S is the set $\{t : 0 \leq t \leq 1000\}$. This is an example of an *optimization* problem in one dimension. The function f that we are trying to optimize is called the *objective function*. The set S may be the entire set of real numbers or a subset thereof because of constraints on the independent variable.

The optimization problems considered in this book can be cast in find-the-max or find-the-min terms. We work with the latter formulation, noting that any max problem can be turned into a min problem simply by negating the objective function.

8.2.1 Graphical Search

If function evaluations are cheap, then plotting can be used to narrow the search for a minimizer. We have designed a simple interactive search environment for this purpose, and its specification is as follows:

```
    function [L,R] = GraphSearch(fname,a,b,Save,nfevals)
%
% Pre:
%    fname is a string that names a function f(x) that is defined
%    on the interval [a,b]. nfevals>=2
%
% Post:
%    Produces sequence of plots of the function f(x). The user specifies
%    the x-ranges by mouseclicks. The fourth argument is used to determine
%    how the plots are saved.
%    If Save is nonzero, then each plot is saved in a separate figure window.
%    If Save is zero or if GraphSearch is called with just three arguments,
%    then only the final plot is saved. [L,R] is the x-range of the final plot.
%    The plots are based on nfevals function evaluations.
%    If GraphSearch is called with less than five arguments, then nfevals is
%    set to 100.
```

Let's apply GraphSearch to the problem of minimizing the following Mercury-to-Earth separation function:

```
    function s = DistMercEarth(t)
%
% Post: The distance from Mercury to Earth.

    xm = -11.9084 +  57.9117*cos(2*pi*t/87.97);
    ym =             56.6741*sin(2*pi*t/87.97);
    xe =  -2.4987 + 149.6041*cos(2*pi*t/365.25);
    ye =            149.5832*sin(2*pi*t/365.25);
    s = sqrt((xe-xm).^2 + (ye-ym).^2);
```

The initial plot is displayed in Fig 8.6. Whenever GraphSearch displays a plot, it prompts

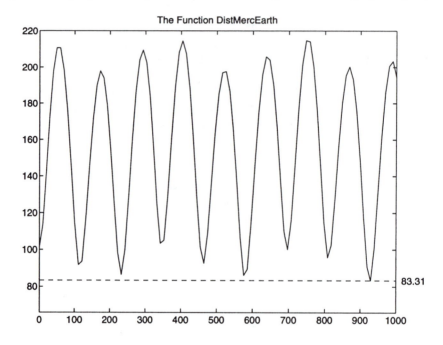

FIGURE 8.6 The GraphSearch environment

for a new, presumably smaller "interval of interest." Left and right endpoints are obtained by mouseclicks. From Fig 8.6 we see that the minimum Mercury-Earth separation is somewhere in the interval [900, 950]. Clicking in the endpoints renders the refined plot shown in Fig 8.7. By repeating the process, we can "zoom" in to relatively short intervals of interest.

Note that our example has many *local minima* across [0, 1000], and this brings up an important aspect of the optimization problem. In some problems, such as the one under consideration, a *global minimum* is sought. In other settings some or all of the *local minima* are of interest. Detailed knowledge of the objective function is usually required to tackle a global min problem. This may come from plotting or from taking into consideration various aspects of the underlying application. The search for a local minimum is easier to automate and is the next item on our agenda.

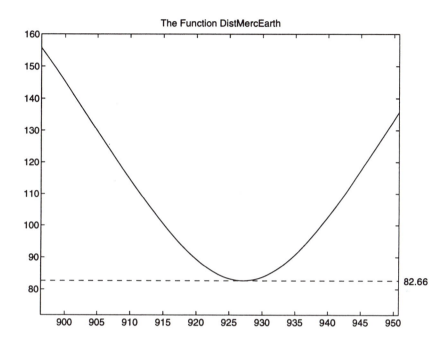

FIGURE 8.7 A refined plot

Problems

P8.2.1 Modify GraphSearch so that on the first plot, an arbitrary number of intervals of interest can be solicited (e.g., $[a_i, b_i]$, $i = 1{:}n$). Each of these should then be graphically searched with the final plot being saved. Thus, upon completion figures 1 through n should contain the final plots for each of the n searches.

8.2.2 Golden Section Search

A function $f(t)$ is *unimodal* on $[a, b]$ if there is a point $t_* \in [a, b]$ with the property that f is strictly monotone decreasing on $[a, t_*]$ and strictly monotone increasing on $[t_*, b]$. This is just a derivative-free way of saying that f has a unique global minimum at t_*. The function DistMercEarth is unimodal on $[900, 950]$.

The method of *golden section search* can be used to find t_*. In this method an interval $[a, b]$ that brackets t_* is maintained from step to step in addition to a pair of "interior" points c and d that satisfy

$$c = a + r(b - a)$$
$$d = a + (1 - r)(b - a)$$

Here, $0 < r < 1/2$ is a real number to be determined in a special way. If $f(c) \geq f(d)$, then from unimodality it follows that f is increasing in $[d, b]$ and so $t_* \in [a, d]$. On the other hand, if $f(c) > f(d)$, then a similar argument tells us that $t_* \in [c, b]$. This suggests the following maneuver to reduce the search space:

```
% Assume:   (i) a < c < d < b where c = a+r*(b-a) and d = a+(1-r)(b-a).
%           (ii) fa = f(a), fc = f(c), fd = f(d), and fb = f(b)

    if fc<=fd
       b = d; fb = fd;
    else
       a = c; fa = fc;
    end
    c = a+r*(b-a); fc = f(c);
    d = a+(1-r)*(b-a); fd = f(d);
```

After this step the length of the search interval is reduced by a factor of $(1 - r)$. Notice that two function evaluations per step are required, and this is where the careful choice of r comes in. If we can choose r so that (1) in the `fc<=fd` case the new d equals the old c and (2) in the `fc>fd` case the new c equals the old d, then the update step can be written as follows:

```
    if fc<=fd
       b = d; fb = fd;
       d = c; fd = fc;
       c = a+r*(b-a); fc = f(c);
    else
       a = c; fa = fc;
       c = d; fc = fd;
       d = a+(1-r)*(b-a); fd = f(d);
    end
```

In this setting only one new function evaluation is required. The value of r sought must satisfy two conditions:

$$a + (1 - r)(d - a) = a + r(b - a)$$
$$c + r(b - c) = a + (1 - r)(b - a)$$

Manipulation of either of these equations shows that

$$\frac{r}{1 - r} = \frac{d - a}{b - a}$$

From the definition of d, we have $d - a = (1 - r)(b - a)$ and so r must satisfy

$$\frac{r}{1 - r} = 1 - r$$

This leads to the quadratic equation $1 - 3r + r^2 = 0$. The root we want is

$$r = \frac{3 - \sqrt{5}}{2}$$

and with this choice the interval length reduction factor is

$$1 - r = 1 - \frac{3 - \sqrt{5}}{2} = \frac{\sqrt{5} - 1}{2} = \frac{1}{r_*} \approx .618$$

where $r_* = (1 + \sqrt{5})/2$ is the well-known *golden ratio*. Packaging these ideas, we obtain

```
    function xmin = Golden(fname,a,b);
%
% Pre:
%    fname    string that names a continuous function  f(x) of
%             a single variable.
%
%
%     a,b     define an interval [a,b]
%
% Post:
%     xmin    approximate local minimizer of f on [a,b].
%
   r = (3 - sqrt(5))/2;
   c = a + r*(b-a);        fc = feval(fname,c);
   d = a + (1-r)*(b-a); fd = feval(fname,d);
   while (d-c) > eps*max(abs(c),abs(d))
      if fc >= fd
         z = c + (1-r)*(b-c);
         % [a c d b ] <--- [c d z b]
         a = c;
         c = d; fc = fd;
         d = z; fd = feval(fname,z);
      else
         z = a + r*(d-a);
         % [a c d b ] <--- [a z c d]
         b = d;
         d = c; fd = fc;
         c = z; fc = feval(fname,z);
      end
   end
   xmin = (c+d)/2;
```

A trace of the (a, c, d, b) when this min-finder is applied to the function DistMercEarth is given in Fig.8.8. The figure shows the location of a, b, c, and d at the beginning of the first five steps. Notice how only one new function evaluation is required per step.

An important observation to make about the termination criteria in Golden is that the square root of the unit roundoff is used. The rationale for this is as follows: If f' and f'' exist, then

$$f(x_* + \delta) \approx f(x_*) + \frac{\delta^2}{2} f''(x_*)$$

since $f'(x_*) = 0$. If there are $O(\text{eps})$ errors in an f-evaluation, then there will be no significant difference between the computed value of f at x_* and the computed value of f at $x_* + \delta$ if $\delta^2 \leq \text{eps}$. Thus there is no point iterating further if the relative spacing of the floating numbers is less than $\sqrt{\text{eps}}$.

If we use Golden to minimize the Mercury-Earth separation function with initial interval $[900, 950]$, then 29 iterations are required, rendering the solution $x_* = 927.1243$

Golden always converges, but if f is not unimodal across the initial interval, then the limit point need not have any bearing on the search for even a local minimizer. Unimodality is to

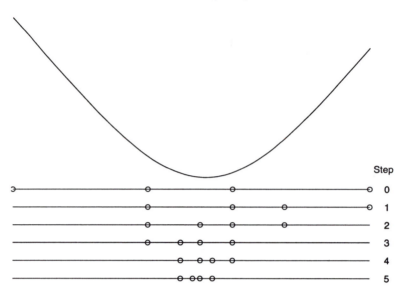

FIGURE 8.8 Golden section search

Golden as the requirement $f(a)f(b) \leq 0$ is to Bisection: Without it we cannot guarantee that anything of interest emerges from the iteration.

8.2.3   The MATLAB fmin Function

The method of golden section search converges linearly, like the method of bisection. To obtain a more rapidly convergent method, we could use the Newton framework to find zeros of the derivative f', assuming that this function and f'' exist. However, this approach requires the encoding of both the first and second derivative functions, and this can be a time-consuming endeavor.

The
Matlab minimizer fmin circumvents this problem but still exhibits quadratic-like convergence. It uses a combination of golden section search and a parabolic fitting method that relies only upon f evaluations. The call

```
tmin = fmin('DistMercEarth',900,950,)
```

applies the method to the Mercury-Earth distance function on the interval $[900, 950]$ and returns the minimizer $t_* = 927.1243$.

At the user's discretion, an "options vector" can be passed to fmin specifying various options. For example,

```
options = [1 .00000001];
tmin = fmin('DistMercEarth',900,950,options)
```

generates a trace because `options(1)` is nonzero and sets the x-tolerance to 10^{-8}, overriding the default value 10^{-4}. The essence of the output is reported in the following table:

k	t_k	$f(t_k)$	Step
1	919.0983	909.95980919060	(initialize)
2	930.9016	846.11648380405	golden
3	938.1966	979.73638285186	golden
4	927.1769	826.56580710972	parabolic
5	927.0499	826.56961589590	parabolic
6	927.1244	826.56196213246	parabolic
7	927.1243	826.56196211235	parabolic
8	927.1243	826.56196211254	parabolic
9	927.1242	826.56196211268	parabolic

A call of the form

```
[tmin options] = fmin(fname,a,b,...)
```

returns the minimizer in `tmin` and iteration information in the vector `options`. In particular, `options(8)` returns $f(t_*)$ and `options(10)` equals the number of f evaluations required. Here is a script that shows how `fmin` can be used to compute the eight local minima of the Mercury-Earth separation function in the interval $[0, 1000]$:

```
% Script File: ShowFmin
%
% Illustrates fmin.

disp('Local minima of the Mercury-Earth separation function.')
disp(sprintf('\ntol = %8.3e\n\n',tol))
disp(' Initial Interval      tmin          f(tmin)     f evals')
disp('-------------------------------------------------------')
options = zeros(18,1);
options(2) = .000001;
for k=1:8
   L = 100+(k-1)*112;
   R = L+112;
   [tmin options] = FMIN('DistMercEarth',L,R,options);
   minfeval = options(8);
   nfevals = options(10);
   disp(sprintf('      [%3.0f,%3.0f]      %10.5f   %10.5f  %6.0f',...
                                 L,R,tmin,minfeval,nfevals))
end
```

Here is the output obtained by running this script file:

Local minima of the Mercury-Earth separation function.
tol = 1.000e-06

Initial Interval	tmin	f(tmin)	f evals
[100,212]	115.42354	89.27527	12
[212,324]	232.09209	86.45270	11
[324,436]	347.86308	100.80500,	10
[436,548]	462.96252	92.21594,	10
[548,660]	579.60462	84.12374,	10
[660,772]	695.69309	99.91281,	9
[772,884]	810.54878	94.96463,	10
[884,996]	927.12431	82.65620,	10

8.2.4   A Note on Objective Functions

We close this section with some remarks that touch on symmetry, dimension, and choice of objective function. We peg the discussion to a nice geometric problem: what is the largest tetrahedron whose vertices P_0, P_1, P_2, and P_3 are on the unit sphere $x^2 + y^2 + z^2 = 1$? A tetrahedron is a polyhedron with four triangular faces.

The problem statement is ambiguous because it assumes that we have some way of measuring the size of a tetrahedron. Let's work with surface area and see what develops. If

$$\left\{ \begin{array}{c} A_{012} \\ A_{023} \\ A_{031} \\ A_{123} \end{array} \right\} \text{ is the area of the face with vertices } \left\{ \begin{array}{c} P_0, P_1, P_2 \\ P_0, P_2, P_3 \\ P_0, P_3, P_1 \\ P_1, P_2, P_3 \end{array} \right\}$$

then our goal is to maximize

$$A = A_{012} + A_{023} + A_{031} + A_{123}$$

Since the four vertices each have a "latitude" θ ($-\pi/2 \le \theta \le \pi/2$) and a "longitude" ϕ ($-\pi < \phi \le \pi$), it looks like this is a problem with eight unknowns since $P_0 = (\theta_0, \phi_0)$, $P_1 = (\theta_1, \phi_1)$, $P_2 = (\theta_2, \phi_2)$, and $P_3 = (\theta_3, \phi_3)$. Moreover, with this formulation there are an infinite number of solutions because any tetrahedron, including the optimal one that we seek, can "slide around" in the sphere. To make the situation less fluid without losing generality, we fix P_0 at the "north pole" and confine P_1 to the zero meridian. With this maneuver, the number of unknowns is reduced to five:

$$\begin{array}{rcl} P_0 & = & (\pi/2, 0) \\ P_1 & = & (\theta_1, 0) \\ P_2 & = & (\theta_2, \phi_2) \\ P_3 & = & (\theta_3, \phi_3) \end{array}$$

A symmetry argument further reduces the number of free parameters if we think of P_0 as the "apex" and the triangle defined by the other three vertices as the "base." It is clear that for the optimum tetrahedron, the base vertices P_1, P_2, and P_3 will all have the same latitude and define an equilateral triangle. (This may be a bit of a stretch if you are not familiar enough with solid geometry, but it can be formally verified.) With these observations, the number of unknowns is reduced to one:

$$
\begin{aligned}
P_0 &= (\pi/2, 0) \\
P_1 &= (\theta, 0) \\
P_2 &= (\theta, 2\pi/3) \\
P_3 &= (\theta, -2\pi/3)
\end{aligned}
$$

Let $T(\theta)$ be the tetrahedron with these vertices. The function that we must optimize is just a function of a single variable:

$$
\begin{aligned}
A(\theta) &= A_{012}(\theta) + A_{134}(\theta) + A_{142}(\theta) + A_{234}(\theta) \\
&= \frac{3\sqrt{3}}{4} \cos(\theta) \left(\sqrt{(3\sin(\theta) - 5)(\sin(\theta) - 1)} + \cos(\theta) \right)
\end{aligned}
$$

which is displayed in Fig 8.9. It is clear that there is a unique maximum somewhere in the interval $[-0.5, 0.0]$.

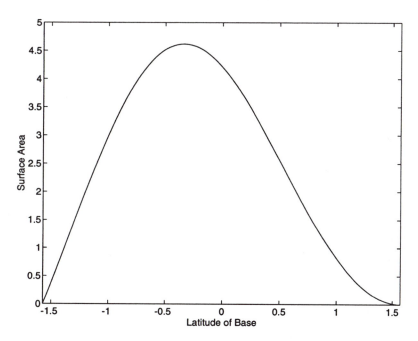

FIGURE 8.9 The surface area of $T(\theta)$

The collapse of the original eight-parameter problem to one with just a single unknown is dramatic and by no means typical of what happens in practice. However, it is a reminder that the exploitation of structure is usually rewarding.

As for finding a solution, we note that maximizing $A(\theta)$ is equivalent to minimizing

$$T_A(\theta) = -\cos(\theta)\left(\sqrt{(3\sin(\theta) - 5)(\sin(\theta) - 1)} + \cos(\theta)\right)$$

which can be handed over to `fmin` for minimization.

Now let's change the problem so that instead of maximizing surface area, we maximize volume. The same symmetry arguments prevail and it can be shown that

$$V(\theta) = \frac{\sqrt{3}}{4}(1 - \sin(\theta)^2)(1 - \sin(\theta))$$

is the volume of $T(\theta)$. The optimizing θ can be obtained by applying `fmin` to

$$T_V(\theta) = (1 - \sin(\theta)^2)(\sin(\theta) - 1)$$

Finally, we could also maximize total edge length:

$$E(\theta) = 3\left(\sqrt{2(1 - \sin(\theta))} + \sqrt{3}\cos(\theta)\right)$$

by minimizing

$$T_E(\theta) = -(\sqrt{(1 - \sin(\theta))} + \sqrt{3/2}\cos(\theta))$$

The script `FindTet` applies `fmin` to these three objective functions, and here are the results:

Objective Function	Time	Theta
Area	1.000	-0.3398369144798928
Volume	1.111	-0.3398369181995862
Edge	1.213	-0.3398369274385691

It turns out that the same tetrahedron solves each problem, and this tetrahedron is *regular*. This means that the faces are identical in area and the edges are equal in length. Notice that accuracy is only good to the square root of the machine precision, confirming the remarks made at the end of §8.2.2.

Because the optimum tetrahedron is regular, we can compute θ a different way. The function

$$\tilde{T}_E(\theta) = \sqrt{1 - \sin(\theta)} - \sqrt{3/2}\cos(\theta)$$

is the difference between the length of edge P_0P_1 and the length of edge P_1P_2. This function is zero at θ_* if $T(\theta_*)$ is the regular, solution tetrahedron. By applying `fzero` to $\tilde{T}_E(\theta)$ we get θ_* to full machine precision: -0.3398369094541218.

The exact solution is given by

$$\theta_* = \arcsin(-1/3)$$

To derive this result, we observe that the function $T_V(\theta)$ is cubic in $s = \sin(\theta)$:

$$p(s) = (1 - s^2)(1 - s)$$

From the equation $p'(s) = 0$, we conclude that $s = -1/3$ is a minimizer and so $\theta_* = \arcsin(-1/3)$.

The point of all this is that the same problem can sometimes be formulated in different ways. The mathematical equivalence of these formulations does *not* necessarily lead to implementations that are equally efficient and accurate.

Problems

P8.2.2 What is the area of the largest triangle whose vertices are situated on the following ellipse

$$\frac{x^2}{4} + \frac{y^2}{9} = 1$$

Solve this problem by applying `Golden`.

8.3   Minimizing Multivariate Functions

The vector-valued function

$$\left[\begin{array}{c} x_1(t) \\ y_1(t) \end{array} \right] = \left[\begin{array}{cc} \cos(\phi_1) & \sin(\phi_1) \\ -\sin(\phi_1) & \cos(\phi_1) \end{array} \right] \left[\begin{array}{c} \frac{P_1 - A_1}{2} + \frac{P_1 + A_1}{2}\cos(t) \\ \sqrt{P_1 A_1}\sin(t) \end{array} \right]$$

describes a "tilted," elliptical orbit having one focus at $(0,0)$. Think of the Sun as being situated at this point. The parameter ϕ_1 is the tilt angle. If $A_1 \geq P_1$, then A_1 and P_1 are the maximum and minimum Sun-to-orbit distances. Let

$$\left[\begin{array}{c} x_2(t) \\ y_2(t) \end{array} \right] = \left[\begin{array}{cc} \cos(\phi_2) & \sin(\phi_2) \\ -\sin(\phi_2) & \cos(\phi_2) \end{array} \right] \left[\begin{array}{c} \frac{P_2 - A_2}{2} + \frac{P_2 + A_2}{2}\cos(t) \\ \sqrt{P_2 A_2}\sin(t) \end{array} \right]$$

be a second such orbit and consider the the display in Fig 8.10. Our goal is to find the minimum distance from a point on the first orbit (A_1, P_1, ϕ_1) to a point on the second orbit (A_2, P_2, ϕ_2). For a distance measure, we use

$$sep(t_1, t_2) = \frac{1}{2}\left[(x_1(t_1) - x_2(t_2))^2 + (y_1(t_1) - y_2(t_2))^2 \right] \tag{8.3.1}$$

This is a function of two variables, and in accordance with terminology established in §8.2, *sep* is the objective function. Note that t_1 selects a point on the outer orbit and t_2 selects a point on the inner orbit. This section is about minimizing multivariate functions such as $sep(t)$.

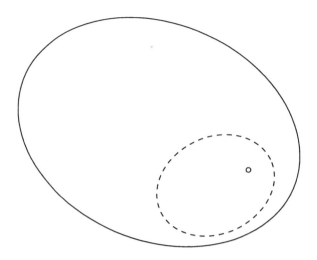

FIGURE 8.10 Orbits $(A_1, P_1, \phi_1) = (10, 2, \pi/8)$ and $(A_2, P_2, \phi_2) = (4, 1, -\pi/7)$

8.3.1   Setting Up the Objective Function

The implementation of $sep(t)$ depends on another function called orbit:

```
function P_loc = Orbit(t,A,P,phi,pcode)
% Pre: t is a row vector of "times" and (A,P,phi) defines an orbit.
%      pcode (optional)  is a valid plot string, e.g., '-', '.', '+', '*', etc.
%
% Post: P_loc = [ x(t); y(t) ] and if nargin==5, these points are plotted.

   c = cos(t); s = sin(t);
   x = ((P-A)/2) + ((P+A)/2)*c;
   y = sqrt(A*P)*s;
   cphi = cos(phi); sphi = sin(phi);
   P_loc = [cphi sphi; -sphi cphi]*[x;y];
   if nargin == 5
      plot(P_loc(1,:),P_loc(2,:),pcode)
   end
```

With this function we can easily generate and plot orbit points. The script

```
Orbit(linspace(0,2*pi),10,2,pi/8,'-');
hold on
Orbit(linspace(0,2*pi),4,1,-pi/7,'-');
Orb1 = Orbit(3,10,2,pi/8);
Orb2 = Orbit(4,1,-pi/7);
plot([Orb1(1) Orb2(1)],[Orb1(2) Orb2(2)])
hold off
```

plots the two orbits in Fig 8.10 and draws a line that connects the "$t_1 = 3$" point on the first orbit with the "$t_2 = 4$" point on the second orbit. The separation between these two points is a function of a 2-vector of independent variables and the parameters A_1, P_1, ϕ_1, A_2, P_2, and ϕ_2:

```
function d = Sep(t,plist)

% Pre:  t is a 2-vector and plist encodes the orbits
%       (A1,P1,phi1) and (A2,P2,phi2).
%
%
% Post: One-half the  square of the distance between the t(1) point
%       on the first orbit and the t(2) point on the second orbit.

A1 = plist(1); P1 = plist(2); phi1 = plist(3);
A2 = plist(4); P2 = plist(5); phi2 = plist(6);

Orb1 = Orbit(t(1),A1,P1,phi1);
Orb2 = Orbit(t(2),A2,P2,phi2);
z   = Orb2-Orb1;
d   = z'*z/2;
```

Note how the the orbit parameters are packaged in a 6-vector. The script

```
plist = [ 10  2  pi/8  4  1  -\pi/7 ];
t = [3 4];
d = sep(t,plist)
```

assigns to d the value $\left((x_1(3) - x_2(4))^2 + ((y_1(3) - y_2(4))^2)\right)/2$ with $A_1 = 10$, $P_1 = 2$, $\phi_1 = \pi/8$, $A_2 = 4$, $P_2 = 1$, and $\phi_2 = -\pi/7$.

As a rule, when solving nonlinear problems in the MATLAB environment, it is good practice not to "hardwire" a function's parameters but to pass them as a second argument. The functions fmin and fmins are designed to handle this flexibility.

It is useful to think of $sep(t)$ as the "elevation" at point $t = (t_1, t_2)$. With this topographical point of view, our goal is to reach the deepest valley point. The terrain in the search area is depicted by the contour plot in Fig 8.11. This map tells us that the minimizing t is somewhere in the vicinity of $(5.5, 4.5)$.

8.3.2   The Gradient

It is foggy, and you are standing on a hillside without a map. To take a step consistent with the goal of reaching the valley bottom, it is reasonable to take that step in the "most downhill" direction. Mathematically, this is in the direction of the negative gradient. Recall that the gradient of a function $f : \mathbb{R}^n \to \mathbb{R}$ at $t = t_c$ is a vector of partial derivatives:

$$\nabla f(t_c) = \begin{bmatrix} \dfrac{\partial f(t_c)}{\partial t_1} \\ \vdots \\ \dfrac{\partial f(t_c)}{\partial t_n} \end{bmatrix}$$

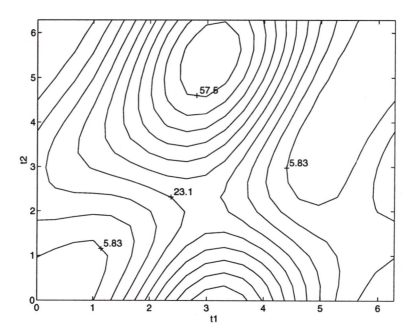

FIGURE 8.11 A contour plot of $sep(t_1, t_2)$.

The gradient always points in the direction of greatest increase and its negation points in the direction of greatest decrease. Taking the t_1 and t_2 partials in (8.3.1), we see that

$$\nabla sep(t) = \begin{bmatrix} \dfrac{\partial sep(t_1, t_2)}{\partial t_1} \\[2mm] \dfrac{\partial sep(t_1, t_2)}{\partial t_1} \end{bmatrix} = \begin{bmatrix} [x_1(t_1) - x_2(t_2)]\dot{x}_1(t_1) & + & [y_1(t_1) - y_2(t_2)]\dot{y}_1(t_1) \\[2mm] -[x_1(t_1) - x_2(t_2)]\dot{x}_1(t_2) & - & [y_1(t_1) - y_2(t_2)]\dot{y}_1(t_2) \end{bmatrix}$$

Substituting the definitions for the component functions, we carefully arrive at the following implementation of the gradient:

```
    function g = gSep(t,plist)
%
% Pre:  t is a 2-vector and plist encodes the orbits
%       (A1,P1,phi1) and (A2,P2,phi2).
%
% Post: The gradient of sep(t,plist).

    A1 = plist(1); P1 = plist(2); phi1 = plist(3);
    A2 = plist(4); P2 = plist(5); phi2 = plist(6);
    alfa1  = (P1-A1)/2; beta1  = (P1+A1)/2; gamma1 = sqrt(P1*A1);
    alfa2  = (P2-A2)/2; beta2  = (P2+A2)/2; gamma2 = sqrt(P2*A2);
```

```
s1      = sin(t(1)); c1      = cos(t(1));
s2      = sin(t(2)); c2      = cos(t(2));
cphi1   = cos(phi1); sphi1   = sin(phi1);
cphi2   = cos(phi2); sphi2   = sin(phi2);
Rot1    = [cphi1 sphi1; -sphi1 cphi1];
Rot2    = [cphi2 sphi2; -sphi2 cphi2];
P1      = Rot1*[alfa1+beta1*c1;gamma1*s1];
P2      = Rot2*[alfa2+beta2*c2;gamma2*s2];
dP1     = Rot1*[-beta1*s1;gamma1*c1];
dP2     = Rot2*[-beta2*s2;gamma2*c2];
g = [-dP1';dP2']*(P2-P1);
```

The derivation not important. But what *is* important is to appreciate that gradient function implementation can be very involved and time-consuming. Our problem is small ($n = 2$) and simple. For high-dimension applications with complicated objective functions, it is often necessary to enlist the services of a symbolic differentiation package.

8.3.3 The Method of Steepest Descent

From multivariable Taylor approximation theory, we know that if λ_c is real and s_c is an n-vector, then

$$f(t_c + \lambda_c s_c) = f(t_c) + \lambda_c \nabla f(t_c)^T s_c + O(\lambda_c^2)$$

Thus we can expect a decrease in the value of f if we set the *step direction* s_c to be

$$s_c = -\nabla f(t_c)$$

and the *step length* parameter λ_c to be small enough. If the gradient is zero, then this argument breaks down. In that case t_c is a critical point, but may not be a local minimum. Our search for the valley bottom can "stall" at such a location.

The practical determination of λ_c is the *line search* problem. Ideally, we would like to minimize $f(t_c + \lambda s_c)$ over all λ. However, an approximate solution to this one-dimensional minimization problem is usually good enough and often essential if f is expensive to evaluate. This brings us to the *steepest descent* framework for computing an improvement t_+ to the current minimizer:

- Compute the gradient $g_c = \nabla f(t_c)$ and set $s_c = -g_c$.

- Inexpensively choose λ_c so $f(t_c + \lambda_c s_c)$ is sufficiently less than $f(t_c)$.

- Set $t_+ = t_c + \lambda_c s_c$.

The design of a reasonable line search strategy is a nontrivial task. To build an appreciation for this point and for the method of steepest descent, we have built a simple environment ShowSD that can be used to minimize the *sep* function. At each step it plots $sep(t_c + \lambda s_c)$ and obtains λ_c by mouseclick. (See Fig 8.12.) After two such graphical line search steps, the ShowSD environment slips into an automated line search mode and performs additional steepest descent steps. In Fig 8.13 the search path is displayed on the contour plot. Figure 8.14 shows how **sep** and the norm of

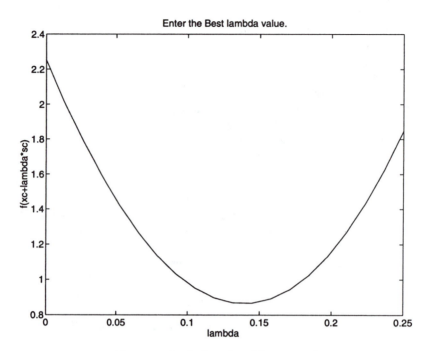

FIGURE 8.12 Graphical line search

its gradient behave during the iteration. The exhibited behavior is typical for the method. The first few iterations are very successful in bringing about a reduction in the value of the objective function. However, the convergence rate for steepest descent (with reasonable line search) is linear and the method invariably starts plodding along with short, ineffective steps.

8.3.4 The MATLAB Fmins Function

The Fmins function can be used to minimize a multivariate function. It uses a simplex search method and does not require gradient function implementation. The following script shows how Fmins is typically invoked:

```
plist = [10  2  pi/8  4  1  -pi/7];
StartingValue = [5;4];
trace = 1;
steptol = .000001;
ftol = .000001;
options = [trace steptol ftol];
[t,options] = Fmins('Sep',StartingValue,options,[],plist);
disp(sprintf('Iterations = %3.0f',options(10)))
disp(sprintf('t(1) = %10.6f,  t(2) = %10.6f',t))
```

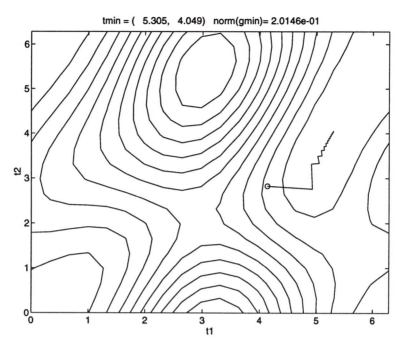

FIGURE 8.13 The steepest descent path

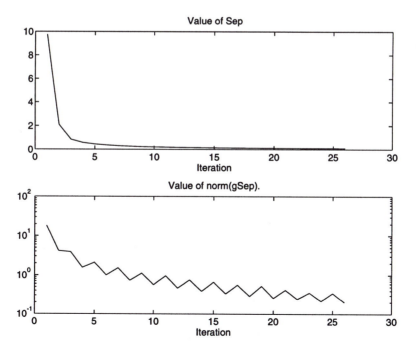

FIGURE 8.14 The descent of $sep(t_1, t_2)$ and $\|\nabla sep(t_1, t_2)\|_2$

The call requires the name of the objective function, a starting value, various option specifications and a list of parameters required by the objective function. Standard input options are used to indicate whether or not a trace of the iteration is desired and how to bring about termination. Setting `trace = 1` in the script means that the results of the `Fmins` iterates are displayed. Setting `trace = 0` turns off this feature. If the iterates are close with respect to steptol and the difference between function evaluations is small with respect to ftol, then the iteration is brought to a halt. An iteration maximum brings about termination if these criteria are too stringent.

Upon return, `t` is assigned the approximate solution and `options(10)` contains the number of required steps.

The function `ShowFmins` can be used to experiment with this minimizer when it is applied to the function `sep`. The starting value is obtained by mouseclicks. The orbit points corresponding to the solution are graphically displayed. (See Fig 8.15.)

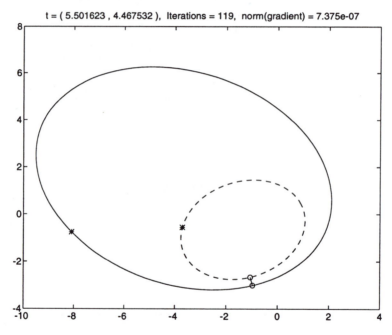

FIGURE 8.15 Solution by `Fmins`.

Problems

P8.3.1 Let $f_c(\lambda) = f(t_c + \lambda s_c)$. Assume that $\lambda_2 > \lambda_1$ and that $f_c(\lambda_2)$ or $f_c(\lambda_1)$ is less than $f_c(0)$. Let λ_c be the minimizer on $[0, \lambda_2]$ of the quadratic that interpolates $(0, f_c(0))$, $(\lambda_1, f_c(\lambda_1))$, and $(\lambda_2, f_c(\lambda_2))$. Modify the line search strategy in the function `SDStep` so that it incorporates this method for choosing λ_c.

P8.3.2 The function *sep* is periodic, and there are an infinite number of minimizers. Use `ShowSD` to find a different solution than the one reported in the text.

P8.3.3 What happens in `ShowSD` if the two orbits intersect? How is the rate of convergence affected if the inner ellipse is very "cigar-like"?

P8.3.4 Modify ShowFmins so that for a given starting value, it reports the number of iterations required when Fmins is run with steptol = ftol = 10^{-d} for $d = 0{:}6$.

8.4 Systems of Nonlinear Equations

Suppose we have a pair of orbits whose intersection points are sought. (See Fig 8.16.) In this

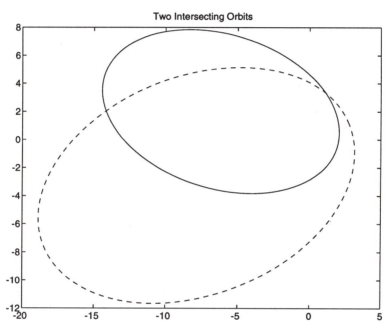

FIGURE 8.16 The orbits $(15, 2, \pi/10)$ and $(20, 3, -\pi/8)$

problem, the unknown is a 2-vector $t = (t_1, t_2)$ having the property that

$$sepV(t) = \left[\begin{array}{c} x_2(t_2) - x_1(t_1) \\ y_2(t_2) - y_1(t_1) \end{array} \right] = \left[\begin{array}{c} 0 \\ 0 \end{array} \right].$$

where

$$\left[\begin{array}{c} x_i(t_i) \\ y_i(t_i) \end{array} \right] = \left[\begin{array}{cc} \cos(\phi_i) & \sin(\phi_i) \\ -\sin(\phi_i) & \cos(\phi_i) \end{array} \right] \left[\begin{array}{c} \dfrac{P_i - A_i}{2} + \dfrac{P_I + A_i}{2} \cos(t_i) \\ \sqrt{P_i A_i} \sin(t_i) \end{array} \right] \qquad i = 1, 2.$$

It is easy to implement this using Orbit:

```
function z = SepV(t,plist)
% Pre:  t is a 2-vector and plist encodes the orbits
%       (A1,P1,phi1) and (A2,P2,phi2).
%
```

```
% Post: The vector from the t(1) point on the first
%        orbit to the t(2) point on the second orbit.

  A1 = plist(1); P1 = plist(2); phi1 = plist(3);
  A2 = plist(4); P2 = plist(5); phi2 = plist(6);

  Orb1 = Orbit(t(1),A1,P1,phi1);
  Orb2 = Orbit(t(2),A2,P2,phi2);
  z    = Orb2 - Orb1;
```

Our goal is to develop a systems version of Newton's method that can zero functions like this.

8.4.1   The Jacobian

In §8.1 we derived Newton's method through the idea of a linear model and we take the same
approach here. Suppose $t_c \in \mathbb{R}^n$ is an approximate zero of the vector-valued function

$$F(t) = \left[\begin{array}{c} F_1(t) \\ \vdots \\ F_n(t) \end{array} \right]$$

To compute an improvement $t_c + s_c$, we build a linear model of F at $t = t_c$. For each component
function we have $F_i(t_c + s_c) \approx F_i(t_c) + \nabla F_i(t_c)^T s_c$ and so

$$F(t_c + s_c) \approx F(t_c) + J(t_c)s_c$$

where $J(t_c) \in \mathbb{R}^{n \times n}$ is a matrix whose ith row equals $\nabla F_i(t_c)^T$. $J(t_c)$ is the *Jacobian* of F at t_c.
The Jacobian of the *sepV(t)* function is given by

$$J(t) = \left[\begin{array}{cc} -\dot{x}_1(t_1) & \dot{x}_2(t_2) \\ -\dot{y}_1(t_1) & \dot{y}_2(t_2) \end{array} \right]$$

and after using the component function definitions we get

```
    function J = JSepV(t,plist)
%
% Pre:   t is a 2-vector and plist encodes the orbits
%        (A1,P1,phi1) and (A2,P2,phi2).
%
% Post: The Jacobian of sepV(t,plist).

  A1 = plist(1); P1 = plist(2); phi1 = plist(3);
  A2 = plist(4); P2 = plist(5); phi2 = plist(6);

  s1 = sin(t(1)); c1 = cos(t(1));
  s2 = sin(t(2)); c2 = cos(t(2));
```

```
beta1 = (P1+A1)/2; gamma1 = sqrt(P1*A1);
beta2 = (P2+A2)/2; gamma2 = sqrt(P2*A2);
cphi1 = cos(phi1); sphi1  = sin(phi1);
cphi2 = cos(phi2); sphi2  = sin(phi2);

Rot1 = [cphi1 sphi1; - sphi1 cphi1];
Rot2 = [cphi2 sphi2; - sphi2 cphi2];

J = [ Rot1*[beta1*s1;-gamma1*c1] Rot2*[-beta2*s2;gamma2*c2]];
```

Do not conclude from this that Jacobian implementation is easy. Our problem is small and simple. Remember that $O(n^2)$ partial derivatives are involved. It does not take a very large n or a very complex F to make Jacobian evaluation a formidable task. An approach using finite differences is discussed in §8.4.3.

8.4.2 The Newton Idea

The Newton idea for systems involves choosing s_c so that the local model of F is zero:

$$F(t_c + s_c) \approx F(t_c) + J(t_c)s_c = 0$$

This involves a function evaluation, a Jacobian evaluation, *and* the solution of a linear system $J(t_c)s = -F(t_c)$, as the following implementation shows:

```
function [tnew,Fnew,Jnew] = NStep(FName,JName,tc,Fc,Jc,plist)
%
% Pre:
%      FName      the name of a function F(t,plist) that accepts  a column
%                 n-vector and returns a column n-vector
%      JName      the name of a function JF(t,plist) that returns the
%                 Jacobian of F at t.
%         tc      a column n-vector, an approximate zero.
%         Fc      F(tc,plist)
%         Jc      JF(tc,plist), assumed nonsingular
%      plist      parameter list
%
% Post:
%       tnew      a column n-vector, an improved approximate obtained
%                 by a single Newton iteration.
%       Fnew      F(tnew,plist)
%       Jnew      JF(tnew,plist)

   sc = -(Jc\Fc);
 tnew = tc + sc;
 Fnew = feval(FName,tnew,plist);
 Jnew = feval(JName,tnew,plist);
```

Repetition of this process produces the *Newton iteration*. Even if we disregard the function and Jacobian evaluation, there are still $O(n^3)$ flops per step, so Newton for systems is a decidedly more intense computation than when the method is applied in the single-variable case.

The function ShowN can be used to explore the Newton iteration behavior when it is applied to the $sepV(t) = 0$ problem. The starting value is obtained by mouseclicking a point on each orbit. The corresponding t-values are computed and form the "initial guess." (See Fig 8.17.) The iterates are displayed graphically, and the result of the first step is shown in Fig 8.18. For this particular starting value, an intersection point is found in eight iterations:

Iteration	tc(1)	tc(2)	norm(Fc)	cond(Jc)
0	2.7688613218079534	2.2363879906910391	7.625e+00	1.368e+01
1	1.3074309674269593	0.6647441388426698	2.062e+00	1.189e+01
2	0.3849611511846585	0.0267749587072792	7.669e-01	1.113e+01
3	0.9221940626755794	0.4980567314409592	5.413e-01	7.458e+00
4	0.6890122854905480	0.3353784868611613	7.822e-02	7.287e+00
5	0.6988770969468836	0.3525102981393918	1.320e-03	6.967e+00
6	0.6982147213897512	0.3519591318238888	7.474e-07	6.968e+00
7	0.6982145171184472	0.3519590450461945	1.191e-13	6.968e+00
8	0.6982145171184865	0.3519590450462368	9.930e-16	6.968e+00

Observe the quadratic convergence that typifies the Newton approach. The condition number of the Jacobian is displayed because if this number is large, then the accuracy of the correction and its possibly large norm could have an adverse effect on convergence. For the problem under consideration, this is not an issue. But if we attempt to find the intersection points of the nearly tangential orbits $(15, 2, \pi/10)$ and $(15, 2.000001, \pi/10)$, then it is a different matter:

Iteration	tc(1)	tc(2)	norm(Fc)	cond(Jc)
0	0.2129893324467656	0.2129893324467656	1.030e-06	3.468e+07
1	-2.9481695167573863	-2.9481687805353460	4.391e-06	1.936e+06
2	-2.9365137654751829	-2.9365138156763226	7.703e-08	1.544e+07
3	-3.0402412755186443	-3.0402413005044142	1.907e-08	2.956e+07
:	:	:	:	:
12	-3.1413980211918338	-3.1413980212404917	7.248e-14	1.501e+10
13	-3.1414973191506728	-3.1414973191745066	2.079e-14	3.064e+10
14	-3.1415554417361111	-3.1415554417454143	3.202e-15	7.850e+10

Intersection = (-14.2659107219684653 , 4.6350610716380816)

Notice that in this case the Jacobian is ill conditioned near the solution and that linear-like convergence is displayed. This corresponds to what we discovered in §8.1 for functions that have a very small derivative near a root.

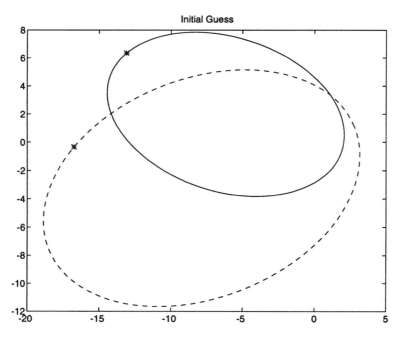

FIGURE 8.17 Initial guess with orbits $(15, 2, \pi/10)$ and $(20, 3, -\pi/8)$

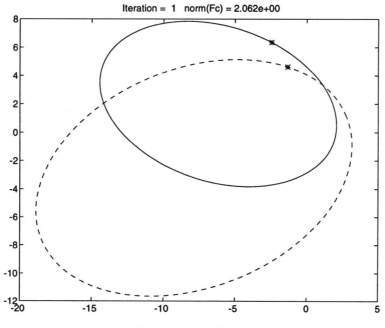

FIGURE 8.18 First step

8.4.3 Finite Difference Jacobians

A fundamental problem with Newton's method is the requirement that the user supply a procedure for Jacobian evaluation. One way around this difficulty is to approximate the partial derivatives with sufficiently accurate divided differences. In this regard, it is helpful to look at the Jacobian from the column point of view. The qth column is given by

$$
\begin{bmatrix} \dfrac{\partial F_1(t_c)}{\partial t_q} \\ \vdots \\ \dfrac{\partial F_n(t_c)}{\partial t_q} \end{bmatrix} = \dfrac{\partial F(t_c)}{\partial t_q}
$$

Thus

$$
\frac{\partial F(t_c)}{\partial t_q} \approx \frac{F(t_1, \ldots, t_q + \delta_k, \ldots, t_n) - F(t_1, \ldots, t_n)}{\delta_k}
$$

where δ_k is a small, carefully chosen real number. Thus an approximate Jacobian can be built up in this fashion column by column. Instead of n^2 partials to encode and evaluate, we have $n + 1$ evaluations of the function F, making the finite difference Newton method very attractive. If the difference parameters δ_k are computed properly, then the quadratic convergence of the exact Newton approach remains in force. The function ShowFDN supports experimentation with the method.

Problems

P8.4.1 Write a function AllInterSect(plist) that returns all the intersection points of two given orbits. Use ShowN.

P8.4.2 How is the convergence rate of the finite difference Newton method affected by the choice of the δ parameters?

P8.4.3 Let C be the unit cube "cornered" in the positive orthant. Assume that the density of the cube at (u, v, w) is given by $d(u, v, w) = e^{-(u-x_1)^2(v-x_2)^2(w-x_3)^2}$, where $x = [x_1, x_2, x_3]^T$ is a vector of parameters to be determined. We wish to choose x so that C's center of gravity is at $(.2, .3, .6)$. This will be the case if each of the nonlinear equations

$$
F_1(x_1, x_2, x_3) = \int_0^1 \int_0^1 \int_0^1 u \cdot d(u, v, w) \, du\,dv\,dw - .2 \int_0^1 \int_0^1 \int_0^1 d(u, v, w) \, du\,dv\,dw
$$

$$
F_2(x_1, x_2, x_3) = \int_0^1 \int_0^1 \int_0^1 v \cdot d(u, v, w) \, du\,dv\,dw - .3 \int_0^1 \int_0^1 \int_0^1 d(u, v, w) \, du\,dv\,dw
$$

$$
F_3(x_1, x_2, x_3) = \int_0^1 \int_0^1 \int_0^1 w \cdot d(u, v, w) \, du\,dv\,dw - .6 \int_0^1 \int_0^1 \int_0^1 d(u, v, w) \, du\,dv\,dw
$$

is zeroed. Solve this problem using Newton's method. Handle the integrals with the trapezoidal rule.

8.4.4 Zeroing the Gradient

If Newton's method is applied to the problem of zeroing the gradient of a function $f : \mathbb{R}^n \to \mathbb{R}^1$, then the Jacobian is special and called the *Hessian*:

$$
J(t_c) = \nabla^2 f(t_c) = H_c = (h_{ij}) \qquad h_{ij} = \frac{\partial^2 f}{\partial t_i \partial t_j}
$$

It is symmetric, and at a local minimum it is also positive definite. In this case a Newton step looks like this:

- Evaluate the gradient $g_c = \nabla f(t_c)$ and the Hessian $H_c = \nabla^2 f(t_c)$.

- Solve the linear system $H_c s_c = -g_c$, possibly using Cholesky.

- Compute the new approximate solution $t_+ = t_c + s_c$.

Hessian computation is more challenging than ordinary Jacobian computation because second derivatives are involved. For the $sep(t)$ function (8.3.1), we find

$$\nabla^2 sep(t) = \begin{bmatrix} (x_1 - x_2)\ddot{x}_1 + (y_1 - y_2)\ddot{y}_1 & -\dot{x}_1\dot{x}_2 - \dot{y}_1\dot{y}_2 \\ -\dot{x}_1\dot{x}_2 - \dot{y}_1\dot{y}_2 & (x_2 - x_1)\ddot{x}_2 + (y_2 - y_1)\ddot{y}_2 \end{bmatrix}$$

whereupon

```
    function [g,H] = gHSep(t,plist)
%
% Pre:  t is a 2-vector and plist encodes the orbits
%       (A1,P1,phi1) and (A2,P2,phi2).
%
% Post: g is the gradient of sep(t,plist) and H is the Hessian.

    A1 = plist(1); P1 = plist(2); phi1 = plist(3);
    A2 = plist(4); P2 = plist(5); phi2 = plist(6);

    alfa1   = (P1-A1)/2; beta1   = (P1+A1)/2; gamma1 = sqrt(P1*A1);
    alfa2   = (P2-A2)/2; beta2   = (P2+A2)/2; gamma2 = sqrt(P2*A2);
    s1      = sin(t(1)); c1      = cos(t(1));
    s2      = sin(t(2)); c2      = cos(t(2));
    cphi1   = cos(phi1); sphi1   = sin(phi1);
    cphi2   = cos(phi2); sphi2   = sin(phi2);
    Rot1    = [cphi1 sphi1; -sphi1 cphi1];
    Rot2    = [cphi2 sphi2; -sphi2 cphi2];
    P1      = Rot1*[alfa1+beta1*c1;gamma1*s1];
    P2      = Rot2*[alfa2+beta2*c2;gamma2*s2];
    dP1     = Rot1*[-beta1*s1;gamma1*c1];
    dP2     = Rot2*[-beta2*s2;gamma2*c2];
    g = [-dP1';dP2']*(P2-P1);

    ddP1    = Rot1*[-beta1*c1;-gamma1*s1];
    ddP2    = Rot2*[-beta2*c2;-gamma2*s2];
    H(1,1) = (P1(1)-P2(1))*ddP1(1) + (P1(2)-P2(2))*ddP1(2) + dP1(1)^2 + dP1(2)^2;
    H(2,2) = (P2(1)-P1(1))*ddP2(1) + (P2(2)-P1(2))*ddP2(2) + dP2(1)^2 + dP2(2)^2;
    H(1,2) = -dP1(1)*dP2(1) - dP1(2)*dP2(2);
    H(2,1) = H(1,2);
```

Because gradient and Hessian computation often involve the evaluation of similar quantities, it usually makes sense to write a single function to evaluate them both.

If we apply Newton's method to the problem of zeroing the gradient of $sep(t)$, then the good properties that we expect of the iteration only apply if the current solution is "close enough." Complementing this behavior is the steepest descent iteration, which tends to perform well in the early stages. Just as we combined bisection and Newton methods for one-dimensional problems in §8.2, we can do the same thing here with steepest descent and Newton applied to the gradient. The function ShowMixed supports experimentation with this idea applied to the $sep(t)$ minimization problem. Here is the result from an experiment in which two steepest descent steps were performed followed by the Newton iteration:

Step	sep	t(1)	t(2)	norm(grad)
0	7.357416	4.281030403596453	2.686289370460838	1.5e+01
1	2.218781	5.030105749935057	2.772440620697425	5.2e+00
2	1.327604	4.855162479526332	3.035849552958382	3.6e+00
3	0.375147	5.068652867195277	3.503617699843756	9.9e-01
4	0.144049	5.217179263949403	3.877813301165616	4.5e-01
5	0.080406	5.351585193520888	4.170987092006141	2.2e-01
6	0.065695	5.454719480756797	4.379814245640807	9.6e-02
7	0.064355	5.497009795477695	4.459321148201743	1.4e-02
8	0.064341	5.501574350842474	4.467446651296436	1.6e-04
9	0.064341	5.501623047366454	4.467532519368982	1.8e-08
10	0.064341	5.501623052934422	4.467532529194003	2.8e-15

The path to the solution is displayed in Fig.8.19. The development of a general-purpose optimizer that combines Newton and steepest descent ideas is very challenging. All we have done here is just give a snapshot of the potential.

Problems

P8.4.4 Automate the switch from steepest descent to Newton in ShowMixed. Cross over when the reduction in $sep(t)$ is less than 1% as a result of a steepest descent step.

8.4.5 Nonlinear Least Squares

A planet moves in the orbit

$$\left[\begin{array}{c} x(t) \\ y(t) \end{array} \right] = \left[\begin{array}{c} \frac{P-A}{2} + \frac{A+P}{2} \cos(t) \\ \sqrt{AP} \sin(t) \end{array} \right]$$

Our goal is to estimate the orbit parameters A and P given that we are able to measure $r(\theta)$, the length of the orbit vector where θ is the angle between that vector and the positive real x-axis. It can be shown that

$$r(\theta) = \frac{2AP}{P(1 - \cos(\theta)) + A(1 + \cos(\theta))}$$

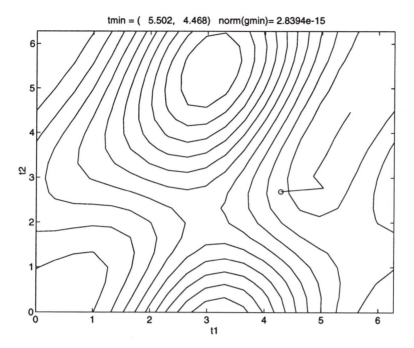

FIGURE 8.19 Steepest Descent/Newton Path

Let r_i be the measured value of this function at $\theta = \theta_i$. For each measurement (θ_i, r_i), define the function

$$F_i(A, P) = r_i - \frac{2AP}{P(1 - \cos(\theta_i)) + A(1 + \cos(\theta_i))}$$

To compute the best choices for A and P from data points $(\theta_1, r_1), \ldots, (\theta_m, r_m)$, we minimize the following function:

$$Orb(A, P) = \frac{1}{2} \sum_{i=1}^{m} F_i(A, P)^2$$

This is an example of the nonlinear least squares problem, a very important optimization problem in which Newton ideas have a key role.

In general, we are given a function

$$F(p) = \begin{bmatrix} F_1(p) \\ F_2(p) \\ \vdots \\ F_m(p) \end{bmatrix}$$

and seek a vector of unknown parameters $p \in \mathbb{R}^n$ so that the sum of squares

$$\rho(p) = \frac{1}{2} \sum_{i=1}^{m} F_i(p)^2 \tag{8.4.1}$$

is as small as possible. We use "p" because nonlinear least squares is often used to resolve parameters in model-fitting problems. A full Newton method involving the step

$$\nabla^2 \rho(p_c) s_c = -\nabla \rho(p_c)$$

entails considerable derivative evaluation since the gradient is given by

$$\nabla \rho(p_c = J(p_c)^T F(p_c)$$

and

$$\nabla^2 \rho(p_c) = J(p_c)^T J(p_c) + \sum_{i=1}^{m} F_i(p_c) \nabla^2 F_i(p_c)$$

prescribes the Hessian. Here, $J(p_c)$ is the Jacobian of the function $F(p)$ evaluated at $p = p_c$:

$$J(p_c) = \left(\frac{\partial F_i(p_c)}{\partial p_j} \right) \qquad 1 \le i \le m,\ 1 \le j \le n$$

Note that once this matrix is found, we have "half" the Hessian. The other half is a linear combination of component function Hessians and very awkward to compute because of all the second derivatives. However, if the model we are trying to determine fits the data well, then near the solution we have $F_i(p) \approx 0$, $i = 1{:}m$, and this part of the Hessian goes away. This leads to the *Gauss-Newton* framework:

- Evaluate $F_c = F(p_c)$ and the Jacobian $J_c = J(p_c)$.

- Solve $(J_c^T J_c) s_c = -J_c^T F_c$.

- Update $p_+ = p_c + s_c$.

Note that s_c is the solution to the normal equations for the least squares problem

$$\min_{s \in \mathbb{R}^m} \ \| J(p_c) s + F(p_c) \|_2$$

Thus in a reliable implementation of the Gauss-Newton step we can use a least squares method to compute the step direction.

The function ShowGN can be used to experiment with this method applied to the problem of minimizing the function $Orb(A, P)$ above. It solicits an exact pair of parameters A_0 and P_0, computes the exact value of $r(\theta)$ over a range of θ, and then perturbs these values with a prescribed amount of noise. (See Fig 8.20.) A (partially) interactive Gauss-Newton iteration with line search is then initiated. Here are some sample results:

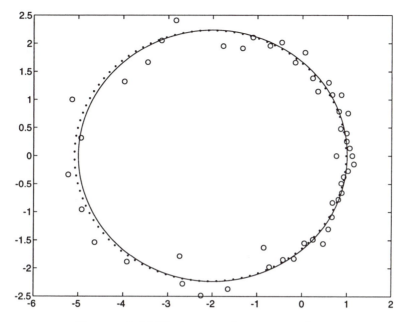

FIGURE 8.20 Fitting an orbit to noisy data

AO = 5.000, PO = 1.000,   Relative Noise = 0.100, Samples = 50

Iteration	wc	A	P	norm(grad)
0	32.5750	3.450226	2.318133	3.429e+01
1	3.6722	4.211857	1.015818	7.519e+00
2	1.0938	5.050354	0.974908	6.534e-01
3	1.0876	5.085463	0.979057	3.037e-03
4	1.0876	5.085633	0.978994	4.492e-05
5	1.0876	5.085630	0.978995	5.565e-07

In this example, the true values of A_0 and P_0 are 5 and 1, respectively. The relative noise in the r values is 10% and the technique does a fairly good job of estimating the parameters from the data.

The Gauss-Newton method works well on small residual problems. On highly nonlinear problems for which the minimum sum of squares is large, the method can be very slow or non-convergent.

Problems

P8.4.5 It is conjectured that the data $(t_1 f_1), \ldots, (t_m, f_m)$ are a sampling of the function

$$F(t) = a_1 e^{\lambda_1 t} + a_2 e^{\lambda_2 t}$$

To determine the *model parameters* a_1 and a_2 (the coefficients) and λ_1 and λ_2 (the time constants), we minimize

$$\phi(a, \lambda) = \sum_{i=1}^{m} [f(t_i) - f_i]^2 = \sum_{i=1}^{m} \left[a_1 e^{\lambda_1 t_i} + a_2 e^{\lambda_2 t_i} - f_i \right]^2 \qquad \lambda = \left[\begin{array}{c} \lambda_1 \\ \lambda_2 \end{array} \right], \quad a = \left[\begin{array}{c} a_1 \\ a_2 \end{array} \right]$$

Defining the m-by-2 matrix E_λ by

$$E_\lambda = \left[\begin{array}{cc} e^{t_1 \lambda_1} & e^{t_1 \lambda_2} \\ e^{t_2 \lambda_1} & e^{t_2 \lambda_2} \\ \vdots & \vdots \\ e^{t_m \lambda_1} & e^{t_m \lambda_2} \end{array} \right]$$

we see that

$$\phi(a, \lambda) = \| E_\lambda a - F \|^2 \qquad f = \left[\begin{array}{c} f_1 \\ f_2 \\ \vdots \\ f_m \end{array} \right]$$

We could apply the MATLAB minimizer **fmins** to this function. However, the four-dimensional search can be reduced to a two-dimensional search by making a simple observation. If $\lambda^{(c)}$ is the current estimate of the time constants and if $E_{\lambda^{(c)}}$ is the corresponding E-matrix, then the best estimate for the coefficients is the vector a that solves

$$\min \| E_{\lambda^{(c)}} a - f \|_2.$$

Let a_λ designate this vector and define

$$\psi(\lambda) = \phi(a_\lambda, \lambda)$$

Apply **fmins** to this function with starting value $\lambda = [0, -1]^T$. To generate data, use

```
function [t,fvals] = GenProb(m,L,R,noise)
rand('normal');
t = linspace(L,R,m)';
fvals = 3*exp(-.2*t) + 5*exp(-4*t) + noise*rand(t);
```

What are the estimated time constants for the problems $(m, L, R, noise) = (10, 0, 2, .001)$, $(10, 1, 3, .001)$, $(10, 2, 4, .001)$, and $(10, 20, 21, .001)$?

M-Files and References

Script Files

ShowMySqrt	Plots relative error associated with MySqrt.
ShowBisection	Illustrates the method of bisection.
ShowNewton	Illustrates the classical Newton iteration.
FindConj	Uses fzero to compute Mercury-Earth conjunctions.
ShowGolden	Illustrates Golden section search.
ShowFmin	Illustrates fmin.
FindTet	Applies fmin to three different objective functions.
ShowFmins	Illustrates Fmins.

Function Files

MySqrt	Canonical square root finder.
Bisection	The method of bisection.
StepIsIn	Checks next Newton step.
GlobalNewton	Newton method with bisection globalizer.
GraphSearch	Interactive graphical search environment.
Golden	Golden section search.
SDStep	Steepest descent step.
ShowSD	Steepest descent test environment.
NStep	Newton step.
ShowN	Newton test environment.
FDNStep	Finite difference Newton step.
ShowFDN	Finite difference Newton test environment.
ShowMixed	Globalized Newton test environment.
GNStep	Gauss-Newton step.
ShowGN	Gauss-Newton test environment.
SineMercEarth	The sine of the Mercury-Sun-Earth angle.
DistMercEarth	Mercury-Earth distance.
TA	Surface area of tetrahedron.
TV	Volume of tetrahedron.
TE	Total edge length of tetrahedron.
TEDiff	Difference between tetrahedron edges.
PtOnOrb	Solicit point on orbit.
Orbit	Generates and plots orbit points.
Sep	Separation between points on two orbits.
gSep	Gradient of sep.
gHSep	Gradient and Hessian of sep.
rho	Residual of orbit fit.
Jrho	Jacobian of rho.

References

J.E. Dennis Jr and R.B. Schnabel (1983). *Numerical Methods for Unconstrained Optimization*, Prentice Hall, Englewood Cliffs, NJ.

P. Gill, W. Murray, and M.H. Wright (1981). *Practical Optimization*, Academic Press, New York.

P. Gill, W. Murray, and M.H. Wright (1991). *Numerical Linear Algebra and Optimization, Vol. 1*, Addison-Wesley, Reading, MA.

Chapter 9

The Initial Value Problem

§**9.1** Basic Concepts

§**9.2** The Runge-Kutta Methods

§**9.3** The Adams Methods

The goal in the *initial value problem* (IVP) is to find a function $y(t)$ given its value at some initial time t_0 and a recipe $f(t, y)$ for its slope:

$$y'(t) = f(t, y(t)) \qquad y(t_0) = y_0$$

In applications we may want to plot an approximation to $y(t)$ over a designated interval of interest $[t_0, t_{max}]$ in an effort to discover qualitative properties of the solution. Or we may require a highly accurate estimate of $y(t)$ at some single, prescribed value $t = T$.

The methods we develop produce a sequence of solution snapshots $(t_1, y_1), (t_2, y_2), \ldots$ that are regarded as approximations to $(t_1, y(t_1)), (t_2, y(t_2)), \ldots \ldots$ All we have at our disposal is the "slope function" $f(t, y)$, best thought of as a MATLAB function f(t,y) that can be called whenever we need information about where $y(t)$ is "headed." IVP solvers differ in how they use the slope function.

In §9.1 we use the Euler methods to introduce the basic ideas associated with approximate IVP solving: discretization, local error, global error, stability, etc. In practice the IVP usually involves a vector of unknown functions, and the treatment of such problems is also covered in §9.1. In this setting the given slope function $f(t, y)$ is a vector of scalar slope functions, and its evaluation tells us how each component in the unknown $y(t)$ vector is changing with t.

The Runge-Kutta and Adams methods are then presented in §9.2 and §9.3 together with the built-in MATLAB IVP solvers ODE23 and ODE45. We also discuss stepsize control, a topic of great practical importance and another occasion to show off the role of calculus-based heuristics in scientific computing.

Quality software for the IVP is very complex. Years of research and development stand behind codes like ODE23 and ODE45. The implementations that we develop in this chapter are designed to build intuition and, if anything, are just the first step in the long journey from textbook formula to production software.

9.1 Basic Concepts

A "family" of functions generally satisfies a differential equation of the form $y'(t) = f(t, y)$. The initial condition $y(t_0) = y_0$ singles out one of these family members for the solution to the IVP. For example, functions of the form $y(t) = ce^{-5t}$ satisfy $y'(t) = -5y(t)$. If we stipulate that $y(0) = 1$, then $y(t) = e^{-5t}$ is the unique solution to the IVP. (See Fig 9.1.) Our goal is to

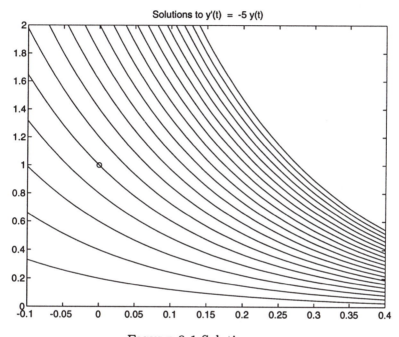

FIGURE 9.1 Solution curves

produce a sequence of points (t_i, y_i) that reasonably track the solution curve as time evolves. The Euler methods that we develop in this section organize this tracking process around a linear model.

9.1.1 Derivation of the Euler Method

From the initial condition, we know that (t_0, y_0) is on the solution curve. At this point the slope of the solution is computable via the function f:

$$f_0 = f(t_0, y_0)$$

To estimate $y(t)$ at some future time $t_1 = t_0 + h_0$ we consider the following Taylor expansion:

$$y(t_0 + h_0) \approx y(t_0) + h_0 y'(t_0) = y_0 + h_0 f(t_0, y_0)$$

This suggests that we use

$$y_1 = y_0 + h_0 f(t_0, y_0)$$

as our approximation to the solution at time t_1. The parameter $h_0 > 0$ is the *step*, and it can be said that with the production of y_1 we have "integrated the IVP forward" to $t = t_1$.

With $y_1 \approx y(t_1)$ in hand, we try to push our knowledge of the solution one step further into the future. Let h_1 be the next step. A Taylor expansion about $t = t_1$ says that

$$y(t_1 + h_1) \approx y(t_1) + h_1 y'(t_1) = y(t_1) + h_1 f(t_1, y(t_1))$$

Note that in this case the right-hand side is not computable because we do not know the exact solution at $t = t_1$. However, if we are willing to use the approximations

$$y_1 \approx y(t_1)$$

and

$$f_1 = f(t_1, y_1) \approx f(t_1, y(t_1))$$

then at time $t_2 = t_1 + h_1$ we have

$$y(t_2) \approx y_2 = y_1 + h_1 f_1$$

The pattern is now clear. At each step we evaluate f at the current approximate solution point (t_n, y_n) and then use that slope information to get y_{n+1}. The key equation is

$$y_{n+1} = y_n + h_n f(t_n, y_n)$$

and its repeated application defines the *Euler method*:

> $n = 0$
> **Repeat:**
> $f_n = f(t_n, y_n)$
>   Determine the step $h_n > 0$.
> $t_{n+1} = t_n + h_n$
> $y_{n+1} = y_n + h_n f_n$.
> $n = n + 1$

The script file ShowEuler solicits the time steps interactively and applies the Euler method to the problem $y' = -5y$, $y(0) = 1$. (See Fig 9.2.) The determination of the step size is crucial.

Our intuition says that we can control error by choosing h_n appropriately. Accuracy should increase with shorter steps. On the other hand, shorter steps mean more f-evaluations as we integrate across the interval of interest. As in the quadrature problem and the nonlinear equation-solving problem, the number of f-evaluations usually determines execution time, and the efficiency analysis of any IVP method must include a tabulation of this statistic. The basic game to be played is to get the required snapshots of $y(t)$ with sufficient accuracy, evaluating $f(t, y)$ as infrequently as possible. To see what we are up against, we need to understand how the errors in the local model compound as we integrate across the time interval of interest.

9.1.2   Local Error, Global Error, and Stability

Assume in the Euler method that y_{n-1} is exact and let $h = h_{n-1}$. By subtracting $y_n = y_{n-1} + h f_{n-1}$ from the Taylor expansion

$$y(t_n) = y_{n-1} + h y'(t_{n-1}) + \frac{h^2}{2} y^{(2)}(\eta) \qquad \eta \in [t_{n-1}, t_n]$$

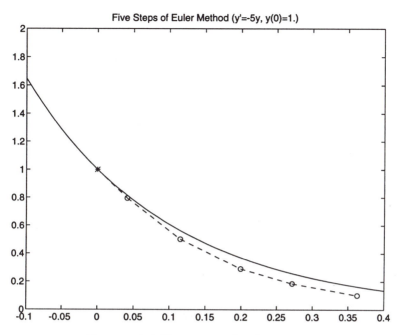

FIGURE 9.2 Five steps of Euler's method

we find that

$$y(t_n) - y_n = \frac{h^2}{2} y^{(2)}(\eta)$$

This is called the *local truncation error* (LTE) In general, the LTE for an IVP method is the error that results when a single step is performed with exact "input data." It is a key attribute of any IVP solver, and the *order* of the method is used to designate its form. A method has order k if its LTE goes to zero like h^{k+1}. Thus the Euler method has order 1. The error in an individual Euler step depends on the square of the step and the behavior of the second derivative. Higher-order methods are pursued in the next two sections.

A good way to visualize the LTE is to recognize that at each step, (t_n, y_n) sits on some solution curve $y_n(t)$ that satisfies the differential equation $y'(t) = f(t, y(t))$. With each step we jump to a new solution curve, and the size of the jump is the LTE. (See Fig 9.3.)

Distinct from the local truncation error is the *global error*. The global error g_n is the actual difference between the $t = t_n$ solution y_n produced by the IVP solver and the *true* IVP solution $y(t_n)$:

$$g_n = y(t_n) - y_n$$

As we have mentioned, the local truncation error in getting y_n is defined by

$$LTE_n = y_{n-1}(t_n) - y_n$$

where $y_{n-1}(t)$ satisfies the IVP

$$y'(t) = f(t, y(t)) \qquad y(t_{n-1}) = y_{n-1}$$

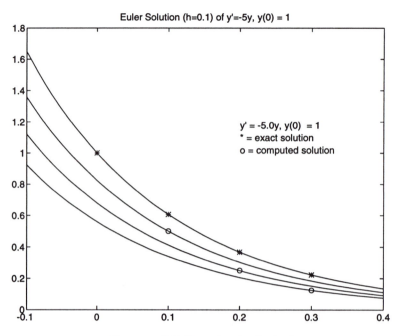

FIGURE 9.3 Jumping trajectories

LTE is tractable analytically and, as we shall see, it can be estimated in practice. However, in applications it is the global error that is usually of interest. It turns out that it is possible to control global error by controlling the individual LTEs if the underlying IVP is *stable*. We discuss this after we prove the following result.

Theorem 9 *Assume that a function $y_n(t)$ exists that solves the IVPs*

$$y'(t) = f(t, y(t)) \qquad y(t_n) = y_n$$

for $n = 0{:}N$, where $(t_0, y_0), \ldots, (t_N, y_N)$ are given and $t_0 < t_1 < \cdots < t_N$. Define the global error by

$$g_n = y_0(t_n) - y_n$$

and the local truncation error by

$$LTE_n = y_{n-1}(t_n) - y_n.$$

If

$$f_y = \frac{\partial f(t, y)}{\partial y} \leq 0$$

for all $t \in [t_0, t_N]$ and none of the trajectories

$$\{(t, y_n(t)) : t_0 \leq t \leq t_N\} \qquad n = 0{:}N$$

intersect, then for $n = 1{:}N$

$$|g_n| \leq \sum_{k=1}^{n} |LTE_k|$$

Proof. If $y_0(t_n) > y_{n-1}(t_n)$, then because f_y is negative we have

$$\int_{t_{n-1}}^{t_n} (f(t, y_0(t)) - f(t, y_{n-1}(t))) \, dt < 0$$

It follows that

$$0 < y_0(t_n) - y_{n-1}(t_n) = (y_0(t_{n-1}) - y_{n-1}(t_{n-1})) + \int_{t_{n-1}}^{t_n} (f(t, y_0(t)) - f(t, y_{n-1}(t))) \, dt$$

$$< (y_0(t_{n-1}) - y_{n-1}(t_{n-1}))$$

and so

$$|y_0(t_n) - y_{n-1}(t_n)| \leq |y_0(t_{n-1}) - y_{n-1}(t_{n-1})| \qquad (9.1.1)$$

Likewise, if $y_0(t_n) < y_{n-1}(t_n)$, then

$$\int_{t_{n-1}}^{t_n} (f(t, y_{n-1}(t)) - f(t, y_0(t))) \, dt < 0$$

and so

$$0 < y_{n-1}(t_n) - y_0(t_n) = (y_{n-1}(t_{n-1}) - y_0(t_{n-1})) + \int_{t_{n-1}}^{t_n} (f(t, y_{n-1}(t)) - f(t, y_0(t))) \, dt$$

$$< y_{n-1}(t_{n-1}) - y_0(t_{n-1})$$

Thus in either case (9.1.1) holds and so

$$|g_n| = |y_0(t_n) - y_n|$$

$$\leq |y_0(t_n) - y_{n-1}(t_n)| + |y_{n-1}(t_n) - y_n|$$

$$< |y_0(t_{n-1}) - y_{n-1}(t_{n-1})| + |y_{n-1}(t_n) - y_n|$$

$$= |g_{n-1}| + |LTE_n|.$$

The theorem follows by induction since $g_1 = LTE_1$. $\square$

The theorem essentially says that if $\partial f / \partial y$ is negative across the interval of interest, then global error at $t = t_n$ is less than the sum of the local errors made by the IVP solver in reaching t_n. The sign of this partial derivative is tied up with the stability IVP. Roughly speaking, if small changes in the initial value induce correspondingly small changes in the IVP solution, then we say that the IVP is *stable*. The concept is much more involved than the condition/stability issues that we talked about in connection with the $Ax = b$ problem. The mathematics is deep and interesting but beyond what we can do here.

So instead we look at the model problem $y'(t) = ay(t), y(0) = c$ and deduce some of the key ideas. In this example, $\partial f/\partial y = a$ and so Theorem 9 applies if $a < 0$. We know that the solution $y(t) = ce^{at}$ decays if and only if a is negative. If $\tilde{y}(t)$ solves the same differential equation with initial value $y(0) = \tilde{c}$, then

$$|\tilde{y}(t) - y(t)| = |\tilde{c} - c|e^{at}$$

showing how "earlier error" is damped out as t increases.

To illustrate how global error might be controlled in practice, consider the problem of computing $y(t_{max})$ to within a tolerance tol, where $y(t)$ solves a stable IVP $y'(t) = f(t, y(t))$, $y(t_0) = y_0$. Assume that a fixed-step Euler method is to be used and that we have a bound M_2 for $|y^{(2)}(t)|$ on the interval $[t_0, t_{max}]$. If $h = (t_{max} - t_0)/N$ is the step size, then from what we know about the local truncation error of the method,

$$|LTE_n| \leq M_2 \frac{h^2}{2}$$

Assuming that Theorem 9 applies,

$$|y(t_{max}) - y_N| \leq \sum_{n=1}^{N} |LTE_n| = M_2 N \frac{h^2}{2} = \frac{t_{max} - t_0}{2} M_2 h$$

Thus to make this upper bound less than a prescribed $tol > 0$, we merely set N to be the smallest integer that satisfies

$$\frac{(t_{max} - t_0)^2}{2N} M_2 \leq tol$$

Here is an implementation of the overall process:

```
    function [tvals,yvals] = FixedEuler(fname,y0,t0,tmax,M2,tol);
%
% Pre:  fname is a string that names a function of the form f(t,y).
%       M2 a bound on the second derivative of the solution to
%                 y' = f(t,y),    y(t0) = y0
%       on the interval [t0,tmax].
%
% Post: determine positive n so that if tvals = linspace(t0,tmax,n), then
%       y(i) is within tol of the true solution y(tvals(i)) for i=1:n.
%
    n = ceil(((tmax-t0)^2*M2)/(2*tol))+1
    h = (tmax-t0)/(n-1);
    yvals = zeros(n,1);
    tvals = linspace(t0,tmax,n)';
    yvals(1) = y0;
    for n=1:n-1
       fval = feval(fname,tvals(n),yvals(n));
       yvals(n+1) = yvals(n)+h*fval;
    end
```

Figure 9.4 shows the error when this solution framework is applied to the model problem $y' = -y$ across the interval $[0, 5]$. The trouble with this approach to global error control is that (1) we

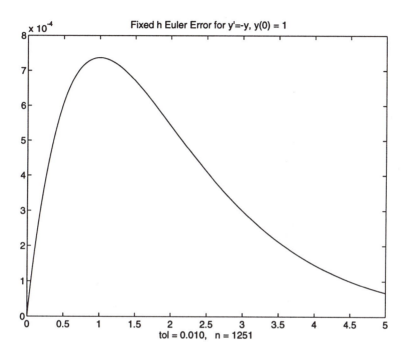

FIGURE 9.4 Error in fixed-step Euler

rarely have good bound information about $|y^{(2)}|$ and (2) it would be better to determine h adaptively so that longer step sizes can be taken in regions where the solution is smooth. This matter is pursued in §9.4.

Rounding errors are also an issue in IVP solving, especially when lots of very short steps are taken. In Fig 9.5 we plot the errors sustained when we solve $y' = -y$, $y(0) = 1$ across $[0, 1]$ with Euler's method in a three-digit floating point environment. The results for steps $h = 1/140$, $1/160$, and $1/180$ are reported. Note that the error gets worse as h gets smaller because the step sizes are in the neighborhood of unit roundoff. However, for the kind of problems that we are looking at, it is the discretization errors that dominate the discussion of accuracy.

Another issue that colors the performance of an IVP solver is the stability of *the method* itself. This is quite distinct from the notion of problem stability discussed earlier. It is possible for a method with a particular h to be unstable when it is applied to a stable IVP. For example, if we apply the Euler method to $y'(t) = -10y(t)$, then the iteration takes the form

$$y_{n+1} = (1 - 10h)y_n$$

To ensure that the errors are not magnified as the iteration progresses, we must insist that

$$|1 - 10h| < 1$$

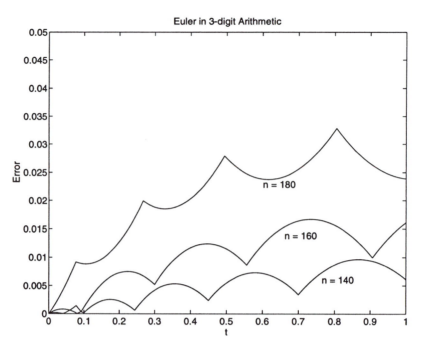

FIGURE 9.5 Roundoff error in fixed-step Euler

(i.e., $h < 2/5$). For h that satisfy this criteria, the method is *stable*. If $h > 2/5$, then any error δ in the initial condition will result in a $(1 - 10h)^n\delta$ contamination of the nth iterate. With this kind of error magnification, we say that the method is *unstable*. Different methods have different "h" restrictions in order to guarantee stability, and sometimes these restrictions force us to choose h much smaller than we would like.

9.1.3 The Backward Euler Method

To clarify this point about method stability, we examine the *backward Euler method*. The (forward) Euler method is derived from a Taylor expansion of the solution $y(t)$ about $t = t_n$. If instead we work with the approximation

$$y(t_{n+1} + h) \approx y(t_{n+1}) + y'(t_{n+1})h = y(t_{n+1}) + f(t_{n+1}, y(t_{n+1}))h$$

and set $h = -h_n = (t_n - t_{n+1})$, then we get

$$y(t_n) \approx y(t_{n+1}) - h_n f(t_{n+1}, y(t_{n+1}))$$

Substituting y_n for $y(t_n)$ and y_{n+1} for $y(t_{n+1})$, we are led to

$$y_{n+1} = y_n + h_n f(t_{n+1}, y_{n+1})$$

and, with repetition, the *backward Euler framework*:

$n = 0$

`Repeat:`

    Determine the step $h_n > 0$.

    $t_{n+1} = t_n + h_n$.

    Let y_{n+1} solve $F(z) = z - h_n f(t_{n+1}, z) = 0$.

    $n = n + 1$

Like the Euler method, the backward Euler method is first order. However, the two techniques differ in a very important aspect. Backward Euler is an *implicit method* because it defines y_{n+1} implicitly. For a simple problem like $y' = ay$ this poses no difficultly:

$$y_{n+1} = y_n + h_n a y_{n+1} = \frac{1}{1 - h_n a} y_n$$

Observe that if $a < 0$, then the method is stable for *all* choices of positive step size. This should be contrasted with the situation in the Euler setting, where $|1 + ah| < 1$ is required for stability.

    Euler's is an example of an *explicit method* because y_{n+1} is defined explicity in terms of quantities already computed. [e.g., y_n, $f(t_n, y_n)$]. Implicit methods tend to have better stability properties than their explicit counterparts. But there is an implementation penalty to be paid because y_{n+1} is defined as a zero of a nonlinear function. In backward Euler, y_{n+1} is a zero of $F(z) = z - h_n f(t_{n+1}, z)$. Fortunately, this does not necessarily require the application of the Chapter 8 root finders. A simpler, more effective approach is presented in §9.3.

9.1.4 Systems

We complete the discussion of IVP solving basics with comments about systems of differential equations. In this case the unknown function $y(t)$ is a vector of unknown functions:

$$y(t) = \begin{bmatrix} z_1(t) \\ \vdots \\ z_d(t) \end{bmatrix}$$

(We name the component functions with a z instead of a y to avoid confusion with earlier notation.) In this case, we are given an initial value for each component function and a recipe for its slope. This recipe generally involves the value of all the component functions:

$$\begin{bmatrix} z_1'(t) \\ \vdots \\ z_d'(t) \end{bmatrix} = \begin{bmatrix} f_1(t, z_1(t), \ldots, z_d(t)) \\ \vdots \\ f_m(t, z_1(t), \ldots, z_d(t)) \end{bmatrix} \qquad \begin{array}{c} z_1(t_0) = z_{10} \\ \vdots \\ z_d(t_0) = z_{d0} \end{array}$$

In vector language, $y'(t) = f(t, y(t)), y(t_0) = y_0$, where the y's are now column d-vectors. Here is a $d = 2$ example:

$$\begin{array}{rcl} u'(t) & = & 2u(t) - .01u(t)v(t) \\ v'(t) & = & -v(t) + .01u(t)v(t) \end{array} \qquad u(0) = u_0, \; v(0) = v_0$$

It describes the density of rabbit and fox populations in a classical predator-prey model. The rate of change of the rabbit density $u(t)$ and the fox density $v(t)$ depend on the current rabbit/fox densities.

Let's see how the derivation of Euler's method proceeds for a systems problem like this. We start with a pair of time-honored Taylor expansions:

$$u(t_{n+1}) \approx u(t_n) + u'(t_n)h_n = u(t_n) + h_n(2u(t_n) - .01u(t_n)v(t_n))$$
$$v(t_{n+1}) \approx v(t_n) + v'(t_n)h_n = v(t_n) + h_n(-v(t_n) + .01u(t_n)v(t_n))$$

Here (as usual), $t_{n+1} = t_n + h_n$. With the definitions

$$y_n = \left[\begin{array}{c} u_n \\ v_n \end{array} \right] \approx \left[\begin{array}{c} u(t_n) \\ v(t_n) \end{array} \right] = y(t_n)$$

and

$$f_n = f(t_n, y_n) = \left[\begin{array}{c} 2u_n - .01u_n v_n \\ -v_n + .01u_n v_n \end{array} \right] \approx \left[\begin{array}{c} 2u(t_n) - .01u(t_n)v(t_n) \\ -v(t_n) + .01u(t_n)v(t_n) \end{array} \right] = f(t_n, y(t_n))$$

we obtain he following vector implementation of the Euler method:

$$\left[\begin{array}{c} u_{n+1} \\ v_{n+1} \end{array} \right] = \left[\begin{array}{c} u_n \\ v_n \end{array} \right] + h_n \left[\begin{array}{c} 2u_n - .01u_n v_n \\ -v_n + .01u_n v_n \end{array} \right]$$

In full vector notation, this can be written as

$$y_{n+1} = y_n + h_n f_n$$

which is exactly the same formula that we developed in the scalar case.

As we go through the next two sections presenting more sophisticated IVP solvers, we shall do so for scalar ($d = 1$) problems, being mindful that all method-defining equations apply at the system level with no modification.

Systems can arise in practice from the conversion of *higher-order* IVPs. In a kth order IVP, we seek a function $y(t)$ that satisfies

$$y^{(k)}(t) = f(t, y(t), y^{(1)}(t), \ldots, y^{(k-1)}(t)) \qquad \text{where} \quad \begin{cases} y(t_0) & = & y_0 \\ y^{(1)}(t_0) & = & y_0^{(1)} \\ & \vdots & \\ y^{(k-1)}(t_0) & = & y_0^{(k-1)} \end{cases}$$

and $y_0, y_0^{(1)}, \ldots, y_0^{(k-1)}$ are given initial values. Higher order IVPs can be solved through conversion to a system of first-order IVPs. For example, to solve

$$v''(t) = 2v(t) + v'(t)\sin(t), \qquad v(0) = \alpha, \; v'(0) = \beta$$

we define $z_1(t) = v(t)$ and $z_2(t) = v'(t)$. The problem then transforms to

$$\begin{array}{l} z_1'(t) = z_2(t) \\ z_2'(t) = 2z_1(t) + z_2(t)\sin(t) \end{array} \qquad z_1(0) = \alpha, \; z_2(0) = \beta$$

Problems

P9.1.1 Produce a plot of the solution to

$$y'(t) = -ty + \frac{1}{y^2} \qquad y(1) = 1$$

across the interval $[1, 2]$. Use the Euler method.

P9.1.2 (a) Compute an approximation to $y(10)$ where

$$x''(t) = (3 - \sin(t))x'(t) + x(t)/(1 + [y(t)]^2)$$

$$y'(t) = -\cos(t)y(t) - x'(t)/(1 + t^2)$$

and $x(0) = 3$, $x'(0) = -1$, and $y(0) = 4$. Use the forward Euler method with fixed step determined so that three significant digits of accuracy are obtained. Hint: Define $z(t) = x'(t)$ and rewrite the recipe for x'' as a function of x, y, and z. This yields a $d = 3$ system.

P9.1.3 Plot the solutions to

$$y'(t) = \begin{bmatrix} -1 & 4 \\ -4 & -1 \end{bmatrix} y(t) \qquad y(0) = \begin{bmatrix} 2 \\ -1 \end{bmatrix}$$

across the interval $[0, 3]$.

9.2 The Runge-Kutta Methods

In a Euler step, we "extrapolate into the future" with only a single sampling of the slope function $f(t, y)$. The method has order 1 because its LTE goes to zero as h^2. Just as we moved beyond the rectangle rule in Chapter 4, so we must now move beyond the Euler framework with more involved models of the slope function. In the Runge-Kutta framework, we sample f at several judiciously chosen spots and use the information to obtain y_{n+1} from y_n with the highest possible order of accuracy.

9.2.1 Derivation

The Euler methods evaluate f once per step and have order 1. Let's sample f twice per step and see if we can obtain a second-order method. We arrange it so that the second evaluation depends on the first:

$$k_1 = hf(t_n, y_n)$$
$$k_2 = hf(t_n + \alpha h, y_n + \beta k_1)$$
$$y_{n+1} = y_n + ak_1 + bk_2$$

Our goal is to choose the parameters α, β, a, and b so that the LTE is $O(h^3)$. From Taylor series we have

$$y(t_{n+1}) = y(t_n) + y^{(1)}(t_n)h + y^{(2)}(t_n)\frac{h^2}{2} + O(h^3)$$

Since

$$y^{(1)}(t_n) = f$$
$$y^{(2)}(t_n) = f_t + f_y f$$

where

$$
\begin{aligned}
f &= f(t_n, y_n) \\
f_t &= \frac{\partial f(t_n, y_n)}{\partial t} \\
f_y &= \frac{\partial f(t_n, y_n)}{\partial y},
\end{aligned}
$$

it follows that

$$
y(t_{n+1}) = y(t_n) + fh + (f_t + f_y f)\frac{h^2}{2} + O(h^3) \tag{9.2.1}
$$

On the other hand,

$$
k_2 = hf(t_n + \alpha h, y_n + \beta k_1) = h\left(f + \alpha h f_t + \beta k_1 f_y + O(h^2)\right)
$$

and so

$$
y_{n+1} = y_n + ak_1 + bk_2 = y_n + (a + b) fh + b\left(\alpha f_t + \beta f f_y\right) h^2 + O(h^3) \tag{9.2.2}
$$

For the LTE to be $O(h^3)$, the equation

$$
y(t_{n+1}) - y_{n+1} = O(h^3)
$$

must hold. To accomplish this, we compare terms in (9.2.1) and (9.2.2) and require

$$
\begin{aligned}
a + b &= 1 \\
2ba &= 1 \\
2b\beta &= 1
\end{aligned}
$$

There are an infinite number of solutions to this system, the canonical one being $a = b = 1$ and $\alpha = \beta = 1/2$. With this choice the LTE is $O(h^3)$, and we obtain a second-order Runge-Kutta method:

$$
\begin{aligned}
k_1 &= hf(t_n, y_n) \\
k_2 &= hf(t_n + h, y_n + k_1) \\
y_{n+1} &= y_n + (k_1 + k_2)/2
\end{aligned}
$$

The actual expression for the LTE is given by

$$
\text{LTE}(RK2) = \frac{h^3}{12}(f_{tt} + 2ff_{ty} + f^2 f_{yy} - 2f_t f_y - 2f f_y^2)
$$

where the partials on the right are evaluated at some point in $[t_n, t_n + h]$. Notice that two f-evaluations are required per step.

The most famous Runge-Kutta method is the classical fourth order method:

$$
\begin{aligned}
k_1 &= hf(t_n, y_n) \\
k_2 &= hf(t_n + \tfrac{h}{2}, y_n + \tfrac{1}{2}k_1) \\
k_3 &= hf(t_n + \tfrac{h}{2}, y_n + \tfrac{1}{2}k_2) \\
k_4 &= hf(t_n + h, y_n + k_3) \\
y_{n+1} &= y_n + \tfrac{1}{6}(k_1 + 2k_2 + 2k_3 + k_4)
\end{aligned}
$$

This can be derived using the same Taylor expansion technique illustrated previously. It requires four *f*-evaluations per step.

The function RKStep can be used to carry out a Runge-Kutta step of prescribed order. Here is its specification along with an abbreviated portion of the implementation:

```
function [tnew,ynew,fnew] = RKstep(fname,tc,yc,fc,h,k)
%
% Pre:
%    fname   string that names a function of the form f(t,y)
%            where t is a scalar and y is a column d-vector.
%       yc   an approximate solution to y'(t) = f(t,y(t)) at t=tc.
%       fc   f(tc,yc).
%        h   the time step.
%        k   the order of the Runge-Kutta method used, 1<=k<=5.
%
% Post:
%     tnew   tc+h, ynew is an approximate solution at t=tnew, and
%     fnew   f(tnew,ynew).

   if k==1
      k1 = h*fc;
      ynew = yc + k1;

   elseif k==2
      k1 = h*fc;
      k2 = h*feval(fname,tc+(h/2),yc+(k1/2));
      ynew  = yc + (k1 + k2)/2;

   elseif k==3
      k1 = h*fc;
      k2 = h*feval(fname,tc+(h/2),yc+(k1/2));
      k3 = h*feval(fname,tc+h,yc-k1+2*k2);
      ynew  = yc + (k1 + 4*k2 + k3)/6;

   elseif k==4
      k1 = h*fc;
      k2 = h*feval(fname,tc+(h/2),yc+(k1/2));
      k3 = h*feval(fname,tc+(h/2),yc+(k2/2));
      k4 = h*feval(fname,tc+h,yc+k3);
      ynew  = yc + (k1 + 2*k2 + 2*k3 + k4)/6;

   elseif k==5
         :
   end
   tnew = tc+h;
   fnew = feval(fname,tnew,ynew);
```

As can be imagined, symbolic algebra tools are useful in the derivation of such an involved sampling and combination of f-values.

Problems

P9.2.1 The RKF45 method produces both a fourth order estimate and a fifth order estimate using six function evaluations:

$$
\begin{aligned}
k_1 &= hf(t_n, y_n) \\
k_2 &= hf(t_n + \tfrac{h}{4}, y_n + \tfrac{1}{4}k_1) \\
k_3 &= hf(t_n + \tfrac{3h}{8}, y_n + \tfrac{3}{32}k_1 + \tfrac{9}{32}k_2) \\
k_4 &= hf(t_n + \tfrac{12h}{13}, y_n + \tfrac{1932}{2197}k_1 - \tfrac{7200}{2197}k_2 + \tfrac{7296}{2197}k_3) \\
k_5 &= hf(t_n + h, y_n + \tfrac{439}{216}k_1 - 8k_2 + \tfrac{3680}{513}k_3 - \tfrac{845}{4104}k_4) \\
k_6 &= hf(t_n + \tfrac{h}{2}, y_n - \tfrac{8}{27}k_1 + 2k_2 - \tfrac{3544}{2565}k_3 + \tfrac{1859}{4104}k_4 - \tfrac{11}{40}k_5) \\
y_{n+1} &= y_n + \tfrac{25}{216}k_1 + \tfrac{1408}{2565}k_3 - \tfrac{2197}{4104}k_4 - \tfrac{1}{5}k_5 \\
z_{n+1} &= y_n + \tfrac{16}{135}k_1 + \tfrac{6656}{12825}k_3 + \tfrac{28561}{56430}k_4 - \tfrac{9}{50}k_5 + \tfrac{2}{55}k_6
\end{aligned}
$$

Write a script that discovers which of y_{n+1} and z_{n+1} is the fourth order estimate and which is the fifth order estimate.

9.2.2 Implementation

Runge-Kutta steps can obviously be repeated, and if we keep the step size fixed, then we obtain the following implementation:

```
    function [tvals,yvals] = FixedRK(fname,t0,y0,h,k,n)
%
% Produces an approximate solution to the initial value problem
% y'(t) = f(t,y(t)), y(t0) = y0 using a strategy that is based upon
% a kth order Runge-Kutta method. Stepsize is fixed.
%
% Pre:  fname = string that names the function f.
%       t0 = initial time.
%       y0 = initial condition vector.
%       h  = stepsize.
%       k  = order of method. (1<=k<=5).
%       n  = number of steps to be taken.
%
% Post: tvals(j) = t0 + (j-1)h, j=1:n+1
%       yvals(:j) = approximate solution at t = tvals(j), j=1:n+1
%
    tc = t0; yc = y0; tvals = tc; yvals = yc; fc = feval(fname,tc,yc);
    for j=1:n
      [tc,yc,fc] = RKstep(fname,tc,yc,fc,h,k);
      yvals = [yvals yc ];
      tvals = [tvals tc];
    end
```

The script file ShowRK can be used to illustrate the performance of the Runge-Kutta methods on the IVP $y' = -y$, $y(0) = 1$. The results are reported in Fig 9.6. All the derivatives of f are

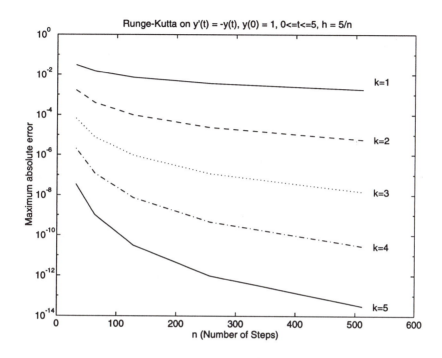

FIGURE 9.6 Runge-Kutta error

"nice," which means that if we increase the order and keep the step size fixed, then the errors should diminish by a factor of h. Thus for $n = 500$, $h = 1/100$ and we find that the error in the kth order method is about 100^{-k}.

Do not conclude from the example that higher-order methods are necessarily more accurate. If the higher derivatives of the solution are badly behaved, then it may well be the case that a lower-order method gives more accurate results. One must also be mindful of the number of f-evaluations that are required to purchase a given level of accuracy. The situation is analogous to what we found in the quadrature unit. Of course, the best situation is for the IVP software to handle the selection of method and step.

Problems

P9.2.2 For $k = 1:5$, how many f-evaluations does the kth-order Runge-Kutta method require to solve $y'(t) = -y(t)$, $y(0) = 1$ with error $\leq 10^{-6}$ across $[0, 1]$?

9.2.3 The MATLAB IVP Solvers ODE23 and ODE45

MATLAB supplies a pair of IVP solvers. ODE23 is based on a pair of second- and third-order Runge-Kutta methods. With two methods for predicting y_{n+1}, it uses the discrepancy of the predictions to determine heuristically whether the current step size is "safe" with respect to a given tolerance. ODE45 is similar, but it is based on a fourth- and fifth-order Runge-Kutta methods. Both codes can be used to solve systems, and to illustrate how they are typically used, we apply them to the following initial value problem:

$$\ddot{x}(t) = -\frac{x(t)}{(x(t)^2 + y(t)^2)^{3/2}} \qquad x(0) = .4 \quad \dot{x}(0) = 0$$

$$\ddot{y}(t) = -\frac{y(t)}{(x(t)^2 + y(t)^2)^{3/2}} \qquad y(0) = 0 \quad \dot{y}(0) = 2$$

These are Newton's equations of motion for the two-body problem. As t ranges from 0 to 2π, $(x(t), y(t))$ defines an ellipse.

Both ODE23 and ODE45 require that we put this problem in the standard $y' = f(t, y)$ form. To that end, we define $u_1(t) = x(t)$, $u_2(t) = \dot{x}(t)$, $u_3(t) = y(t)$, $u_4(t) = \dot{y}(t)$. The given IVP problem transforms to

$$\dot{u}_1(t) = u_2(t) \qquad\qquad u_1(0) = .4$$
$$\dot{u}_2(t) = -u_1(t)/(u_1(t)^2 + u_3(t)^2)^{3/2} \qquad u_2(0) = 0$$
$$\dot{u}_3(t) = u_4(t) \qquad\qquad u_3(0) = 0$$
$$\dot{u}_4(t) = -u_3(t)/(u_1(t)^2 + u_3(t)^2)^{3/2} \qquad u_4(0) = 2$$

We then write the following function, which returns the derivative of the u vector:

```
    function up = Kepler(t,u)
%
% Pre:  t (time) is a scalar and u is a 4-vector whose components
%       have the property that
%
%
%              u(1) = x(t)        u(2) = (d/dt)x(t)
%              u(3) = y(t)        u(4) = (d/dt)y(t)
%
%
%       where (x(t),y(t)) are the equations of motion in
%       the 2-body problem.
%
% Post:up is a 4-vector with the property that it is
%       the derivative of u at time t.
%

    r3 = (u(1)^2 + u(3)^2)^1.5;
    up(1) =   u(2);
    up(2) = -u(1)/r3;
    up(3) =   u(4);
    up(4) = -u(3)/r3;
```

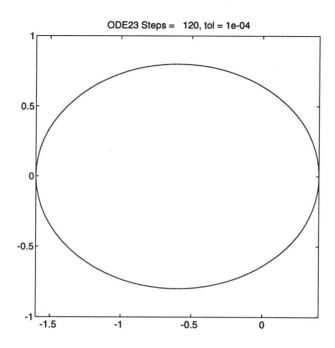

FIGURE 9.7 The ODE23 orbit

With this function available, we can call ODE23 and plot the results:

```
tInitial = 0;
tFinal   = 2*pi;
tol = .0001;
uInitial = [ .4; 0 ; 0 ; 2];
trace = 0;

[t, u] = ODE23('Kepler', tInitial, tFinal, uInitial, tol,trace);

figure
plot(u(:,1),u(:,3))
axis('square','equal')
title(sprintf('ODE23 Steps = %5.0f, tol = %5.0e',length(t),tol))
```

The call requires that we supply the name of the slope function, the initial and final times of the integration, the initial value, a tolerance, and an indication of whether or not we want to see a trace of the solution process. The output consists of a column vector t of times across the interval where ODE23 obtained solutions and a matrix u which contains the solutions at these times. In our problem there are four unknown functions, so u has four columns. Going back to the transform of the original IVP, we know that the first and third functions correspond to $x(t)$

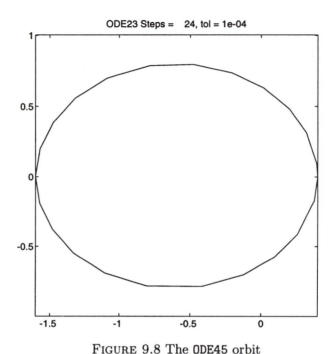

FIGURE 9.8 The ODE45 orbit

and $y(t)$ and that is why we plot u(:,1) versus u(:,3). (See Fig 9.7.) Instead of ODE23, we could call ODE45:

```
[t, u] = ODE45('Kepler', tInitial, tFinal, uInitial, tol,trace)
```

The calling sequence is identical. In this particular problem, ODE45 requires fewer function evaluations. (See Fig 9.8.) However, the time steps are so long that the resulting plot needs refinement. One possibility is to use splines:

```
tvals = linspace(tInitial,tFinal);
xvals = spline(t,u(:,1),tvals);
yvals = spline(t,u(:,3),tvals);
figure
plot(xvals,yvals)
axis('square','equal')
title(sprintf('ODE45 Steps = %5.0f, tol = %5.0e',length(t),tol))
xlabel('(With Spline Smoothing)')
```

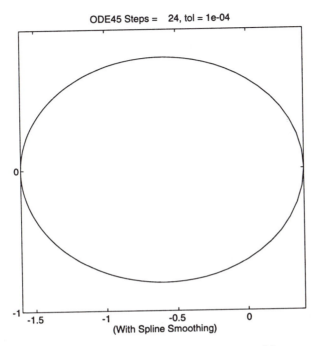

ODE45 Steps = 24, tol = 1e-04

(With Spline Smoothing)

FIGURE 9.9 The smoothed ODE45 orbit

(See Fig 9.9.) To force output at prescribed points, we can use a sequence of calls to ODE45 (or ODE23):

```
        :
h = (tFinal-tInitial)/n;
t0 = tInitial;
for k=1:n
    [t, u] = ODE45('Kepler', t0, t0+h, uInitial, tol,trace);
    k= length(t)
    disp(sprintf(' %10.3f    %10.3f ',u(k,1),u(k,3)));
    t0 = t0+h;
    uInitial = u(k,:)
end
```

Notice how the final u value in one call is the initial condition for the next call.

Problems

P9.2.3 Solve the following IVP problem using ODE23 and ODE45:

$$\ddot{x}(t) = 2\dot{y}(t) + x(t) - \frac{\mu_*(x(t)+\mu)}{r_1^3} - \frac{\mu(x(t)-\mu_*)}{r_2^3} \qquad x(0) = 1.2 \quad \dot{x}(0) = 0$$

$$\ddot{y}(t) = -2\dot{x}(t) + y(t) - \frac{\mu_* y(t)}{r_1^3} - \frac{\mu y(t)}{r_2^3} \qquad y(0) = 0 \qquad \dot{y}(0) = -1.0493575$$

where $\mu = 1/82.45$, $\mu_* = 1 - \mu$, and

$$r_1 = \sqrt{((x(t) + \mu)^2 + y(t)^2}$$
$$r_2 = \sqrt{((x(t) - \mu_*)^2 + y(t)^2}$$

Examine the effect of tol on the number of required steps.

9.3   The Adams Methods

From the fundamental theorem of calculus, we have

$$y(t_{n+1}) = y(t_n) + \int_{t_n}^{t_{n+1}} y'(t)dt$$

and so

$$y(t_{n+1}) = y(t_n) + \int_{t_n}^{t_{n+1}} f(t, y(t))dt$$

The Adams methods are based on the idea of replacing the integrand with a polynomial that interpolates $f(t, y)$ at selected solution points (t_j, y_j). The kth order Adams-Bashforth method is explicit and uses the current point (t_n, y_n) and $k - 1$ "historical" points. The kth order Adams-Moulton method is implicit and uses the future point (t_{n+1}, y_{n+1}), the current point, and $k - 2$ historical points. The implementation and properties of these two IVP solution frameworks are presented in this section.

9.3.1   Derivation of the Adams-Bashforth Methods

In the kth order *Adams-Bashforth* (AB) method, we set

$$y_{n+1} = y_n + \int_{t_n}^{t_{n+1}} p_{k-1}(t)dt \tag{9.3.1}$$

where $p_{k-1}(t)$ interpolates $f(t, y)$ at (t_{n-j}, y_{n-j}), $j = 0{:}k - 1$. We are concerned with the first five members of this family:

Order	Interpolant	AB Interpolation Points
1st	constant	(t_n, f_n)
2nd	linear	$(t_n, f_n), (t_{n-1}, f_{n-1})$
3rd	quadratic	$(t_n, f_n), (t_{n-1}, f_{n-1}), (t_{n-2}, f_{n-2})$
4th	cubic	$(t_n, f_n), (t_{n-1}, f_{n-1}), (t_{n-2}, f_{n-2}), (t_{n-3}, f_{n-3})$
5th	quartic	$(t_n, f_n), (t_{n-1}, f_{n-1}), (t_{n-2}, f_{n-2}), (t_{n-3}, f_{n-3}), (t_{n-3}, f_{n-3})$

If $k = 1$, then the one-point Newton-Cotes rule is applied and we get

$$y_{n+1} = y_n + h_n f(t_n, y_n) \qquad h_n = t_{n+1} - t_n$$

Thus the first-order AB method is the Euler method.

In the second-order Adams-Bashforth method, we set

$$p_{k-1}(t) = f_{n-1} + \frac{f_n - f_{n-1}}{h_{n-1}}(t - t_{n-1})$$

in (9.3.1). This is the linear interpolant of (t_{n-1}, f_{n-1}) and (t_n, f_n), and we obtain

$$\int_{t_n}^{t_{n+1}} f(t, y(t))dt \approx \int_{t_n}^{t_{n+1}} \left(f_{n-1} + \frac{f_n - f_{n-1}}{h_{n-1}}(t - t_{n-1}) \right) dt$$

$$= \frac{h_n}{2} \left(\frac{h_n + 2h_{n-1}}{h_{n-1}} f_n - \frac{h_n}{h_{n-1}} f_{n-1} \right).$$

If $h_n = h_{n-1} = h$, then from (9.3.1)

$$y_{n+1} = y_n + \frac{h}{2}(3f_n - f_{n-1})$$

The derivation of higher-order AB methods is analogous. A table of the first five Adams-Bashforth methods along with their respective local truncation errors is given in Fig. 9.10 The

Order	Step	LTE
1	$y_{n+1} = y_n + hf_n$	$\frac{h^2}{2} y^{(2)}(\eta)$
2	$y_{n+1} = y_n + \frac{h}{2}(3f_n - f_{n-1})$	$\frac{5h^3}{12} y^{(3)}(\eta)$
3	$y_{n+1} = y_n + \frac{h}{12}(23f_n - 16f_{n-1} + 5f_{n-2})$	$\frac{3h^4}{8} y^{(4)}(\eta)$
4	$y_{n+1} = y_n + \frac{h}{24}(55f_n - 59f_{n-1} + 37f_{n-2} - 9f_{n-3})$	$\frac{251h^5}{720} y^{(5)}(\eta)$
5	$y_{n+1} = y_n + \frac{h}{720}(1901f_n - 2774f_{n-1} + 2616f_{n-2} - 1274f_{n-3} + 251f_{n-4})$	$\frac{95h^6}{288} y^{(6)}(\eta)$

FIGURE 9.10 Adams-Bashforth family

derivation of the LTEs for the AB methods is a straightforward computation that involves the Newton-Cotes error:

$$y(t_{n+1}) - y_n = \int_{t_n}^{t_{n+1}} (f(t, y_n(t)) - p_{k-1}(t))dt$$

9.3.2 Implementation

To facilitate experimentation with the AB method, we have written a function that can carry out any of the methods specified in Fig 9.10:

```
      function [tnew,ynew,fnew] = ABstep(fname,tc,yc,fvals,h,k)
%
% Pre:
%   fname    string that names a function of the form f(t,y)
%            where t is a scalar and y is a column d-vector.
%
%       yc   approximate solution to y'(t) = f(t,y(t)) at t=tc.
%
%   fvals    d-by-k matrix where fvals(:,i) is an approximation
%            to f(t,y) at t = tc +(1-i)h, i=1:k
%
%        h   time step.
%
%        k   order of the AB method used, 1<=k<=5.
%
% Post:
%     tnew   tc+h
%
%     ynew   approximate solution at t=tnew
%
%     fnew   f(tnew,ynew).

    if k==1
       ynew = yc + h*fvals;
    elseif k==2
       ynew = yc + (h/2)*(fvals*[3;-1]);
    elseif k==3
       ynew = yc + (h/12)*(fvals*[23;-16;5]);
    elseif k==4
       ynew = yc + (h/24)*(fvals*[55;-59;37;-9]);
    elseif k==5
       ynew = yc + (h/720)*(fvals*[1901;-2774;2616;-1274;251]);
    end
    tnew = tc+h;
    fnew = feval(fname,tnew,ynew);
```

In the systems case, fval is a matrix and ynew is yc plus a matrix-vector product.

Note that k snapshots of $f(t, y)$ are required, and this is why Adams methods are called *multistep* methods. Because of this there is a "start-up" issue with the Adams-Bashforth method: How do we perform the first step when there is no "history"? There are several approaches to this, and care must be taken to ensure that the accuracy of the generated start-up values is consistent with the overall accuracy aims. For a kth order Adams framework we could use a kth order Runge-Kutta method, to get $f_j = f(t_j, y_j)$, $j = 1{:}k - 1$. The function StartAB does just that:

```
    function [tvals,yvals,fvals] = StartAB(fname,t0,y0,h,k)
%
% Uses k-th order Runge-Kutta to generate approximate
% solutions to
%                   y'(t) = f(t,y(t))    y(t0) = y0
%
% at t = t0, t0+h, ... , t0 + (k-1)h.
%
% Pre:
%        fname is a string that names the function f.
%        t0 is the initial time.
%        y0 is the initial value.
%        h is the step size.
%        k is the order of the RK method used.
%
% Post:
%        tvals = [ t0, t0+h, ... , t0 + (k-1)h].
%        For j =1:k, yvals(:,j) = y(tvals(j))  (approximately).
%        For j =1:k, fvals(:,j) = f(tvals(j),yvals(j)).
%
    tc = t0;    yc = y0;    fc = feval(fname,tc,yc);
    tvals = tc; yvals = yc; fvals = fc;

    for j=1:k-1
       [tc,yc,fc] = RKstep(fname,tc,yc,fc,h,k);
       tvals = [tvals tc];
       yvals = [yvals yc];
       fvals = [fc fvals];
    end
```

With this function available, we are able to formulate a fixed-step Adams-Bashforth solver:

```
    function [tvals,yvals] = FixedAB(fname,t0,y0,h,k,n)
%
% Produces an approximate sol'n to the initial value problem y'(t) = f(t,y(t)),
% y(t0) = y0 using a strategy that is based upon a k-th order Adams-Bashforth
% method. Stepsize is fixed.
%
% Pre:
%        fname    string that names the function f.
%          t0     initial time.
%          y0     initial condition vector.
%           h     stepsize.
%           k     order of method. (1<=k<=5).
%           n     number of steps to be taken,
%
```

```
% Post:
%    tvals(j)      t0 + (j-1)h, j=1:n+1
%    yvals(:j)     approximate solution at t = tvals(j), j=1:n+1
%

    [tvals,yvals,fvals] = StartAB(fname,t0,y0,h,k);
    tc = tvals(k);
    yc = yvals(:,k);
    fc = fvals(:,k);

    for j=k:n
        % Take a step and then update.
        [tc,yc,fc] = ABstep(fname,tc,yc,fvals,h,k);
        tvals = [tvals tc];
        yvals = [yvals yc];
        fvals = [fc fvals(:,1:k-1)];
    end
```

If we apply this algorithm to the model problem, $y' = -y, y(0) = 1$. (See Fig 9.11.) Notice that for the kth-order method, the error goes to zero as h^k, where $h = 1/n$.

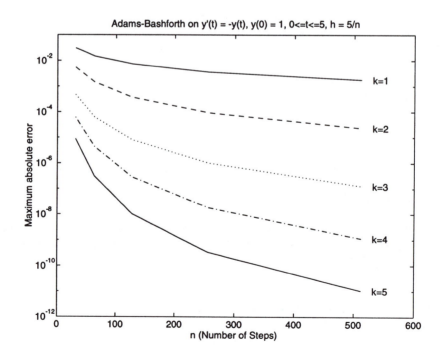

FIGURE 9.11 kth order Adams-Bashforth error

9.3.3   The Adams-Moulton Methods

The kth order *Adams-Moulton* (AM) method is just like the kth-order Adams-Bashforth method, but the points at which we interpolate the integrand in

$$y_{n+1} = y_n + \int_{t_n}^{t_{n+1}} p_{k-1}(t)dt$$

are "shifted" one time step into the future. In particular, the kth-order Adams-Moulton method uses a degree $k-1$ interpolant of the points (t_{n+1-j}, f_{n+1-j}), $j = 0{:}k-1$:

Order	Interpolant	AM Interpolation Points
1st	constant	(t_{n+1}, f_{n+1})
2nd	linear	$(t_{n+1}, f_{n+1}), (t_n, f_n)$
3rd	quadratic	$(t_{n+1}, f_{n+1}), (t_n, f_n), (t_{n-1}, f_{n-1})$
4th	cubic	$(t_{n+1}, f_{n+1}), (t_n, f_n), (t_{n-1}, f_{n-1}), (t_{n-2}, f_{n-2})$
5th	quartic	$(t_{n+1}, f_{n+1}), (t_n, f_n), (t_{n-1}, f_{n-1}), (t_{n-2}, f_{n-2}), (t_{n-3}, f_{n-3})$

For example, in the second-order Adams-Moulton method we set

$$p_{k-1}(t) = f_n + \frac{f_{n+1} - f_n}{h_n}(t - t_n)$$

the linear interpolant of (t_n, f_n) and (t_{n+1}, f_{n+1}). We then obtain the approximation

$$\int_{t_n}^{t_{n+1}} f(t, y(t))dt \approx \int_{t_n}^{t_{n+1}} \left(f_n + \frac{f_{n+1} - f_n}{h_n}(t - t_n)\right) = \frac{h_n}{2}(f_n + f_{n+1})$$

and thus

$$y_{n+1} = y_n + \frac{h_n}{2}(f(t, y_{n+1}) + f_n)$$

As in the backward Euler method, which is just the first order Adams-Moulton method, y_{n+1} is specified implicitly through a nonlinear equation. The higher-order Adams-Moulton methods are derived similarly, and in Fig 9.12 we specify the first five members in the family.

The LTE coefficient for any AM method is slightly smaller than the LTE coefficients for the corresponding AB method. Analogous to ABstep, we have

```
    function [tnew,ynew,fnew] = AMstep(fname,tc,yc,fvals,h,k)
%
% Pre:
%     fname     string that names a function of the form f(t,y)
%               where t is a scalar and y is a column d-vector.
```

Order	Step	LTE
1	$y_{n+1} = y_n + hf(t_{n+1}, y_{n+1})$	$-\dfrac{h^2}{2}y^{(2)}(\eta)$
2	$y_{n+1} = y_n + \dfrac{h}{2}\left(f(t_{n+1}, y_{n+1}) + f_n\right)$	$-\dfrac{h^3}{12}y^{(3)}(\eta)$
3	$y_{n+1} = y_n + \dfrac{h}{12}\left(5f(t_{n+1}, y_{n+1}) + 8f_n - f_{n-1}\right)$	$-\dfrac{h^4}{24}y^{(4)}(\eta)$
4	$y_{n+1} = y_n + \dfrac{h}{24}\left(9f(t_{n+1}, y_{n+1}) + 19f_n - 5f_{n-1} + f_{n-2}\right)$	$-\dfrac{19h^5}{720}y^{(5)}(\eta)$
5	$y_{n+1} = y_n + \dfrac{h}{720}\left(251f(t_{n+1}, y_{n+1}) + 646f_n - 264f_{n-1} + 106f_{n-2} - 19f_{n-3}\right)$	$-\dfrac{3h^6}{160}y^{(6)}(\eta)$

FIGURE 9.12 The Adams-Moulton methods

```
%        yc    an approximate solution to y'(t) = f(t,y(t)) at t=tc.
%     fvals    d-by-k matrix where fvals(:,i) is an approximation
%              to f(t,y) at t = tc +(2-i)h, i=1:k
%         h    time step.
%         k    order of the AM method used, 1<=k<=5.
% Post:
%      tnew    tc+h
%      ynew    an approximate solution at t=tnew
%      fnew    f(tnew,ynew).

   if k==1
      ynew = yc + h*fvals;
   elseif k==2
      ynew = yc + (h/2)*(fvals*[1;1]);
   elseif k==3
      ynew = yc + (h/12)*(fvals*[5;8;-1]);
   elseif k==4
      ynew = yc + (h/24)*(fvals*[9;19;-5;1]);
   elseif k==5
      ynew = yc + (h/720)*(fvals*[251;646;-264;106;-19]);
   end
   tnew = tc+h;
   fnew = feval(fname,tnew,ynew);
```

We could discuss methods for the solution of the nonlinear $F(z) = 0$ that defines y_{n+1}. However, we have other plans for the Adams-Moulton methods that circumvent this problem.

9.3.4   The Predictor-Corrector Idea

A very important framework for solving IVPs results when we couple an Adams-Bashforth method with an Adams-Moulton method of the same order. The idea is to *predict* y_{n+1} using an Adams-Bashforth method and then to *correct* its value using the corresponding Adams-Moulton method. In the second-order case, AB2 gives

$$y_{n+1}^{(P)} = y_n + \frac{h}{2}(f_n + f_{n-1})$$

which then is used in the right-hand side of the AM2 recipe to render

$$y_{n+1}^{(C)} = y_n + \frac{h}{2}\left(f(t_{n+1}, y_{n+1}^{(P)}) + f_n\right)$$

For general order we have the following implementation in terms of `ABstep` and `AMstep`:

```
   function [tnew,yPred,fPred,yCorr,fCorr] = PCstep(fname,tc,yc,fvals,h,k)
%
% Pre:
%      fname    string that names a function of the form f(t,y)
%               where t is a scalar and y is a column d-vector.
%
%         yc    approximate solution to y'(t) = f(t,y(t)) at t=tc.
%
%      fvals    a d-by-k matrix where fvals(:,i) is an approximation
%               to f(t,y) at t = tc +(1-i)h, i=1:k
%
%          h    time step.
%
%          k    order of the Runge-Kutta method used, 1<=k<=5.
%
% Post:
%       tnew    tc+h,
%
%      yPred    the predicted solution at t=tnew
%
%      fPred    f(tnew,yPred)
%
%      yCorr    the corrected solution at t=tnew
%
%      fCorr    f(tnew,yCorr).

      [tnew,yPred,fPred] = ABstep(fname,tc,yc,fvals,h,k);
      [tnew,yCorr,fCorr] = AMstep(fname,tc,yc,[fPred fvals(:,1:k-1)],h,k);
```

The repeated application of this defines a fixed-step predictor-corrector framework:

```
    function [tvals,yvals] = FixedPC(fname,t0,y0,h,k,n)
%
% Produces an approximate sol'n to the initial value problem y'(t) = f(t,y(t)),
% y(t0) = y0 using a strategy that is based on a k-th order
% Adams PC method. Stepsize is fixed.
%
% Pre:
%     fname     string that names the function f.
%        t0     initial time.
%        y0     initial condition vector.
%         h     stepsize.
%         k     order of method. (1<=k<=5).
%         n     number of steps to be taken.
%
% Post:
%   tvals(j)    t0 + (j-1)h, j=1:n+1
% yvals(:j)     approximate solution at t = tvals(j), j=1:n+1
%

    [tvals,yvals,fvals] = StartAB(fname,t0,y0,h,k);
    tc = tvals(k);
    yc = yvals(:,k);
    fc = fvals(:,k);

    for j=k:n
       % Take a step and then update.
       [tc,yPred,fPred,yc,fc] = PCstep(fname,tc,yc,fvals,h,k);
       tvals = [tvals tc];
       yvals = [yvals yc];
       fvals = [fc fvals(:,1:k-1)];
    end
```

The error associated with this method when applied to the model problem is given in Fig 9.13.

Problems

P9.3.1 Write functions

```
        [tvals,yvals] = AFixedAB(A,t0,y0,h,k,n)
        [tvals,yvals] = AFixedAM(A,t0,y0,h,k,n)
```

that can be used to solve the IVP $y'(t) = Ay(t)$, $y(t_0 = y_0$, where A is a d-by-d matrix. In AFixedAM a linear system will have to be solved at each step. Get the factorization "out of the loop."

P9.3.2 Use FixedABand FixedPC to solve the IVP described in problem P9.2.3. Explore the connections between step size, order, and the number of required function evaluations.

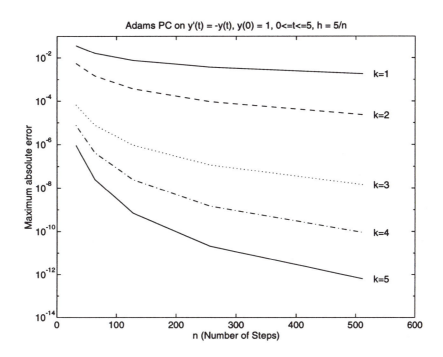

FIGURE 9.13 kth Order Predictor-Corrector Error

9.3.5 Stepsize Control

The idea behind error estimation in adaptive quadrature is to compute the integral in question in two ways and then accept or reject the better estimate based on the observed discrepancies. The predictor-corrector framework presents us with a similar opportunity. The quality of $y_{n+1}^{(C)}$ can be estimated from $|y_{n+1}^{(C)} - y_{n+1}^{(P)}|$. If the error is too large, we can reduce the step. If the error it is too small, then we can lengthen the step. Properly handled, we can use this mechanism to integrate the IVP across the interval of interest with steps that are as long as possible given a prescribed error tolerance. In this way we can compute the required solution, more or less minimizing the number of f evaluations. The MATLAB IVP solvers ODE23 and ODE45 are Runge-Kutta based and do just that. We develop a second-order adaptive step solver based on the second-order AB and AM methods.

Do we accept $y^{(C)}$ as our chosen y_{n+1}? If

$$\Delta = |y_{n+1}^{(P)} - y_{n+1}^{(C)}|$$

is small, then our intuition tells us that $y_{n+1}^{(C)}$ is probably fairly good and worth accepting as our approximation to $y(t_{n+1})$. If not, there are two possibilities:

- Refine $y_{n+1}^{(C)}$ through repeated application of the AM2 formula:

$$y_{n+1} = y_{n+1}^{(C)}$$
Repeat:
$$y_{n+1} = y_n + \frac{h}{2}\left(f(t_{n+1}, y_{n+1}) + f_n\right)$$

A reasonable termination criterion might be to quit as soon as two successive iterates differ by a small amount. The goal of the iteration is to produce a solution to the AM2 equation.

- Halve h and try another predict/correct step [i.e., produce an estimate y_{n+1} of $y(t_n + h/2)$].

The latter approach is more constructive because it addresses the primary reason for discrepancy between the predicted and corrected value: an overly long step h.

To implement a practical step size control process, we need to develop a heuristic for estimating the error in $y_{n+1}^{(c)}$ based on the discrepancy between it and $y_{n+1}^{(P)}$. The idea is to manipulate the LTE expressions

$$y(t_{n+1}) \;=\; y_{n+1}^{(P)} + \frac{5}{12}h^3 y^{(3)}(\eta_1) \qquad \eta_1 \in [t_n, t_n + h]$$

$$y(t_{n+1}) \;=\; y_{n+1}^{(C)} - \frac{1}{12}h^3 y^{(3)}(\eta_2) \qquad \eta_2 \in [t_n, t_n + h]$$

We make the assumption that $y^{(3)}$ does *not* vary much across $[t_n, t_n + h]$. Subtracting the first equation from the second leads to approximation

$$|y_{n+1}^{(C)} - y_{n+1}^{(P)}| \;\approx\; \frac{1}{2}h^3 |y^{(3)}(\eta)| \qquad \eta \in [t_n, t_n + h]$$

and so

$$|y_{n+1}^{(C)} - y(t_{n+1})| \;\approx\; \frac{1}{6}|y_{n+1}^{(C)} - y_{n+1}^{(P)}|$$

This leads to the following framework for a second-order predictor-corrector scheme:

$$y_{n+1}^{(P)} = y_n + \frac{h}{2}(f_n + f_{n-1})$$
$$y_{n+1}^{(C)} = y_n + \frac{h}{2}\left(f(t_{n+1}, y_{n+1}^{(P)}) + f_n\right)$$
$$\epsilon = \frac{1}{6}|y_{n+1}^{(C)} - y_{n+1}^{(P)}|$$

If ϵ is too big, then

   reduce h and try again.

Else if ϵ is about right, then

   set $y_{n+1} = y_{n+1}^{(C)}$ and keep h.

Else if ϵ is too small, then

   set $y_{n+1} = y_{n+1}^{(C)}$ and increase h.

The definitions of "too big," "about right," and "too small" are central. Here is one approach. Suppose we want the global error in the solution snapshots across $[t_0, t_{max}]$ to be less than δ. If it takes n_{max} steps to integrate across $[t_0, t_{max}]$, then we can heuristically guarantee this if

$$\sum_{n=1}^{n_{max}} \text{LTE}_n \leq \delta$$

Thus if h_n is the length of the nth step, and

$$|\text{LTE}_n| \leq \frac{h_n \delta}{t_{max} - t_0}$$

then

$$\sum_{n=1}^{n_{max}} \text{LTE}_n \leq \sum_{n=1}^{n_{max}} \frac{h_n \delta}{t_{max} - t_0} \leq \delta$$

This tells us when to accept a step. But if the estimated LTE is considerably smaller than the threshold, say

$$\epsilon \leq \frac{1}{10} \frac{\delta h}{t_{max} - t_0}$$

then it might be worth doubling h.

If the ϵ is too big, then our strategy is to halve h. But to carry out the predictor step with this step size, we need $f(t_n - h/2, y_{n-1/2})$ where $y_{n-1/2}$ is an estimate of $y(t_n - h/2)$. "Missing" values in in this setting can be generated by interpolation or by using (for example) an appropriate Runge Kutta estimate.

Problems

P9.3.1 Derive an estimate for $|y_{n+1}^{(C)} - y(t_{n+1})|$ for the third-, fourth- and fifth-order predictor-corrector pairs.

M-Files and References

Script Files

ShowTraj	Shows family of solutions.
ShowEuler	Illustrates Euler method.
ShowTrunc	Shows effect of truncation error.
EulerRoundoff	Illustrates Euler in three-digit floating point.
ShowAB	Illustrates FixedAB.
ShowPC	Illustrates FixedPC.
ShowRK	Illustrates FixedRK.
ShowKepler	Illustrates ODE23 and ODE45 on a system.

Function Files

`StartAB`	Gets starting values for Adams methods.
`ABStep`	Adams-Bashforth step (order $<=5$).
`FixedAB`	Fixed step size Adams-Bashforth.
`AMStep`	Adams-Moulton step (order $<=5$).
`PCStep`	AB-AM predictor-corrector Step (order $<=5$).
`FixedPC`	Fixed stepsize AB-AM predictor-corrector.
`RKStep`	Runge-Kutta step (order $<=5$).
`FixedRK`	Fixed step size Runge-Kutta.
`Kepler`	For solving two-body IVP.

References

C.W. Gear (1971). *Numerical Initial Value Problems in Ordinary Differential Equations*, Prentice Hall, Englewood Cliffs, NJ.

J. Lambert (1973). *Computational Methods in Ordinary Differential Equations*, John Wiley, New York.

J. Ortega and W. Poole (1981). *An Introduction to Numerical Methods for Differential Equations*, Pitman, Marshfield, MA.

L. Shampine and M. Gordon (1975). *Computer Solution of Ordinary Differential Equations: the Initial Value Problem*, Freeman, San Francisco.

Bibliography

R. Bartels, J. Beatty, and B. Barsky (1987). *An Introduction to Splines for Use in Computer Graphics and Geometric Modeling*, Morgan Kaufmann, Los Altos, CA.

Å. Björck (1996). *Numerical Methods for Least Squares Problems*, SIAM Publications, Philadelphia, PA.

T.F. Coleman and C.F. Van Loan (1988). *Handbook for Matrix Computations*, SIAM Publications, Philadelphia, PA.

S.D. Conte and C. de Boor (1980). *Elementary Numerical Analysis: An Algorithmic Approach, Third Edition*, McGraw-Hill, New York.

P. Davis (1963). *Interpolation and Approximation*, Blaisdell, New York.

P. Davis and P. Rabinowitz (1984). *Methods of Numerical Integration, 2nd Ed.*, Academic Press, New York.

C. de Boor (1978). *A Practical Guide to Splines*, Springer, Berlin.

J.E. Dennis Jr and R.B. Schnabel (1983). *Numerical Methods for Unconstrained Optimization*, Prentice Hall, Englewood Cliffs, NJ.

J. Dongarra, I. Duff, D. Sorensen, and H. van der Vorst (1990). *Solving Linear Systems on Vector and Shared Memory Computers*, SIAM Publications, Philadelphia, PA.

G.E. Forsythe and C. Moler (1967). *Computer Solution of Linear Algebraic Systems*, Prentice-Hall, Englewood Cliffs, NJ.

C.W. Gear (1971). *Numerical Initial Value Problems in Ordinary Differential Equations*, Prentice Hall, Englewood Cliffs, NJ.

P. Gill, W. Murray, and M.H. Wright (1981). *Practical Optimization*, Academic Press, New York.

P. Gill, W. Murray, and M.H. Wright (1991). *Numerical Linear Algebra and Optimization, Vol. 1*, Addison-Wesley, Reading, MA.

G.H. Golub and J.M. Ortega (1993). *Scientific Computing: An Introduction with Parallel Computing*, Academic Press, Boston.

G.H. Golub and C.F. Van Loan (1996). *Matrix Computations, Third Edition*, Johns Hopkins University Press, Baltimore, MD.

W.W. Hager (1988). *Applied Numerical Linear Algebra*, Prentice-Hall, Englewood Cliffs, NJ.

D. Kahaner, C.B. Moler, and S. Nash (1989). *Numerical Methods and Software*, Prentice Hall, Englewood Cliffs, NJ.

J. Lambert (1973). *Computational Methods in Ordinary Differential Equations*, John Wiley, New York.

C.L. Lawson and R.J. Hanson (1996). *Solving Least Squares Problems*, SIAM Publications, Philadelphia, PA.

M. Marcus (1993). *Matrices and* MATLAB*: A Tutorial*, Prentice Hall, Upper Saddle River, NJ.

J. Ortega and W. Poole (1981). *An Introduction to Numerical Methods for Differential Equations*, Pitman, Marshfield, MA.

R. Pratap (1995). *Getting Started with* MATLAB, Saunders College Publishing, Fort Worth, TX.

L. Shampine and M. Gordon (1975). *Computer Solution of Ordinary Differential Equations: the Initial Value Problem*, Freeman, San Francisco.

G.W. Stewart (1973). *Introduction to Matrix Computations*, Academic Press, New York.

A. Stroud (1972). *Approximate Calculation of Multiple Integrals*, Prentice Hall, Englewood Cliffs, NJ.

The Student Edition of MATLAB, *Users Guide*, The MathWorks Inc., Natick, Massachusetts.

C.F. Van Loan (1992). *Computational Frameworks for the Fast Fourier Transfrom*, SIAM Publications, Philadelphia, PA.

Index

abs, 57
Abscissas (quadrature), 124
Adams methods, 328
Adams-Bashforth methods, 328
   error 332
   implementation, 330
   start-up, 331
Adams-Moulton methods, 333
   error 334
Adaptive quadrature, 138
all, 41
And operation (&), 31
any, 41, 57
axis, 17, 21, 43

Back substitution, 197
Backslash (\), 27
Backward Euler, 316
Backward Euler, 316
Backward stability, 220
Band matrices, 159
Banded systems, 200
Bandwidth (matrix), 159
Barrier synchronization, 150
Benchmarking, 55
Bidiagonal solving, 202
Big O notation, 83
Binary search, 97
Bisection, 262
Block structure, 161
Boolean operations, 31
Breakpoint continuity, 113

ceil, 38
chol, 249
Circulant matrix, 158
Cholesky factorization, 236
   block, 250
   dot product, 244
   gaxpy, 246
   recursive 247

   saxpy, 244
   scalar, 243
   shared memory, 254
   tridiagonal, 240
clc, 22
clock, 56
close all, 22
Colon notation, 6, 25, 26
Column vector, 3
Column vs. row, 71
Comments (%), 12
Complete spline, 115
Complex numbers, 4
Composite rules, 133
cond, 220
condest, 220
Condition number, 219
Continuation, 24
contour, 175
Contour plots, 175, 290, 293, 303
Critical section, 153
Cubic Hermite interpolation, 105
Cubic splines, 112

delete, 35
Determinant vs. condition, 222
diag, 160
Diagonal dominance, 236
diary, 35
Differentiation, 58
disp, 12
Divided differences, 86

elfun, 9
Ellipses (...), 24
Empty Vector, 6
eps, 50
Error, 44
   absolute, 45
   composite rules, 136
   finite difference, 60

interpolating polynomial, 84
Newton-Cotes, 129
relative, 45
etime, 56
Euler's method, 309
error, 314
fixed step, 314
stability, 315
eval, 61
Explicit IVP solver, 316
Exponentiation of vector, 52

Fast Fourier transform, 179
feval, 61
find, 33
Finite difference Jacobian, 300
Finite difference Newton, 272
Floating point system
arithmetic, 50
exponent, 49
mantissa, 49
normalized, 50
floor, 38
Flop counting, 83
flops, 58
fmin, 282
Fmins, 292
for-loop, 4
format, 8
Forward substitution, 195
Fourth order Runge-Kutta, 320
Function Evaluation, 11, 104, 264
Functions, 52
user-defined 53
vector arguments, 56

Gaussian Elimination, 207
Gauss-Legendre rule, 144
Gauss-Newton framework, 304
Gauss quadrature, 143
Gaxpy operation , 245
ginput, 42
global, 140, 274
Global error, 311
Global error control, 314
Global max and min, 278
Globalizing Newton's method, 269
Golden section search, 279
Gradient, 289
Graphical Search, 277

help, 5
Hermite interpolation, 105
Hessenberg systems, 204
Higher order IVPs, 318
hilbert, 156
hist, 36
hold, 19
home, 45
Horner's rule, 73

if-then-else, 30
imag, 180
Implicit IVP solver, 316
inf, 51
Initial guess, 261
Initial value problems, 308
input, 30
Interconnection network, 186
Interpolation, 68
inverse, 90
two-dimensional, 90
piecewise cubic Hermite, 105
piecewise linear, 94
Interpolating polynomial error, 84
Inverse interpolation, 90

Jacobian, 296
finite difference, 300

Least square fitting, 224
length, 5
Least squares sensitivity, 235
Level-3 operations, 250
Linear systems, 194
linspace, 7
Load balancing, 254
Local truncation error (LTE), 311
logspace, 7
long, 8
LU factorization, 208
use of, 216

.m suffix, 12
Machine precision, 50
Matrices
banded, 159
block, 161
circulant, 158
symmetric, 156
Toeplitz, 159

Matrix-vector products, 23, 164
Matrix-matrix multiplication, 26, 167
   two-processor, 188
Matrix set-up, 25, 155, 173
max, 32
mean, 36
median, 36
Memory manager, 4,5
Mesh, 187
Messages
   send 186
   receive, 186
Midpoint rule, 132
mkpp, 120
Monte Carlo, 41
Multiplier matrices, 205
Multiple right hand sides, 197
Multiprocessors
   Shared memory, 146
   Distributed memory, 186
Multivariate minimization, 287

nargin, 61
Natural spline, 116
Nested multiplication, 73, 80
Newton-Cotes error, 129
Newton-Cotes rules, 125
Newton's method, 266
   globalizing, 269
   finite difference, 272
   systems, 295
Node program, 148
Nonlinear least squares, 302
Nonlinear problems, 258
norm, 171
Normal distribution, 36
Norms, 170
Not operation ($\tilde{\ }$), 32
Not-a-knot spline, 117
num2str, 61

Objective function, 288
ODE23, 323
ODE45, 323
ones, 56
Open rules, 132
Or operation (|), 32
Orthogonal matrices, 229
Orthonormal basis, 230
Outer product, 169

Overflow, 51, 231

path, 54
Piecewise cubic Hermite interpolation, 105
Piecewise polynomial, 94
Piecewise linear interpolation
   adaptive, 101
   static, 100
Pivoting, 214
plot, 2, 9, 19, 26, 27, 33
polyfit, 92, 227
Polygons, 20
Polygon smoothing, 41
Polynomial
   Interpolation, 68
   representation, 69
   Newton, 76
polyval, 92, 227
Post-condition, 53
pp-representation, 119
ppval, 120
Pre-condition, 53
Predictor-corrector framework, 335
Prompt (>>), 3

QR factorization, 229, 232
quad, 141
quad8, 141
Quadratic convergence, 267
Quadrature
   Gauss, 143
   Newton-Cotes, 125
   rule, 124
   spline, 146
   two-dimensional, 176

rand, 36
randn, 36
random processes, 36
real, 180
Recursion
   adaptive piecewise linear interpolation, 102
   adaptive quadrature, 138
   Cholesky, 247
   fast Fourier transform, 183
   Newton representation, 79
   Strassen, 183
Recv, 186, 190
rem, 31
Residual (least squares), 235

Residual vs. error, 218
Ring, 186
Root finding, 259
Rotations, 229, 231
Roundoff error, 51
Row interchanges, 213
Row/column orientation, 165, 195
Row vector, 3
Runge phenomena, 84
Runge-Kutta methods, 319
    error, 323
    step, 321

Saxpy operation 165
Script files, 4,12
Search
    binary, 97
    golden section, 279
    graphical, 277
    linear, 96
Secant method, 272
Second order Runge-Kutta, 319
Semicolon, 7
semilog, 47
Send, 186, 190
Shared memory Cholesky 254
Shared memory multiprocessor, 146
short, 8
Significant digits, 45
Simpson's rule, 126
size, 25
sort, 33
Speed-up, 192
Splines, 112
    complete, 115
    end conditions, 119
    natural, 116
    not-a-knot, 117
    quadrature, 146
    smoothing, 326
sprintf, 12, 32
Square root computation, 259
Stability of IVP solver, 312
Stability and pivoting, 212
Static scheduling, 149
std, 36
Steepest descent, 291
Stepsize control, 337
Stirling's formula, 45
Stride, 6

String manipulation, 62
Strings, 12
Structure exploitation, 166
Subinterval location, 98
subplot, 22
Subvector, 9
sum, 32
Superpositioning, 19
Systems
    differential equations, 317
    linear equations, 194
    nonlinear, 295
Symmetric positive definite systems, 236
Symmetry, 156

Taylor approximation, 46
text, 43
Theorem 1 (floating point), 50
Theorem 2 (interpolation error), 84
Theorem 3 (cubic Hermite error), 107
Theorem 4, (4-point Newton-Cotes error), 129
Theorem 5, (stored matrix error), 171
Theorem 6, ($Ax = b$ sensitivity), 220
Theorem 7, (positive definiteness), 238
Theorem 8, (quadratic convergence), 268
Theorem 9, (local and global error), 312
Three-digit arithmetic, 62
Transpose, 4
Trapezoidal rule, 126
Triangular systems, 194
Tridiagonal Cholesky, 240
Tridiagonal systems, 201
tril, 160
triu, 160
Two-point boundary value problem, 237
type, 12

Underflow, 51
Uniform distribution, 36
Unimodal function, 279
Unit lower triangular, 208
unmkpp, 120

Vandermonde approach, 70
Vandermonde matrix, 71, 157
Vector multiply, 72
Vectors
    concatenation, 21
    display of, 3
    multiplication of, 72

setting up, 3
transposing, 4
empty, 6
Vectorization 9, 15, 74

Weights (quadrature), 124
`while`-loop, 30

`xor`, 32

`zeros`, 5, 25